Data and Knowledge in a Changing World

Springer

Berlin
Heidelberg
New York
Barcelona
Budapest
Hong Kong
London
Milan
Paris
Santa Clara
Singapore
Tokyo

C. Bardinet · J.-J. Royer (Eds.)

Geosciences and Water Resources: *Environmental Data Modeling*

With 95 Figures and 35 Tables

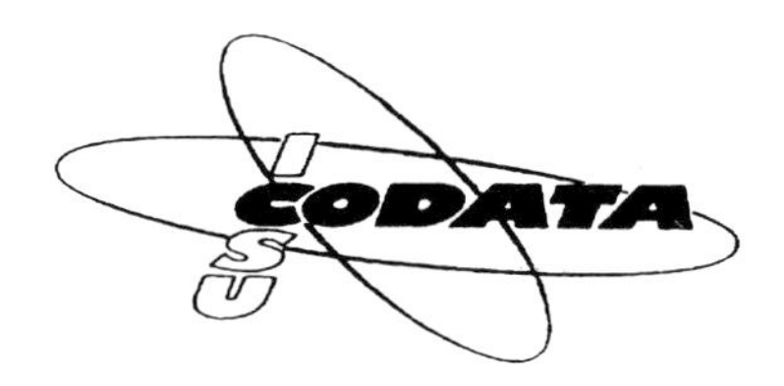

CODATA Secretariat
Phyllis Glaeser, Exec. Director
51, Boulevard de Montmorency
F - 75016 Paris

Editors:

Dr. Claude Bardinet
E.N.S
45, rue d'Ulm
75230 Paris Cedex 05
France

Dr. Jean-Jacques Royer
CNRS - CRPG
15, rue ND des Pauvres, B.P. 20
54501 Vandoeuvre-Lès-Nancy Cedex
France

The image on the front cover comes from an animation which shows worldwide Internet traffic. The color and height of the arcs between the countries encode the data-packet counts and destinations, while the "skyscraper" glyphs (or icons) encode total traffic volume at any site. This image was generated by Stephen G. Eick at the AT&T Bell Laboratories.

ISBN 3-540-61947-x Springer-Verlag Berlin Heidelberg New York

Cataloging-in-publication Data applied for

Die Deutsche Bibliothek – Cip-Einheitsaufnahme
Geosciences and water resources : environmental data modeling ; with 35 tables / ICSU CODATA. C. Bardinet ; J.-J. Royer (ed.). - Berlin ; Heidelberg ; New York ; Barcelona ; Budapest ; Hong Kong ; London ; Milan ; Paris , Santa Clara ; Sigapore ; Tokyo : Springer, 1997
(Data and knowledge in a changing world)
ISBN 3-540-61947-X
NE: Bardinet, Claude [Hrsg.]; International Council of Scientific Unions / Committee on Data for Science and Technology

Typesetting: Camera-ready by editors
SPIN:10555340 51/3020-5 4 3 2 1 0 - Printed on acid-free paper

FOREWORD

This volume contains selected up-to-date professional papers prepared by specialists from various disciplines related to geosciences and water resources. Thirty papers discuss different aspects of environmental data modeling. It provides a forum bringing together contributions, both theoretical and applied, with special attention to *Water in Ecosystems*, *Global Atmospheric Evolution*, *Space and Earth Remote Sensing*, *Regional Environmental Changes*, *Accessing Geoenvironmental Data* and *Ecotoxicological Issues*.

"*Geosciences and Water Resources* : *Environmental Data Modeling*" is now the fourth volume in the Series "*Data and Knowledge in a Changing World*".

Launched by **CODATA** after the 14th International Conference of the Committee on Data for Sciences and Technology, in Chambéry, the purpose of this new Series is to collect from widely varying fields a wealth of information pertaining to the intelligent exploitation of data in science and technology and to make that information available to a multidisciplinary community.

The present series encompasses a broad range of contributions, including computer-related handling and visualization of data, to the major scientific, technical, medical and social fields. The titles of the previous published volumes are:

- The Information Revolution : Impact on Science and Technology.
- Modeling Complex Data for Creating Information.
- Industrial Information and Design Issues.

These titles, edited by J. -E. Dubois and N. Gershon, demonstrate the objectives of the series on Data and Knowledge which is open to contributions of various kinds, namely those defined by E. Fluck:

- fostering the improvement, not only of the quality and accessibility of quantitative and qualitative data, but of the classical and ground-breaking methods by which numeric and symbolic data are acquired, analyzed and managed;

- presenting new data and knowledge interfaces designed to optimize interoperability and thereby increase the potential of sharing data among databases and networks;

- intensifying international cooperation in communication and data sharing. This implies work dealing with standardization, data quality agreements and conceptual data descriptions (metadata, syntactic and semantic approaches), the evolution of Internet-based facilities, other forms of worldwide communications and electronic publishing;

- developing creative designs in the field of engineering, including cognitive aspects critical to data-based decision making.

In view of the changing nature of the information required to study complex systems, CODATA's scope now includes data activities and interdisciplinary modeling such as that encountered in far-reaching projects (e.g. Global Change, various Genome projects, environmental and biodiversity issues, etc.) or in certain medical information systems. Most of the above will therefore be highlighted in this Series.

Moreover, the present Communication Revolution intensifies the role of data and information, thus challenging our ability to adapt our quality control to the new dimensions of the emerging network systems on local, regional and international levels.

As dozens of new sites appear daily on the Internet, driving on the information highway becomes more and more difficult. Paradoxically, the increasing power of the network may lead to a giant information traffic jam: *too much information may kill information*. Now is the time to structure the access and to cross-validate information systems and databases. This requires a lot of work but is necessary to ensure reliable access to pertinent scientific data. The 12th International CODATA Conference, held in Columbus in 1990, was entitled "*Data for Discovery*". This slogan is still alive today and could be extended by "*Internet for Discovery*". More than ever, today's powerful emerging tools should help in the production and use of information, the future resource for the society of worldwide knowledge that the coming century will witness.

In this evolving world where, in many ways, data are becoming essential resources, this Series aims to present innovative concepts ultimately leading to new information paradigms. In so doing, it should become an intensive, thought-provoking forum.

J.-E. Dubois
President of CODATA

CONTENTS

CHAPTER 3: SPACE AND EARTH REMOTE SENSING GIS APPLICATION TO ENVIRONMENTAL IMPACT STUDIES

CHAPTER 4: REGIONAL ENVIRONMENTAL CHANGES APPLICATION TO SOIL EROSION MODELING

Chapter 5: Accessing Geoenvironmental Data From On-line to Expert Systems

Chapter 6: Ecotoxicological Issues

AUTHORS

ABOUJAOUDE Adel
Laboratoire d'Hydraulique de France - 6, Rue de Lorraine, 38130 Echirolles, France

ALLÉ Paul
C.R.P.G. - C.N.R.S - 15, rue Notre-Dame des Pauvres, B.P. 20,
54501 Vandoeuvre-Les-Nancy Cedex, France

AMATYA K. M.
Core Consultancy Ltd - P.O. Box 720, Kathmandu, Nepal

ANKLEY Gerald T.
U.S. Environmental Protection Agency, Environmental Research Laboratory Duluth, 6201 Congdon Boulevard, Duluth, MN 55804 USA

ANTONINETTI Massimo
CNR Istituto di Ricerca sul Rischio Sismico - Via Ampere 56, 20131 Milano, Italy

ARFI Catherine
Laboratoire de Chimie Analytique - Faculté de Pharmacie, 27 Bd Jean Moulin,
13385 Marseille Cedex 5, France

BAKEER Reda M.
Ecole Centrale de Paris, Lab. MSS, 92295 Chatenay Malabry Cedex, France
Department of Civil and Environmental Engineering, Tulane University,
New Orleans, Louisiana, USA

BARDINET Claude
Ecole Normale Supérieure - Laboratoire G-SAT - 45, rue d'Ulm,
75230 Paris Cedex 05, France, E-mail: bardinet@mercator.ens.fr

BARNIER Michel
Ministre Délégué aux Affaires Européennes, 37 quai d'Orsay, 75700 Paris
France

BERAUD Jean-François
B.U.R.G.E.A.P. - 70, rue Mademoiselle, 75015 Paris, France

BERGER M.
CEA/Département de Recherche Fondamentale sur la Matière Condensée,
SCIB/LAN, F-38054 Grenoble Cedex 9, France

BLAKE Gérard
Laboratoire de Biologie et Biochimie Appliquées, École Supérieure d'Ingénieurs
de Chambéry, Université de Savoie, 73376 Le Bourget du Lac, France

BOCCARDO Piero
Dipartimento di Georisorse e Territorio - Politecnico di Torino
C. so Duca degli Abruzzi, 24, 10129 Torino, Italy

BORTOLAMI Giancarlo
Dipartimento di Scienze della Terra - Universita di Torino
Via Accademia delle Scienze 5, 10123 Torino, Italy

BOUILLÉ François
L.I.S.T. - Boîte 1000 - Université Pierre et Marie Curie - 4 place Jussieu,
75252 Paris Cedex 05, France

BOURNAY E.
I.C.I.M.O.D. - P.O. Box 3226, Kathmandu, Népal

BOUSCAREN Remy
CITEPA - 10 rue du Fg Poissonnière, 75010 Paris, France

BRIVIO Pietro Alessandro
MTV - Space Applications Institute, Joint Research Centre - TP 440,
Via Fermi 1, I-21020 Ispra (Va), Italy, E-mail: alessandro.brivio@jrc.it

CADET J.
CEA/Département de Recherche Fondamentale sur la Matière Condensée,
SCIB/LAN, F-38054 Grenoble Cedex 9, France

CALL Daniel J.
Lake Superior Research Institute, University of Wisconsin-Superior
Superior,WI 54880 USA

CHANG J. P.
CITEPA - 10 rue du Fg Poissonnière, 75010 Paris, France

DE VITO Claudia
Dipartimento di Scienze della Terra - Universita di Torino
Via Accademia delle Scienze 5, 10123 Torino, Italy

DUBOST Michel
International Centre for Alpine Environments - ICALPE
Office for Mediterranean Mountains,
Casa pastureccia, Riventosa, 20250 Corte, France

FAVRE Jean-Louis
Ecole Centrale de Paris, Lab. MSS, 92295 Chatenay Malabry Cedex, France

FONTELLE J. P.
CITEPA - 10 rue du Fg Poissonnière, 75010 Paris, France

GABERT Gottfried
CODATA Germany, Hauptstr. 21, Insernhagen (FB) D-30916, Germany

GIRAULT I.
CEA/Département de Recherche Fondamentale sur la Matière Condensée,
SCIB/LAN, F-38054 Grenoble Cedex 9, France

GREENBERG Arthur
Department of Chemistry, University of North Carolina at Charlotte,
Charlotte, NC 28223 USA

GREGOIRE J. M.
MTV - Inst. Remote Sensing Applications, JRC-CEC, 21020 Ispra (VA), Italy

HOPPE P.
B.R.G. - Stilleweg 2, 30655 Hannover, Germany

ABICHINO Giorgio
CNR - Centro di Studio per i Problemi Minerari - C/o Politecnico di Torino,
Corso Duca degli Abruzzi 24, 10129 Torino, Italy

INCARDONA M.-F.
CEA/Département de Recherche Fondamentale sur la Matière Condensée,
SCIB/LAN, F-38054 Grenoble Cedex 9, France

JIMENEZ-ESPINOSA Rosario
Dpt. Geologia - University of Jaén, Escuela Politécnica, Avda. de Madrid, 35
23071 Jaén, Spain

KHREIM Jean-François
CNRS-CEFE - Route de Mende, B.P. 5051, 34033 Montpellier, France

KOFFI B.
MTV - Inst. Remote Sensing Applications, JRC-CEC, 21020 Ispra (VA), Italy

LACAZE Bernard
CNRS-CEFE - Route de Mende, B.P. 5051, 34033 Montpellier, France
E-mail : lacaze@srvlinux.cepe.cnrs-mop.fr

LECOMTE Paul
B.R.G.M. - Avenue de Concyr, B.P. 6009, 45060 Orléans Cedex 2, France

MASCLET Pierre
Laboratoire d'Études des Systèmes Atmosphériques Multiphasiques,
Université de Savoie, Campus Scientifique,
73376 Le Bourget du Lac, France

MATMATTE Lyes
Ecole Centrale de Paris, Lab. MSS, 92295 Chatenay Malabry Cedex, France
Department of Civil and Environmental Engineering, Tulane University,
New Orleans, LA, USA

MEKENYAN Ovanes G.
Bourgas University of Technology, Department of Physical Chemistry
8010 Bourgas, Bulgaria
Lake Superior Research Institute, University of Wisconsin-Superior
Superior,WI 54880 USA

MOLINA-SANCHEZ L.
Dpt. Hidrogeologia y Quimica Analitica, University of Almeria,
La Canada 04120, Almeria, Spain

MOLKO D.
CEA/Département de Recherche Fondamentale sur la Matière Condensée,
SCIB/LAN, F-38054 Grenoble Cedex 9, France

MORIN B.
CEA/Département de Recherche Fondamentale sur la Matière Condensée,
SCIB/LAN, F-38054 Grenoble Cedex 9, France

NAVARRETE F.
Dpt. Hidrogeologia y Quimica Analitica, University of Almeria,
La Canada 04120, Almeria, Spain

NECHITAILENKO Vitaly A.
Geophysical Center, Russian Academy of Sciences, Molodezhnaya Str. 3,
Moscow 117296, Russia, E-mail: vitaly@wdcb.rssi.ru

OBER G.
Remote Sensing Dept IRRS-CNR - 56 Via Ampère, 20131 Milano, Italy

PASTOR J.
Laboratoire de Chimie Analytique - Faculté de Pharmacie, 27 Bd Jean Moulin,
13385 Marseille Cedex 5, France

PAULI A. M.
Laboratoire de Chimie Analytique - Faculté de Pharmacie, 27 Bd Jean Moulin,
13385 Marseille Cedex 5, France

PEPE Monica
CNR Istituto di Ricerca sul Rischio Sismico - Via Ampere 56, 20131 Milano, Italy

PEREZ Alain
ELF - Tour ELF, 92078 Paris La Défense Cedex 45, France

PFENDT P.
Institute of Chemistry - Technology and Metallurgy, P.O. Box 550,
Belgrade 11001, Yougoslavia

PHILIMONOV D. A.
National Research Center for Biologically Active Compounds, 23 Kirova Str.,
Staraya Kupavna, Moscow Region, 142450, Russia

PHILIMONOVA E. A.
Institute for High Temperatures, Russian Academy of Sciences, Russia

POLIC Predrag
Faculty of Chemistry - University of Belgrade, Akademski trg 16, P.O. Box 158,
Belgrade 11001, Yougoslavia

POLVERELLI M.
CEA/Département de Recherche Fondamentale sur la Matière Condensée,
SCIB/LAN, F-38054 Grenoble Cedex 9, France

PORTUGAL H.
Laboratoire de Chimie Analytique - Faculté de Pharmacie, 27 Bd Jean Moulin,
13385 Marseille Cedex 5, France

PRADHAN P.
I.C.I.M.O.D. - P.O. Box 3226, Kathmandu, Népal

PURIC M.
Biological Institute - Podgorica, Yougoslavia

QUAGLINO Alberto
Dipartimento di Georisorse e Territorio - Politecnico di Torino,
C.so Duca degli Abruzzi 24, 10129 Torino, Italy

RAO Narasimha K. L.
Professor of Geology, S.V. University, Tirupati - 517 502,
Andhra Pradesh, India

RAOUL S.
CEA/Département de Recherche Fondamentale sur la Matière Condensée,
SCIB/LAN, F-38054 Grenoble Cedex 9, France

RAVANAT J.-L.
CEA/Département de Recherche Fondamentale sur la Matière Condensée,
SCIB/LAN, F-38054 Grenoble Cedex 9, France

REDDY D. K. Samarasimha
Minister for Panchayat Raj and Rural Development
Government of Andhra Pradesh, Hyderabad 500 001, Andhra Pradesh, India

RENACCO E.
Laboratoire de Chimie Analytique - Faculté de Pharmacie, 27 Bd Jean Moulin,
13385 Marseille Cedex 5, France

RODDA John C.
President of the International Association of Hydrological Sciences
Institut of Hydrology, Wallingford, Oxfordshire, 0X10 8BB, U.K.

ROTIVAL-LIBERT C.
Laboratoire de Chimie Analytique - Faculté de Pharmacie, 27 Bd Jean Moulin,
13385 Marseille Cedex 5, France

ROYER Jean-Jacques
C.R.P.G. - C.N.R.S - 15, rue Notre-Dame des Pauvres, B.P. 20,
54501 Vandoeuvre-Les-Nancy Cedex, France
E-mail : royer@crpg.cnrs-nancy.fr

SHTUKA Arben
LIAD - ENSG, B.P. 40, Rue du Doyen Marcel Roubault,
54501 Vandoeuvre-Les-Nancy Cedex, France
E-mail: shtuka@extreme.ensg.u-nancy.fr

TARTARI Gabriele A.
CNR Istituto Italiano di Idrobiologia, Largo Tonolli 50/52,
28048 Verbania-Pallanza, Italy

TARTARI Gianni
Istituto di Ricerca Sulle Acque - CNR, 20047 Brugherio, Italy
E-mail: tartari@irsa1.irsa.rm.cnr.it

THIEMAN James R.
NASA - Goddard Space Flight Center, National Space Science Data Center, Code 633.2, Greenbelt, MD 20771, USA
E-mail: thieman@nssdca.gsfc.nasa.gov

USSEGLIO-POLATERA Jean-Marc
Laboratoire d'Hydraulique de France - 6, rue de Lorraine, 38130 Echirolles, France

VALSECCHI Sara
Istituto di Ricerca Sulle Acque - CNR, 20047 Brughero, Italy

VANDEVELDE Thierry
Compagnie Générale des Eaux - 32, place Ronde, 92982 Paris La Défense, France

VEDINA Olga
Institute of Plant Physiology, Moldavian Academy of Sciences, Padurii 22, Kishinev 277002, Moldavia

VEITH Gilman D.
U.S. Environmental Protection Agency, Environmental Research Laboratory Duluth, 6201 Congdon Boulevard, Duluth, MN 55804 USA

ZHELEZNIAK M. B.
National Research Center for Biologically Active Compounds, 23 Kirova Str., Staraya Kupavna, Moscow Region, 142450, Russia

PREFACE

C. Bardinet and J.- J. Royer

This book focuses specifically on such urgent topics as: water in the ecosystem, environment and toxicity, space and earth remote sensing and environmental changes. Many of the chapters were inspired by work prepared for the 14th CODATA Conference held in Chambéry, France, and updated since then. A major theme is the crucial need for reliable sources of information and observation on a world wide scale together with better global methods of investigation. One of the conclusions was that additional work and research is necessary in the area of geoscience and water resources evaluation.

During the past twenty years, enormous efforts have been made to gather, standardize, validate and analyze information data for science and technology. Several attempts were made to coordinate these efforts at an international level. CODATA, among other organisms, was one of the pioneers in these fields contributing to most current problems such as global warming, the effect of CFC on the ozone hole, the world-wide water resources shortage, the role of anthropic activity on natural cycles.

The place of data in environmental research is now recognized as essential, and one can say that the main questions are related to the availability and quality of data and data access. Concerning Earth's environment, CODATA is participating in the survey on data availability on the following themes:

- Risk data bases in eco-toxicology of industrial materials including solid, liquid and gaseous phases;
- Hazard data bases in geno-toxicology;
- Thermodynamic, chemical and diffusive properties of hazardous components with a special focus on cleaning techniques, reassessment processes and substitute technology;

- Legislative data bases on ecology including international and national aspects;
- Biodiversity data bases including taxonomy, inventories of main species, industrial and social aspects;
- World wide preservation of water (surface and underground) quality data bases including mapping, pollution levels, cleaning processes;
- World wide preservation of uncontaminated soil data bases, in agriculture, national parks, industrial and urban wastes, including acid rainfall and radiation aspects;
- Software information systems for modeling and simulating ecological risk analysis;
- Retention level of ecological information including air, soil and water pollution.

Several International Scientific Programs are involved in these areas. All of them need validated data and information on a world scale. Lack of controlled data may lead to misunderstanding or neglect some aspects of important problems such as anthropic impacts on the Earth's system or the evaluation of natural resources. The paper presented by Dr. Rodda as an introduction to the Chapter "Water in Ecosystems : a non renewable resource" of this volume illustrates perfectly the problem of unreliable data. On the resource evaluation of water at the word scale, this author concludes "*It is very strange that at the time when global demand for water is growing faster than ever, knowledge of the world's water resources is declining ...* ".

CODATA contributes to various international actions in the Global Change Research Program. CODATA is a Committee of ICSU specializing in data for the Global Change Programme (IGBP - International Geosphere Biosphere Programme) whose purpose is to provide an answer to the debated question of possible human responsibility in the observed warming of the Earth. Is the peculiar inertia of the Earth's system taken into account in present speculation on global warming? Do the models developed by different groups account for the ability of interference and retroactive actions of every component of the Global Change System? Are models completely effective? For example, it is admitted nowadays that when observing the land degradation of the Sahel one has to specify the causes resulting from natural cycles of the climate and those resulting from human activities (land cover degradation). Two chapters of this volume "Global atmospheric evolution: impact of anthropic activities" and "Regional environmental changes: application to soil erosion modeling" are devoted to these important questions.

IGBP uses data bases, maps and atlases on different scales in order to diffuse images and data on the state of the Earth via the multimedia highway. IGBP needs various studies from geophysics to human settlement pollution, together with many other topics dealing with the 2-D or 3-D coordinates (including the relief) and even 4-D (including time).

This is why the GIS (Geographic Information Systems) are so important in the IGBP program. The concept of GIS, born in the 1970's, was really only accepted 15 years later.

New ways have been introduced for taking into account the dynamics of geographical problems on different scales, and tools have been developed recently integrating new systems in data acquisition, transfer and exchange, data analysis, integration and modeling. In associating 2-D and 3-D graphical representation of multi thematic approaches, a GIS gives the representation of the dynamics of the Earth through images, data and texts.

During the 80's in the field of IGBP, it appeared absolutely necessary to promote the combined use of data collected from ground, marine and aircraft stations together with remotely sensed data. This is now widely used as illustrated in the chapter "Space and Earth remote sensing: GIS application to environmental impact studies".

This implies the need to locate existing data and information, not an easy task in the world of databases. Since the beginning of the 90's, INTERNET has improved data exchange and integration. Such an approach supposes the combined use of computer techniques and multimedia tools. These aspects are discussed in the chapter "Accessing Geoenvironmental data: from on-line to expert systems".

To protect our environment, it is necessary to clean waste gases reducing the emission of toxic impurities. Conventional industry and eco-industry have to reduce their impact on the environment, while the Offices for Environment protection must control the levels of pollution and the evolution of technical methods for promoting a new era of sustainable development. Ecotoxicological research has to be developed on the basis of valid data. The chapter on "Eco-toxicological issues" discusses these aspects.

Consideration of the Earth's environment means studying the issue of sustainable development at different geographic and subject levels in order to include all its multidisciplinary aspects in terms of research and education, research and industrial development, administration and policy. CODATA which is specialized in collecting, validating and standardizing information exchange, can play a major role in this debate.

For instance, the sea level survey is one of the important measurements in the Global warming survey. A database on the evolution of sea water density, sea temperature and ice sheet melting is absolutely necessary in this field. But all the measurements relating to sea level rise are based on a very poor network of tidalmeter stations around the world. Measurement of the ozone hole mechanisms is another important contribution to the global warming and greenhouse effect survey. Human activities modify the atmospheric carbon dioxide rate and contribute to an additional greenhouse effect. One question is what is and what could be the role of the biosphere in the global ecological balance. The taxonomic bases are the key for understanding the question of biodiversity in relation to climatic changes (see chapter "Global atmospheric evolution: impact of anthropic activities").

It is particularly important to consider data access and data processing for studying the Global environment and the natural hazards risks (desertification, flooding areas, volcanic risks, seismic risks, high mountain risks, deforestation and erosion, sea level variations, atmospheric pollution, paleoenvironment level and variation etc.). In these matters explanations and predictions depend on the accessibility and the quality of data and methodology, i.e.:

- the diffusion of higher-level methodology (GIS hardware and software) in image processing,
- the accessibility of higher-level multisensor data products (stereo, VIS, NIR, MIR, TIR, Radar),
- and the evolution of quality control concerning the integration of in situ data and remotely sensed data.

To answer these questions, computer and geoscience research are based on both ground station and remote sensing (RS) data.

One important task is to assure access to these data sets, at all the required geographical levels, by scientific users in all parts of the world, in the international computer networks.

CODATA considers that the scientific community must solve the questions of multiscale analysis, data quality and control and the accessibility of pertinent data series for all subjects. This could help to answer the following questions:

- *Question 1*

  Concerning satellite remote sensing data, - *what kind of data are and will be available now and in the near future*? - Is there an international data exchange market and an data policy for international scientific programs such as IGBP?

  Concerning space and earth remote sensing research and GIS applications - *are we sure that the data are available on the requested scale and in time*? If we care about the volume of the existing data and the volume of data really available, we have to care about on-line access of geo-environmental data.

- *Question 2*

  Concerning operational data for disaster monitoring and hazard mapping - *is it always possible to use, in time, optical and radar multitemporal satellite data, for the control of landslide, volcanic terrain, and flooded zones*?

- *Question 3*

  Concerning the mapping of global and regional scale image data - *is it always possible to receive on-line access to weather satellite data, upper atmospheric data on pollution, and all the other data required*?

- *Question 4*

  Concerning the monitoring of forest ecosystems - *is it possible to establish a reference database* i.e.: - calibration of the scanner and radar instrument in order to promote multitemporal and multisensor analysis, and an evaluation of vegetation indices; - relationships between infrared radiance and biophysical properties of various zonal types of forest, and all the other data required?

- *Question 5*

  Concerning water in ecosystems - *are we sure that water could ever be considered as a renewable resource*, particularly if we consider the needs of water in agriculture and in urban and industrial areas in relation to population growth, and if we consider the effect of pollution on water supply?

  We have to use models based on validated data. For example, for modeling toxicity in aquatic ecosystems, data bases have to include chemical data describing the structure, as well as data describing the spatial distribution of the sources of pollutants.

  This is the case for the photo-induced toxicity of the polycyclic aromatic hydrocarbons (PAHs) which has to be studied at the structure level of the parent chemicals, and at the emission level.

  Question 6

  Concerning the global atmospheric evolution - *are we sure we know the impact of anthropic activities, and are all the required data available*?

In environmental research, the issue of the availability of all existing data has to be solved. Here, the case of international CSCE " Open Sky Treaty " signed in Helsinki on March 24, 1992, is of great interest. This treaty decided a tool should be built for conflict prevention and crisis management, and that this tool could be used to control environmental protection.

For example, in using overflight data registration, it could be possible to control the security of nuclear power installations and the dissemination of nuclear products, but also to facilitate the exchange of information concerning the environment. This could concern sensitive industrial areas, as well as the main ecosystems and the natural biotopes.

From ecosystems to ecotoxicology, and from the equilibrium of atmospheric components to the equilibrium of water cycles, the survey of the environment is first of all a question of data, data accessibility and data interpretation.

The various topics discussed in this volume on geosciences and water resources demonstrate the interdisciplinarity of science and technology - the environment is no exception. This strengthens the need for scientific conferences where scientists of different fields can communicate and discuss their results. The editors hope that this volume will contribute significantly to widening the dialogue between chemists, physicists, computer specialists, biologists, geoscientists and environmental specialists.

Acknowledgments: The editors would like to express their thanks to all the authors of this book for their contributions and hard work. They also wish to thank Mrs. Bernice Dubois for her valuable help in reviewing the proofs and sometimes finalizing the English version of the papers. The final version of this book was made possible with the help of a number of people among whom are Mrs. Phyllis Glaeser, Executive Director of CODATA, for her advice in editing, Mrs. Ewa Niemirowicz for help in translating abstracts into French and Mrs. Françoise Kern for assembling the different contributions and editing the book.

Several organisations have also contributed to make this book possible: the Laboratoire National d'Essais (L.N.E.) for their never-failing and valuable assistance; Aidel for providing us with some of their excellent computer tools, in particular the database software SUPERDOC; the Ecole Nationale Supérieure - Laboratoire G-SAT for giving material support and the Centre de Recherches Pétrographiques et Géochimiques, (C.R.P.G.), research laboratory of the Centre National de la Recherche Scientifique (C.N.R.S.) for providing editing and image processing facilities.

Note : CODATA is connected to the Internet (*http://www.cisti.nrc.ca*). Additional information, especially abstracts of papers presented in this and previous volumes can also be accessed (*http://mercator.ens.fr/home/bardinet/codata.html*).

INTRODUCTION

THE QUEST FOR A HEALTHIER ENVIRONMENT

Michel BARNIER

French Minister of the Environment, 1993-1995
Président du Conseil Général de la Savoie
Ministre Délégué aux Affaires Européennes,
37 quai d'Orsay, 75700 Paris, France

Ladies and Gentlemen,

I am particularly happy to be present at this 14th International Conference on the use of data of whose importance and renown I am fully aware.

I am especially touched by its being held in Chambéry, after Beijing in 1992 and just before Tsukuba in 1996, and I am honored that the Savoie was chosen to host this International Conference.

The Committee on Data for Science and Technology, working under the aegis of ICSU, has indeed an essential role in promoting relations between science and industry for the use and distribution of interdisciplinary data with particular regard to the Environment.

I seize this opportunity to mention that, at the start of 1994, I inaugurated the offices of the French Institute of the Environment whose mission it is to establish a statistical system on the Environment and to publish a "State of the Environment" report. On the European level, and particularly for the European Environment Agency just set up in Copenhagen, it will also supply data essential for following the evolution of effects of environmental policies of the various countries comprising the European Union.

You know how important to me is the dissemination of environmental information. It is indeed indispensable to understand in order to act, to understand before acting so as to avoid the double pitfall of indifference and "catastrophism". This seems to me the best way to carry out a coherent and rational policy.

I know that in a few minutes you will participate in an important event. I refer to the Ceremony granting the "Albert Einstein" science Award to the well-known American scientist, Professor F. Sherwood Rowland and the "José Vasconcelos" education Award to the great Australian mathematician, P. Joseph O'Halloran.

It is a great honor for all of us to be privileged to welcome to this Conference Professor Rowland, one of the very first to have studied and pointed out the reduction of the ozone layer. His work again illustrates the importance and the need for obtaining numerous and reliable statistical data to apprehend the ever-more complex phenomena we have to face.

Indeed today, environmental problems result more and more from the formidable development of industrialization and urbanization and after the Cairo Conference on Population and Development, we are reminded that this mutation also is inseparable from the demographic explosion.

Just one figure to illustrate this phenomenon: where it took humanity more than two million years to reach a population of five billion people (in 1987), it will take only 44 years to go from five to ten billion at the present growth rate.

The great challenge awaiting us is indeed to find a balance between man and his Environment. This balance concerns us all. It is, to my mind, one of the major issues at the start of the 21st century, and the Rio Conference stressed the importance of what we call sustainable development.

This global framework is particularly justified by the fact that the *environment knows no borders*. I should like to give you two examples: global pollution and the pluridisciplinary approach required by the Environment.

1 Environment: A Global Problem

Environmental problems have an increasingly global impact, in particular with regard to questions concerning the general balance of our planet such as climate, in its broad sense, i.e. exchanges between earth, atmosphere and oceans or again, biodiversity.

I have the feeling that in only a few years we have gone from a local view of pollution in a static milieu to a far more dynamic version of the Environment. Satellite observation has enabled us to understand that air masses, like water masses, circulate all over the Earth. We must, therefore, consider the *Earth System* as a whole and have, for certain polluting factors, a global approach.

1.1 Climate Change

The systematic construction of climatology is recent and far from complete. Instruments for analyzing and measuring the climate are being forged gradually. Paleoclimatology, or the knowledge of past climates by the study of sediments and glaciers developed during the 19th century. Later, in the 1950s, stricter techniques and methodologies were perfected, in particular by great depth ice samples.

Today large computer models and the use of satellites aim at grasping climate complexity. This latter results from the numerous parameters involving the entire Earth and its atmosphere and also from the fact that climate is in a state of perpetual change. The disturbing element in this gigantic mechanism is, of course, the impact of an uncontrolled development of human action on forests, water resources, soil quality. With regard to global warming, this is caused by the spectacular increase of gas production giving rise to the hothouse effect.

Some experts estimate that the average earth temperature will rise from 3 to 7 °C between now and the year 2100 which would have serious consequences (some of them as yet unknown) on ecosystems. Much remains to be done concerning knowledge and prediction.

France, for its part, is determined to fulfill the commitments made at the Rio Conference, and it has ratified the two Conventions adopted there. The "Climate" Convention was ratified unanimously by both houses of Parliament early this year, and it constitutes substantial progress in planning for planetary cooperation, particularly as a principle of solidarity between Northern and Southern countries. France is a partner in, among others, the international Geosphere-Biosphere program dealing with the evolution of problems on a planetary scale.

France is further committed to defending a similar approach in the European debate. Her efforts, and those of the European Commission, have been useful. The European Union decided, for example, that the commitment made by signing the United Nations Framework Convention, i.e. the return by the year 2000 to the CO_2 emission levels of 1990, would be fulfilled by the whole of the Union.

It is my conviction that a common objective achieved by common tools constitutes the best guarantee for a fair sharing of the economic burden caused by the required adjustment.

On this issue, I am persuaded that a fiscal approach remains the best instrument from the standpoint of global economic efficiency, given the extreme dispersion of dioxide that concerns every one of us in our daily lives.

I intend to try to convince my partners of the validity of these proposals that differ a good deal from the Commission's present ecotax project concerning primarily the competition among our heads of industry. I shall endeavor to promote this project during the French Presidency of the European Union.

1.2 The Ozone Layer

The other global problem occupying a central position in debates on the Environment is the deteriorating ozone layer.

The capacity of CFC molecules to destroy ozone was shown in 1974, and in 1979 we were able to observe what is called the *hole in the ozone layer*. The latest results presented by Professor Rowland show us that the ozone layer has been diminishing from the 1970s on by from 5 to 10% each year. It is with great pleasure that I learned in 1995 that Professors Sherwood Rowland (USA), Paul Crutzen (Netherlands) and Mario Molina (Mexico) had received the Nobel Prize for Chemistry for their discovery concerning the deteriorating ozone layer by chemical agents, products of our industrial civilization.

Today it is felt that the harm has already been largely done, since most of the CFC emitted by human action has not yet reached the higher atmosphere and has, therefore, not yet produced its devastating effects on the ozone layer.

Nonetheless, having identified the evil, man can at least do his best to stop it from spreading and from getting worse. This example shows that the harm caused to the Environment by human action is often revealed late and, granted the inertia of the climatological system, our information must be as exact as possible.

1.3 Biodiversity

I do want to take this opportunity to pay homage to the scientists who first developed the idea of the diversity of genetic patrimony and whose vision was particularly relevant and ambitious in light of the importance this concept has today. Indeed in less than a decade, it resulted in the signing of a World Convention on biodiversity.

Certainly France regrets that this Convention goes no farther along the road to conservation and does not establish world lists, for example. However, the text accords great importance to national strategies regarding the identification of the components of biological diversity and to the scientific "watch", to measures and to programs for protection as well as to setting up protected zones, and France ratified this Convention.
May I recall here the significance accorded by France to the DIVERSITAS program carried out on UNESCO initiative and whose aim it is to acquire an in-depth scientific knowledge in the area of biological diversity.

2 THE ENVIRONMENT: A PLURIDISCIPLINARY FIELD

2.1 Pluridisciplinary aspects

For some years now, environment problems have changed in scale, in nature, in urgency, and their complexity, that ranges from the local to the global, leads to a growing demand for scientific data and databases, for prediction models and to developing technological and industrial research in order to control the ecological and economic blockages caused by this change.

If some of these can be controlled in the medium range, others can lead to dead ends, even more because of the many uncertainties involved. Anticipation is, therefore, an essential function and sometimes the lack of scientific data makes it impossible to provide an immediate answer to important questions.

The work I decided to undertake on polluted sites highlights, for example, the difficulties involved in hierarchizing sites, in knowing the impact of polluted materials on the medium. Still today, doubt subsists as to the risks presented by certain substances for the Environment or for human health.

This complexity and this diversity lead to integrating many fields and actors from ever more varying horizons: scientists, of course, such as biologists, chemists and physicists, but also sociologists, lawyers and economists.

2.2 Environment and Ethics

We have just evoked France's ratification of the Biodiversity Convention. To justify action in favor of nature conservation, utilitarian arguments are often resorted to, such as: "Natural species must be protected because they can be useful one day if such and such a molecule were to be discovered".

This approach counts, of course, but it is not the most important in my eyes. I am among those who believe that wild species and natural habitats constitute a patrimony in themselves. Patrimony is, by definition, what we received in heritage and what we are to transmit to future generations. This patrimony is not only economic; it is also cultural. It involves a moral dimension that, it seems to me, essential not to lose sight of and that goes far beyond the scientific dimension. In this context I am happy to note the current and continued CODATA actions to manage a worldwide inventory of the vegetable and animal kingdoms. Many of these projects are facilitated by their computerization on Internet.

2.3 Environment and Rights

Robert Poujade said "Environmental policy is expressed globally; it involves all policies but it is not limited to any one of them".

The Environment contributes to the good of all, whether by protecting nature, preserving species, fighting pollution and harmful effects or defending the framework of life itself.

Recognizing this has contributed, for example, to working out, over the past 20 years, a specific right to the environment, a right still young and changing rapidly in line with the new problems pinpointed by science. This pluridisciplinary nature of the Environment renders such a right very complex indeed; I therefore decided to launch a broad project of renovation and modernization of environmental legislation. This code of the Environment aims at clarifying and regrouping certain texts currently included in legislation as diverse as the rural code, the city planning code or the mining code. It should be completed by the end of 1994 and presented during next Spring's parliamentary session.

2.4 Environment and Sociology

The weight of questions linked to the Environment in our societies is also measured by the antagonism or conflicts it arouses.

The man of the street is often in contradiction with himself, simultaneously driver and pedestrian, consuming more and more synthetic products, yet demanding they be ever purer and more natural. He agrees to the necessity of installing plants for treating waste products but local egotism results in each one wanting to see the project carried out in his neighbor's backyard.

It no longer suffices to work out increasingly severe technical requirements; it is necessary to discuss and explain. To bring a project to fruition today involves carrying out a real strategy going far beyond the purely technical aspects. Indeed a whole teaching process must be set up to change age-old daily behavior habits.

Cooperation and information exchanges among elected representatives, environmental protection associations and local populations are more and more important for success in completing a project.

3 CONCLUSION

As we can indeed see, the Environment no longer has any borders. Global pollution problems confirm the need for reaching agreement on a planetary scale.

We already knew that Environment was a transversal science, requiring various technical competencies; henceforth its social dimension must also be considered.

If today I want the Ministry of the Environment to be more respected, consulted, listened to in political life, it is in truth because the Environment is so close to us and so deeply concerns us all.

Thank you.

Chambéry, Monday 19 September 1994.

Since this lecture, given then as Minister of the Environment, the relevance of the issues discussed is recognized by all. In this book on many aspects of the environment studied during the Chambéry Conference, CODATA presents updates taking into account current results of the struggle to preserve our environment.

KEYWORDS

Environment, climate, ozone, ethics, rights, sociology.

A LA RECHERCHE D'UN ENVIRONNEMENT MEILLEUR

Michel BARNIER

Ministre français de l'Environnement, 1993-1995
Président du Conseil Général de la Savoie
Ministre Délégué aux Affaires Européennes,
37 quai d'Orsay, 75700 Paris, France

Mesdames, Messieurs,

Je suis particulièrement heureux de pouvoir être présent à cette 14ème conférence internationale sur l'utilisation des données dont je connais l'importance et le retentissement.

Je suis particulièrement touché de sa tenue à Chambéry, après Pékin en 1992 et avant Tsukuba en 1996, et je suis très honoré que la Savoie ait été choisie pour accueillir cette conférence internationale.

CODATA, sous l'égide de l'ICSU, joue un rôle essentiel en promouvant des relations entre la science et l'industrie afin d'utiliser et de diffuser des données interdisciplinaires en particulier dans l'environnement.

Je saisis cette occasion pour dire que début 1994 j'ai inauguré les bureaux de l'Institut français de l'environnement qui doit établir un système statistique sur l'environnement et publier un Rapport sur la Situation de l'Environnement. Au niveau européen, il fournira aussi des données nécessaires pour suivre l'évolution des effets des politiques sur l'environnement des pays de l'Union européenne.

Vous savez combien importante est pour moi la diffusion d'informations sur l'environnement. Il est, en effet, indispensable de comprendre pour agir, de comprendre avant d'agir en évitant le double écueil de l'indifférence et du catastrophisme. C'est, me semble-t-il, la meilleure façon de conduire une politique cohérente et rationnelle.

Je sais que vous allez dans quelques instants procéder à un événement important. Il s'agit de la remise des prix de la science "Albert Einstein" attribué au Professeur Rowland, éminent scientifique américain, et de l'éducation "José Vasconcelos" remis au grand mathématicien australien, le Professeur Monsieur O'Halloran.

C'est un grand honneur pour nous tous d'avoir le privilège d'accueillir à cette conférence Monsieur Rowland, qui est l'un des premiers a avoir mis l'accent sur la question de la réduction de la couche d'ozone. Ses travaux illustrent encore l'importance et la nécessité d'obtenir des données statistiques nombreuses et fiables pour appréhender les phénomènes de plus en plus complexes auxquels nous devons faire face.

En effet, aujourd'hui, les problèmes d'environnement résultent de plus en plus du formidable développement de l'industrialisation et de l'urbanisation et à l'heure ou se termine la conférence du Caire sur la population et le développement, cela nous rappelle que cette mutation est aussi indissociable de l'explosion démographique. Un seul chiffre pour illustrer ce phénomène: alors qu'il aura fallu à l'humanité plus de 2 millions d'années pour atteindre 5 milliards d'individus (en 1987), il lui suffirait de 44 ans pour passer de 5 à 10 milliards, au taux actuel de croissance.

Le grand défi qui nous attend est bien de trouver l'équilibre entre l'homme et son environnement. Cet équilibre est l'affaire de tous. Il s'agit , me semble-t-il, d'un des enjeux majeurs du début du 21ème siècle et la Conférence de Rio a souligné l'importance de ce qu'on appelle le développement durable.

Ce cadre global se justifie, en particulier, du fait que l'*environnement ne connaît pas de frontières*. Je voudrais brièvement l'illustrer de deux façons devant vous : les pollutions globales et l'approche pluridisciplinaire que nécessite l'environnement.

1 L'ENVIRONNEMENT : UN PROBLÈME GLOBAL

Les problèmes environnementaux ont, de plus en plus, un impact global notamment dès qu'il s'agit de problèmes touchant l'équilibre général de notre planète comme le climat au sens large du terme, c'est à dire les échanges entre la terre, l'atmosphère et les océans, ou comme la biodiversité.

J'ai le sentiment qu'on est passé en quelques années d'une vision locale des pollutions sur un milieu statique à une version beaucoup plus dynamique de l'environnement.

L'observation par satellite a permis de comprendre que les masses d'air comme les masses d'eau circulaient sur l'ensemble de la Terre. Il faut donc considérer le *Système Terre* dans son ensemble et avoir pour certains polluants une approche globale.

1.1 Le changement du climat

La construction systématique de la climatologie est récente et loin d'être achevée. Les instruments d'analyse et de mesure du climat se forgent progressivement. Au XIXe siècle, s'est constituée la paléoclimatologie, c'est à dire la connaissance des climats du passé par l'étude des sédiments et des glaciers. Puis à partir des années 50, des techniques et des méthodologies plus rigoureuses sont mises au point notamment par des prélèvements de glace à de grandes profondeurs.

Aujourd'hui, de grands modèles informatiques et l'utilisation des satellites visent à saisir la complexité des climats. Celle-ci résulte en effet à la fois du nombre de paramètres impliquant la totalité de la Terre et de son atmosphère et également du fait que les climats sont en perpétuel changement. L'élément perturbateur dans cette mécanique gigantesque est, bien sûr, l'impact d'un développement non maîtrisé des activités humaines sur les forêts, les ressources en eau, la qualité des sols. Dans le cas du réchauffement planétaire, c'est l'augmentation spectaculaire de la production de gaz qui concoure à l'effet de serre qui en est la cause.

Certains experts estiment que la température moyenne de la Terre s'élèverait de 3 à 7°C d'ici 2100 ce qui entraînerait des conséquences pour certaines encore inconnues sur les écosystèmes. Beaucoup reste cependant à faire pour la connaissance et la prévision.

La France, pour sa part, a tenu à ce que les engagements pris à la conférence de RIO soient tenus et, elle a ratifié les deux conventions adoptées à cette occasion. La convention "climat" dont le Parlement français a autorisé la ratification à l'unanimité des deux chambres en début d'année, constitue une avancée substantielle dans l'ébauche d'une coopération planétaire en instaurant notamment un principe de solidarité entre les pays du nord et du sud.

La France participe notamment au programme international Géosphère-Biosphère qui traite de l'évolution des problèmes à l'échelle de la Terre. De la même façon, la France s'attache à défendre une approche similaire dans le débat européen. Ses efforts, aussi bien que ceux de la Commission européenne, ont été utiles. La Communauté a, par exemple, formellement décidé que l'engagement pris lors de la signature de la convention cadre des Nations Unies, c'est à dire le retour en l'an 2000 des émissions de CO_2 à leur niveau de 1990 serait accompli par la Communauté dans son ensemble.

Je suis, à cet égard, convaincu qu'un objectif commun atteint à partir d'outils communs, constitue la meilleure garantie d'un partage équitable du poids économique de l'ajustement nécessaire.

Je reste, sur ce sujet, persuadé qu'une approche fiscale demeure le meilleur instrument d'un point de vue de l'efficacité économique globale, compte tenu de l'extrême dispersion des émissions de gaz carbonique qui concernent chacun d'entre nous dans sa vie quotidienne.

Je vais m'attacher à essayer de convaincre mes partenaires du bien-fondé de ces propositions sensiblement différentes du projet actuel d'écotaxe de la Commission respectant notamment la compétitivité de nos industriels. Je m'attacherai à promouvoir ce dossier dans le cadre de la présidence française de l'Union européenne.

1.2 La couche d'Ozone

L'autre problème global qui occupe une position centrale dans les débats sur l'environnement est la détérioration de la couche d'ozone.

La faculté des molécules de CFC de détruire l'ozone a été démontrée en 1974, et c'est en 1979 que l'on a pu observer ce qu'on a appelé le *trou dans la couche d'ozone*. Les derniers résultats présentés par Monsieur Rowland nous montrent que la couche d'ozone s'est réduite depuis les années 70 de 5 à 10 % par an. C'est d'ailleurs avec plaisir que j'ai pris connaissance en 1995 de l'attribution du Prix Nobel de chimie aux Professeurs Sherwood Rowland (États-Unis), Paul Crutzen (Pays-Bas) et Mario Molina (Mexique) pour leur découverte relative à la détérioration de la couche d'ozone sous l'effet d'agents chimiques, produits de notre civilisation industrielle.

Aujourd'hui, on considère que le mal a déjà été en grande partie commis, puisque l'essentiel des CFC émis par les activités humaines n'a pas encore rejoint la haute atmosphère et donc ils n'ont pas encore produit leurs effets dévastateurs sur la couche d'ozone.

Toutefois, le mal étant identifié, l'homme peut au moins s'efforcer d'arrêter sa propagation et son aggravation. Cet exemple montre que les dommages que causent les activités humaines à l'environnement se manifestent souvent avec un certain retard et, compte tenu de l'inertie du système climatologique, il faut posséder une information aussi précise que possible.

1.3 La biodiversité

Je voudrais profiter de ma venue ici pour rendre hommage aux scientifiques qui sont à l'origine du développement de la notion de diversité du patrimoine génétique, et qui ont eu une vision particulièrement pertinente et ambitieuse comme en témoigne l'importance qu'a pris ce concept aujourd'hui. En effet, en moins d'une décennie, il a abouti à la signature d'une convention mondiale sur la biodiversité.

Certes, la France a regretté que la convention n'aille pas plus loin dans la voie de la conservation par l'instauration notamment de listes mondiales. Cependant le texte donne une large place aux stratégies nationales en ce qui concerne l'identification des composantes de la diversité biologique et à la veille scientifique, les mesures et programmes de protection ainsi que l'établissement de zones protégées, et la France a accepté de ratifier cette convention.

Je voudrais rappeler ici, notamment toute l'attention que la France accorde au programme DIVERSITAS mené sur l'initiative de l'UNESCO et qui vise à approfondir les connaissances scientifiques dans le domaine de la diversité biologique.

2 L'Environnement : un Domaine Pluridisciplinaire

2.1 Environnement et science

Les problèmes d'environnement ont, depuis quelques années, changé d'échelle, de nature, d'urgence et leur complexité qui s'étend du local au global conduit à un développement de la recherche technologique et industrielle afin de maîtriser les impasses écologiques et économiques causées par cette évolution.

Si certains d'entre eux sont maîtrisables à moyen terme, d'autres peuvent conduire à des butoirs, d'autant que les incertitudes sont grandes. L'anticipation est donc une fonction essentielle et parfois le manque de données scientifiques ne permet pas de répondre immédiatement à des questions importantes.

Le travail que j'ai décidé d'engager sur les sites pollués, met par exemple, en évidence, des difficultés pour hiérarchiser les sites, connaître l'impact des polluants sur le milieu. Aujourd'hui encore, des incertitudes demeurent quant aux risques que peuvent présenter certaines substances pour l'environnement ou la santé humaine.

Cette complexité et cette diversité conduisent, pour répondre à ces questions, à intégrer de nombreuses disciplines, des acteurs d'horizons de plus en plus différents : des scientifiques bien sûr, tels que biologistes, chimistes, physiciens, mais aussi des sociologues, des juristes, des économistes.

2.2 Environnement et éthique

Nous venons d'évoquer la ratification par la France de la Convention sur la biodiversité. Pour justifier les actions de conservation de la nature on a souvent recours à des arguments utilitaires tels que : "il faut protéger les espèces naturelles car elles pourront nous être utiles un jour si l'on découvre telle ou telle molécule".

Cette approche compte beaucoup, mais ce n'est pas la plus importante à mes yeux. Je suis de ceux qui pensent que les espèces sauvages et les milieux naturels constituent un patrimoine. Le patrimoine est par définition ce que nous avons reçu en héritage et ce que nous allons transmettre aux générations futures. Ce patrimoine n'est pas seulement économique, il est aussi culturel. Il y a là, une dimension morale qu'il me paraît essentiel de ne pas perdre de vue et qui dépasse largement toute notion scientifique. Dans ce domaine, je note avec satisfaction l'intérêt des actions actuelles et continues menées par CODATA pour gérer un inventaire mondial des règnes végétal et animal. De nombreux projets sont facilités par leur informatisation sur Internet.

2.3 Environnement et droit

Robert Poujade disait "la politique de l'environnement s'exprime globalement, elle implique toutes les politiques mais ne se réduit à aucune d'elles".

L'environnement participe à la satisfaction contemporaine de l'intérêt général, qu'il s'agisse de protection de la nature, de la préservation des espèces, de la lutte contre les pollutions et les nuisances ou encore de la défense du cadre de vie.

Cette reconnaissance contribue par exemple depuis une vingtaine d'années à l'élaboration d'un droit particulier à l'environnement, encore jeune et qui évolue rapidement au fur et à mesure que la science met en évidence des problèmes nouveaux. Cette pluridisciplinarité de l'environnement a eu pour conséquence de rendre complexe ce droit ; j'ai donc décidé de lancer un grand chantier de rénovation et de modernisation de cette législation de l'environnement. Ce code de l'environnement a pour but de clarifier et de regrouper certains textes actuellement inclus dans des législations aussi diverses que le code minier. Il devrait être achevé à la fin de 1994 et présenté lors de la session parlementaire du printemps prochain.

2.4 Environnement et sociologie

Le poids des interrogations liées à l'environnement dans nos sociétés se mesure également aux antagonismes ou aux conflits qu'il suscite.

L'homme de la rue est souvent en contradiction avec lui même, à la fois automobiliste et piéton, consommant de plus en plus de produits de synthèse et les réclamant plus purs et naturels. Il est d'accord pour reconnaître la nécessité d'installations de traitement de déchets, mais les égoïsmes localisés font que chacun préfère que le projet se réalise chez l'autre, dans le jardin du voisin.

Il ne suffit plus d'élaborer des prescriptions techniques toujours plus sévères, mais il faut savoir dialoguer, expliquer. Mener à terme un projet résulte, aujourd'hui, d'une véritable stratégie qui dépasse largement l'aspect purement technique. Il faut mettre en œuvre une véritable pédagogie pour changer des comportements quotidiens ancestraux.

La concertation, l'information avec les élus, les associations de protection de l'environnement, les populations locales prend de plus en plus d'importance dans la réussite de l'aboutissement d'un projet.

3 Conclusion

On le voit donc bien, l'environnement n'a plus de frontière. Les problèmes de pollutions globales confirment la nécessité de trouver des accords à une échelle planétaire.

On savait que l'environnement était une science transversale, nécessitant des compétences techniques variées, il faut désormais prendre en considération sa composante sociale.

Si je souhaite que le ministère de l'environnement soit aujourd'hui plus respecté, consulté, écouté dans la vie politique, c'est bien parce que l'environnement est proche de nous et qu'il nous concerne tous.

Merci

Chambéry, Lundi 19 Septembre 1994.

Depuis cette conférence que je donnais alors en tant que Ministre de l'Environnement, les sujets abordés continuent à être d'une actualité reconnue par tous. Dans ce livre sur de nombreux aspects de l'environnement étudiés lors de la Conférence de Chambéry, CODATA présente des bilans recalés compte tenu des résultats actuels de la lutte menée pour préserver notre environnement.

Chapter 1

WATER IN ECOSYSTEMS
A non Renewable Resource

ON THE PROBLEMS OF ASSESSING THE WORLD'S WATER RESOURCES

John C. RODDA

President of the International Association of Hydrological Sciences, Institute of Hydrology, Wallingford , UK
Hon. Professor, Institute of Earth Studies, The University of Wales, Aberystwyth , UK
Formerly Director, Hydrology and Water Resources Department, World Meteorological Organization, Geneva, Switzerland

Institut of Hydrology, Wallingford, Oxfordshire, 0X10 8BB, U.K.

ABSTRACT

It is very strange that at the time when global demand for water is growing faster than ever, knowledge of the world's water resources is declining. How this knowledge has been and is being acquired is described. The systems that provide this knowledge are considered along with the errors involved. Work on reducing these errors, particularly the introduction of total quality management, is discussed. It is concluded that reliable assessment of the world's water resources is not possible currently.

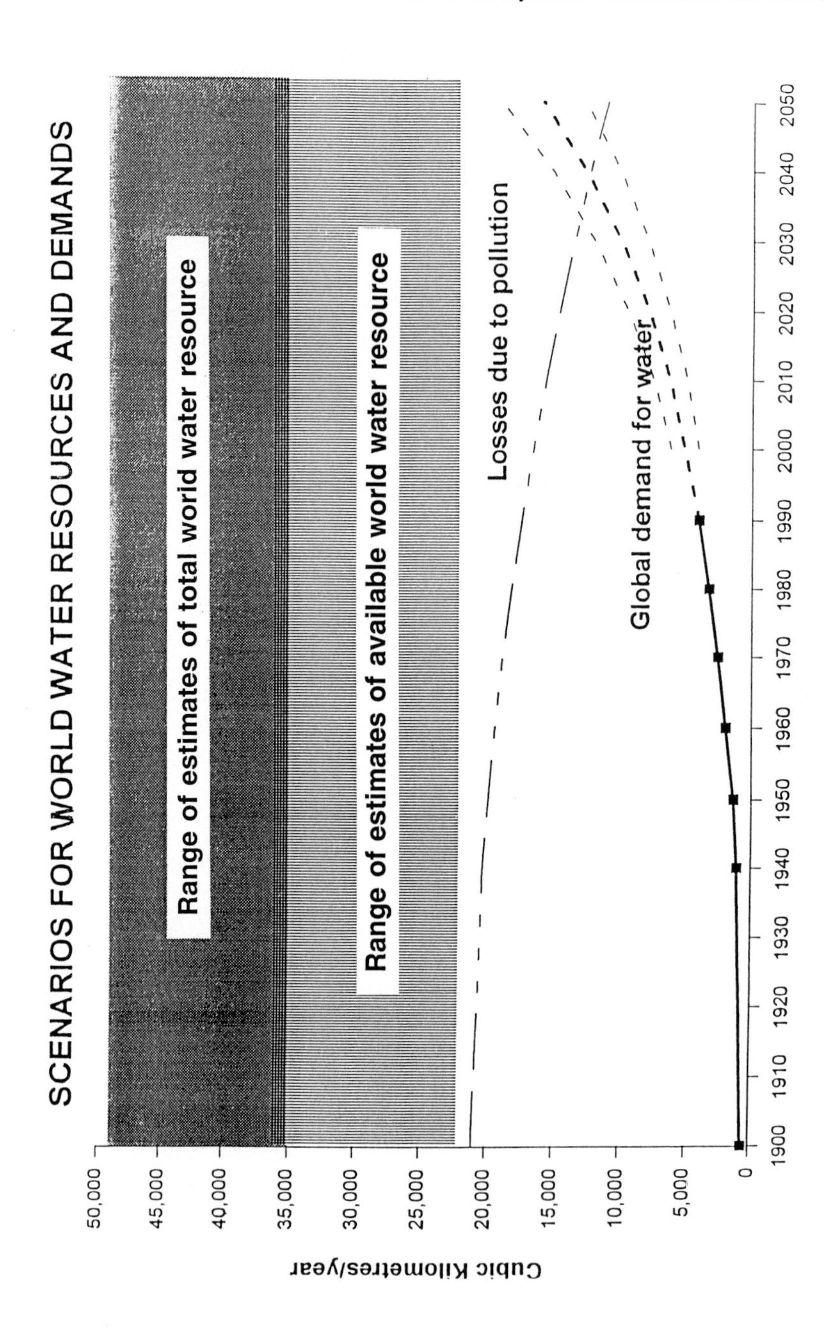

Figure 1. Scenarios for world water resources and demands

This situation presents problems to the bodies involved in their development and management and to those organizations who wish to provide the Commission for Sustainable Development and the United Nations General Assembly with an assessment of global water resources for their sessions in 1997.

Sur les Problèmes d'Evaluation des Ressources Mondiales en Eaux

RÉSUMÉ

Il est surprenant qu'à une époque où la demande globale en eau est en train de croître plus rapidement que jamais, les connaissances sur les ressources mondiales en eau sont en train de diminuer. On décrit comment ces connaissances ont été et sont actuellement acquises. Les systèmes qui produisent ces connaissances, sont particulièrement étudiés y compris les erreurs d'acquisition. Les travaux sur la réduction de ces erreurs, en particulier l'introduction de la qualité au niveau de la gestion, sont discutés. En conclusion, il est observé que l'estimation fiable des ressources en eau au niveau mondial n'est pas possible à l'heure actuelle. Cette situation pose problème aux organismes impliqués dans le développement et la gestion des ressources en eau, et aux organisations qui souhaitent fournir un bilan global des ressources en eau à la Commission du Développement Durable et à l'Assemblée Générale des Nations Unies pour leurs sessions en 1997.

1 INTRODUCTION

The demand for water for drinking, irrigation, power and for a host of other purposes continues to increase, bringing greater pressures on the world's freshwater resources ([1]). Some regions are currently experiencing considerable shortages principally because the margin between the finite resource and the demand is very small or non-existent. Droughts, such as those that recur in parts of Africa, reinforce these conditions, emphasizing the dangers of over-utilization of a dwindling resource.

This situation must worsen as the world population accelerates, consumption per head increases, pollution of surface water and groundwater worsens and climate change threatens. There are predictions that, with the doubling of the global population by the middle of the next century and the soaring demand, a world water crisis will develop ([2], [3]). Figure 1 shows how this scenario may arise, making the world's freshwater resources even more precious than today. Is the current situation in the basin of the Aral Sea a foretaste of things to come? What is worse, water resources could easily become a source of conflict in some of the globe's 200 international river basins. Then, in contrast to this scarcity, from time to time most parts of the world are subject to floods. Indeed, the toll of death and destruction due to floods is the largest of any of the different natural disasters ([4]). With the increasing population there will be an even bigger future target for floods, avalanches and landslides.

Variability and scarcity of water resources can handicap the progress of a developing country towards the sustainable development that the UN Conference on Environment and Development (UNCED) and Agenda 21 ([5]) aim to promote.

Table I. Approximate quantities of water in the various parts of the hydrological cycle with replacement periods. (from Young et al. [60]).

Category	Total volume (km^3 10^3)	% of total	% of fresh	Annual volume recycled (km^3)	Replacement period
Oceans	1338000	96.5		505000	2654 y
Groundwater to 2000m	23400	1.7		16700	1400 y
Predominantly fresh groundwater	10530	0.76	30.1	—	—
Soil Moisture	16.5	0.001	0.005	16500	1 y
Glaciers and permanent snow	24064.1	1.74	68.7	—	—
Antarctica	21600	1.56	61.7	—	—
Greenland	2340	0.17	6.68	2477	9700 y
Arctic Islands	83.5	0.006	0.24	—	—
Other Mountain areas	40.6	0.003	0.12	25	1600 y
Ground ice (permafrost)	300	0.022	0.86	30	10000 y
Lakes	176.4	0.013		10376	17 y
Freshwater lakes	91	0.007	0.26	—	—
Salt water lakes	85.4	0.006		—	—
Marshes	11.47	0.0008	0.03	2294	5 y
Rivers	2.12	0.0002	0.006	49400	16 d
Biological water	1.12	0.0001	0.003	—	—
Atmospheric water	12.9	0.001	0.04	600000	8 d
Total water	1385984.61	100[a]			
Total freshwater	35029.21	2.53	100[a]		

[a] Some duplication in categories and sub-categories.

They can also hinder developed countries in pursuit of this goal. But, of course, water encapsulates the dilemma of development or environment. Every scheme to provide a better water supply, to reduce the risk of flooding or to meet other human needs, modifies the hydrological cycle. They alter the transport of materials about the earth through the geochemical cycles. They also impact on the various living organisms that depend on the aquatic environment. They may also be changing the global patterns of transport of energy.
With these worries, it might be assumed that reliable information on freshwater resources and on the use of water would be readily available on a national basis, regionally and globally. But this is not so. For many nations, countrywide knowledge of the hydrological cycle is at best imprecise, while data on water use are in an even worse case. There are a number of reasons for this. Precipitation, evaporation, river flows and the storage of water in the soils, aquifers, lakes, glaciers and seasonal snow cover are not easy to measure. Neither are their chemical constituents. The acquisition and analysis of these data are rudimentary in many areas. The necessary government institutions are lacking and are badly equipped to meet their responsibilities in many others. In addition, there is to date no world-wide system for collecting and exchanging these data for forecasting and other purposes as there is for meteorological data through WMO's World Weather Watch. And while a few countries have working systems for collecting data on water use, most do not have them and figures for water use are unreliable. The consequence is that the statements of world water resources (Table I) and water use (Table II) are seriously in error.

Of course, at the level of the individual, these errors must be the concern of everyone who has or aspires to a wholesome and reliable water supply. At the global level, they concern the Commission for Sustainable Development and those bodies who will try to provide an assessment of global water resources to the Commission and the UN General Assembly for their sessions in 1997.
In the face of a coming water crisis, it seems imperative to improve knowledge of water resources. And without a proper assessment of water resources defined as the determination of the sources, extent, dependability of water resources for their utilization and control ([7]) there can be no integrated river basin planning and management while the pursuit of sustainable development will remain elusive. What is required is that governments accept the holistic approach to freshwater which they supported at the International Conference on Water and the Environment (Dublin, January 1992) ([6]) and at UNCED. The luxury of an 'à la carte' approach to water resources development and management is no longer tenable ([8]); the full menu of integrated river basin management must be taken.

2 Capturing Knowledge Of The Hydrological Cycle

Several theories are to be found in the early literature to explain how springs rise and rain forms. But with the birth of scientific hydrology ([9]) in 1674, following the publication of "De l'origine des fontaines" by Pierre Perrault ([10]), there was the first scientific evidence that atmospheric transport provides precipitation in sufficient quantities to cause rivers to flow. Perrault's study of the Coquille at Aignay le Duc, a headwater tributary of the Seine, and the contemporary work of Mariotte, on the basin of the Seine to Paris ([11]), (Table III) are the earliest assessments of the water balance of any basin.

Table II. Dynamics of water consumption in the world according to various kinds of human activity (Km^3/Years) (from Shiklomanov ([23])).

Water Uses	1900	1940	1950	1960	1970	1975	1980	1990	2000
Irrigated lands	47.3	75.8	101	142	173	192	217	272	347
Agriculture									
A	525	893	1 130	1 550	1 850	2 050	2 290 (68.9)	2 680	3 250 (62.6)
B	409	679	859	1 180	1 400	1 570	1 730 (88.7)	2 050	2 500 (86.2)
Industry									
A	37.2	124	178	330	540	612	710 (21.4)	973	1 280 (4.7)
B	3.5	9.7	14.5	24.9	38.0	47.2	61.9 (3.1)	88.5	117 (4.0)
Municipal supply									
A	16.1	36.3	52.0	82.0	130	161	200 (6.1)	300	441 (8.5)
B	4.0	9.0	14	20.3	29.2	4.2	41.1 (2.1)	52.4	64.5 (2.2)
Reservoirs									
A	0.3	3.7	6.5	23.0	66.0	103	120 (3.6)	170	220 (4.2)
B	0.3	3.7	6.5	23.0	66.0	103	120 (6.1)	170	220 (7.6)
TOTALS (rounded)									
A	579	1 060	1 360	1 990	2 590	2 930	3 320 (100)	4 130	5 190 (100)
B	417	701	894	1 250	1 540	1 760	1 950 (100)	2 360	2 900 (100)

Note A: Total water consumption B: Irretrievable water losses.
Percentage figures in parentheses.

Table III. Seventeenth century water budgets from the Perrault and Mariotte studies in the Seine basin (from Tixeront ([11])).

	Perrault - Aignay-le-Duc		Mariotte - Paris	
	From the Author	Present Real Values	From the Author	Present Real Values
Catchment Area (km^2)	118.8	93	53 500	44 320
P = rain per year (mm)	518	900	459	750
R = runoff per year:				
millions of m^3	9.5	31.5	3 574	8 600
mm / year	80	340	67	194
Deficit : P - R (mm)	438	560	392	556

Lingering doubts over the ability of the atmosphere to transport sufficient moisture were finally scotched by Dalton ([12]). He combined lysimeter-made measurements of evaporation with discharge estimates based on measurements carried out on the Thames and with a large number of records of rainfall, to determine the water balance for England and Wales.

During the 19th century there was an increase in the ability to measure the different hydrological variables and a gradual expansion of countrywide instrument networks. The skills of collecting and analysing hydrometeorological data were developed together with the organization of the earliest national hydrological and meteorological services. A few of these analyses were for purely scientific purposes, but an increasing number were aimed at the design and operation of various structures and schemes to cope with the increasing demand for water, to provide drainage and to alleviate floods. Water balance studies were important to a number of these purposes and to questions which began to be asked about the impact of human activities on the water balance, such as about deforestation. These studies involved the range of scales ([13]) pertinent to water as a resource; at one extreme, studies of the global water balance, at the other, studies of experimental plots and small basins.

3 Making A Water Balance

Probably the earliest example of a water balance study on the smallest scale started in the 1890s in the basins of the Sperbelgraben and Rappengraben in Switzerland ([14]). This investigation into the hydrological differences between forest and pasture was the first in a long series of paired basin studies ([15], [16]), such as Waggon Wheel Gap (USA), Coweeta (USA), Valdai (Russia), Hupselse Beek (Netherlands), and Plynlimon (UK), aimed primarily at determining the hydrological impact of land use differences and changes.

Table IV. Material transport from the continents to the oceans (from Walling ([30])).

CONTINENT	SUSPENDED SEDIMENT		DISSOLVED		SEDIMENT / DISSOLVED RATIO
	$(10^6 t\ y^{-1})$	$(t\ km^{-2} y^{-1})$	$(10^6 t\ y^{-1})$	$(t\ km^{-2} y^{-1})$	
Africa	530	35	201	13	2.6
Asia*	6 433	229	1 592	57	4.0
Europe	230	50	425	92	0.5
North and Central America	1 462	84	758	43	1.9
Oceania/Pacific Islands+	3 062	589	293	56	10.5
South America	1 788	100	603	34	3.0

*Mainland Asia, includes Eurasian Arctic.
+Includes Australia and the large Pacific Islands.

These studies were promoted by UNESCO's initial International Hydrological Decade (IHD) and its subsequent International Hydrological Programme (IHP), a number of the results being employed to assist in the development of the comparative approach to hydrology ([17]). More recently, these basin studies and the application of the results from them have been promoted in Europe by the IHP Friend Project ([18]) and in several other parts of the world by the series of similar projects the Friend concept has fostered. Utilizing the regional data bases it develops, Friend has two main objectives. The first is the application of these data bases to water resources problems, such as the estimation of extreme events for sites with no records. The second is region-wide assessment of the impact of human activities on the hydrological cycle.

Studies of the world water balance also seem to have commenced in the late nineteenth century: a series of examples being listed by Lvovitch ([19], [20]) and Baumgartner and Richel ([21]). More recently Korzun ([22]) and Shiklomanov ([23]) have assessed the global budget and its regional patterns, while the variations of its components have been summarized on a number of occasions, such as by Street-Perrott et al. ([24]). In recent years, knowledge of the processes that determine the water balance has been improving because of detailed experiments within basin studies and because of the series of large scale experiments on land/atmosphere interactions being conducted in different parts of the world ([25], [26]).

Now some of these experiments come under the International Geosphere Biosphere Programme (IGBP) core project on the Biological Aspects of the Hydrological Cycle ([27]). Others fall within the World Climate Research Programme, Global Energy and Water Cycle Experiment (GEWEX) ([28]), with its Continental Scale International Project, focusing on the Mississippi Basin ([29]).

The hydrological cycle provides the power to transport the materials in the different geochemical cycles; consequently measurements of the fluxes in the water balance globally and basin-wide are essential in determining the budgets of these different materials. Knowledge of runoff is needed to estimate the transported loads of sediment ([30]), carbon ([32]) and other determinants ([33]); Table IV shows estimates of the transport of material in suspension and solution from the land mass to the world ocean. Precipitation amounts must be known, in addition to dry deposition, in order to determine the loads being deposited from the atmosphere ([33]). A summary of the movement of water and material about the globe has been made by Berner and Berner ([34]).

4 Collecting Hydrological Data

Data on the water balance at the global scale, or for the smallest headwater basin, must be determined from measurements. Traditionally these measurements have been derived from networks of ground-based instruments, but now data are available in an increasing amount from weather radars and from satellite imagery. Unfortunately only a few countries employ these data routinely in assessments of water resources. This situation is not likely to improve as most future satellites are not designed with hydrological applications in mind.

Table V provides a summary of the statistics for the global hydrological instrument network, compiled from statistics on national networks ([35]), which in some cases, include the networks employed for research purposes, such as representative basin studies. These instruments and methods of observation are operated on a routine basis by the world's Hydrological Services who collect, analyze and apply the data from them. For many parts of the world and for certain hydrological variables, coverage is poor. For more limited areas, the networks are dense and most of the hydrological variables are measured, while much of the data produced appear to have the desirable characteristics of reliability, continuity and representativeness. However, the contrasts between the data rich and data poor areas of the world, and particularly the fact that over 70% of the globe, (namely the oceans), lacks measurements of precipitation and evaporation, rarely feature in comments on published global water budgets. Error, accuracy and precision are words that seem to be absent from most of these discussions. However, they appear more frequently in the reports of small basin studies. They should be, of course, the concern of those who operate these instrument networks and manage the data obtained from them on a regular basis. Agency-wide, national and international programmes aid this effort to assure quality. Certain initiatives on the international level aim to assist in quality assurance, as well as in making international data sets more readily available. There are, for example, the World Glacier Monitoring Service in Zurich, the Global Runoff Data Centre in Koblenz, the Global Precipitation Climatology Centre at Offenbach and the Collaborating Centre for Surface and Groundwater Quality at Burlington.

Table V. The global hydrological network (from Infohydro Manual, 2nd edition, ([35])).

VARIABLE	TYPE OF STATION	NUMBER OF STATIONS						
		AFRICA (RA I) (1)	ASIA (RA II) (2)	S.AMERICA (RA III) (3)	N.& C.AMERICA (RA IV) (4)	S.W. PACIFIC (RA V) (5)	EUROPE (RA VI) (6)	GLOBAL TOTAL
PRECIPITATION	Non-recording	17 036	39 456	19 247	19 973	15 276	40 367	151 355
	Recording	2 639	18 864	4 124	5 280	3 332	8 422	42 661
	Telemetry	8	1 916	211	1 023	515	459	4 132
	Radar	9	56	3	82	8	35	193
EVAPORATION	Pan	1 508	3 686	2 031	2 716	1 120	1 499	12 560
	Indirect method	374	7	40	11	1 049	488	1 969
DISCHARGE	Total*	5 703	11 543	7 924	13 211	5 838	19 798	64 017
	Non-recording	3 045	8 479	5 691	2 080	2 043	6 137	27 475
	Recording	1 856	3 064	2 233	11 128	3 795	13 661	35 737
	Telemetry	39	2 033	158	3 613	1 075	2 561	9 479
STAGE (WATER LEVEL)	Total*	3 410	6 405	5 872	11 274	1 167	10 474	38 602
	Non-recording	2 244	3 800	4 244	1 725	522	5 826	18 361
	Recording	877	2 300	1 628	9 549	642	4 599	19 595
	Telemetry	15	1 257	194	1 734	192	1 768	5 160
SEDIMENT DISCHARGE	Suspended	859	3 820	1 561	5 217	619	3 712	15 788
	Bedload	6	685	505	0	1	549	1 746
WATER QUALITY		5 297	5 045	2 752	31 462	1 690	55 379	101 625
GROUNDWATER	Water Level - Observation wells - Production wells	 4 884 31 804	 16 657 63 705	 1 133 14 150	 19 818 14 099	 18 585 13 504	 85 075 38 452	 146 152 175 714
	Temperature - Observation wells - Production wells	 287 243	 2 541 88	 5 200 5 539	 21 097 21 501	 4 888 888	 18 967 1 641	 52 980 29 900
	Water Quality - Observation wells - Production wells	 4 898 5 674	 1 964 45 187	 320 3 416	 13 757 14 825	 7 935 3 172	 14 889 23 711	 43 763 95 985

*The total includes stations not distinguished as "recording" and "non-recording"

Each of these global centres quality controls the data it acquires before archiving them and, in addition, as part of the GEMS Water Quality Programme ([36]), a considerable amount of help is provided to national services to improve and maintain the performance and standards of their analytical services for water quality. Unfortunately, the data held by these centres only cover some countries. The most recent are frequently two or three years old and time series are often short.

Lack of a readily accessible and reliable body of hydrological data for the globe has led WMO to initiate the establishment of a World Hydrological Cycle Observing System (WHYCOS), which would consist initially of about 1,000 stations located around the world on the major rivers ([37]). Each station would monitor about 15 variables, including flow and physico-chemical determinants of water quality, which would be transmitted via one of the geostationary satellites, such as Meteosat, to national, regional and global centres. These centres would build up archives of data over the period of operation of WHYCOS (at least 20 years) and process it to create tools for decision making, as well as for science.

These archives would be extensions to existing national archives and to those compiled for Friend purposes, as well as to the existing global data centres, such as the Global Runoff Data Centre. WHYCOS would also contribute to the Global Climate Observing System (GCOS) and to the Global Terrestrial Observing System (GTOS). More importantly, WHYCOS would seek to build up the capabilities of the Hydrological Services in those countries where networks, staff levels and facilities are in decline. This decline has been revealed in a number of recent studies, a decline which is most marked in Africa ([38], [39]). Because of this decline, an increase in the errors surrounding water balance estimates is to be expected. These errors would apply to most scales, basin-wide to global. Indeed, it is something of an enigma that at the time when global demand for water is rising faster than ever before, the errors in assessing just how much water is available for use are generally increasing. There seems to be the expectation that there will always be enough water in the well or to fill the reservoir.

In research on the water balance of small basins and for the process studies within them, the instrument systems are normally more advanced and more complete than for national networks. Toebes and Ouryvaev ([40]) provided an overview of observational and other practices for representative and experimental basins and there are many more recent reviews of experience in individual basin studies ([41], [42], [43]). Large-scale studies of hydrological processes such as FIFE, HAPEX-MOBIHY and BOREAS are relatively recent. Essentially, they couple simultaneous measurements of a number of variables on different scales in intensive field campaigns, measurements that may have previously been approached through different disciplines. These studies are described extensively in the literature, such as by Dozier ([44]) who considers their experimental design using ground-based measurements, aircraft and satellites.

5 THE MARGIN OF ERROR

Unfortunately for hydrologists, meteorologists and others involved, determining water resources needs measurements to be made in the natural environment where conditions are continually changing with time and where human activities now impose further modifications.

Table VI. Problems of assessing the water balance from point precipitation measurements

1.	Spatial coverage is often incomplete
2.	Temporal coverage is often incomplete
3.	There are at least 54 different types of standard gauge in use in 136 countries covering about 90% of the land area of the globe and in addition a large number of different types of rain recorders
4.	Errors of measurement have not been determined for each gauge type
5.	Installation of gauges and their sites may not meet the required practice
6.	Changes have occurred in gauge exposure
7.	Gauges have been moved
8.	New types of gauge have been introduced without comparisons with old versions
9.	Observer practice has altered
10.	Station histories are not documented

To sample these changes, and particularly the extremes they involve, the measurements are best made over long-time periods, the presence of the sensor should not alter the variable being observed and the instruments should be located at sites which are properly representative of the area or basin being sampled. Table VI indicates some of the problems of point precipitation measurement. For out-of-river variables, the representativeness problem has been eased somewhat by the advent of weather radars and satellites, through the images they provide of the fields of several variables. However, these images need measurements from strategically placed ground-based instruments for their calibration and interpretation. For the measurement of other variables, for example in taking soil water content and water quality measurements, the representativeness problem largely remains. Concentrations of constituents vary vertically and horizontally in a water body, and of course with time. Depth-integrated samples go some way to overcoming these difficulties and obviously the more samples taken, the more nearly the samples represent the whole ([45]). These difficulties have often led to very dense networks being established initially, to be later reduced as hydrological patterns become established.

These and other problems, such as those concerned with the storage and analysis of the data captured from the field, introduce errors into the measurements of the hydrological variables. Of course the error is strictly the difference between the result of a measurement and the true value of the quantity being measured ([46]). But for most hydrological variables, as for many other environmental variables, the true value is unknown and even a best approximation, such as might be obtained in the laboratory from a series of repeated measurements, is not often available. This has led to a preference for the use of the term uncertainty - the interval about the measurement within which the true value of the quantity can be expected to be with a stated probability ([46]). There are many different sources of error in hydrological measurements and Fig. 2 shows how these may arise in the practice of flow measurement with a weir or a flume. The overall uncertainty of measurement X_Q ([47]) depends on the standard of construction of the gauging structure, the correct application of the design specifications and a number of other factors and X_Q will also vary with discharge.

Herschy ([48], [49]) discussed the errors of the various methods of flow measurement drawing on material published by the International Organization for Standardization (ISO) and also by WMO. He concluded that if ISO standards are followed, then flow measurements made at a single gauge are expected to have the following upper limits of uncertainties at the 95% confidence limits:

Single determination of discharge	7%
15 minute average of discharge	5%
Daily mean, monthly mean or annual discharge	5%

It would be most unwise to assume that river flows generally are measured to these limits: many countries are not able to meet ISO Standards ([50]), nor the requirements of WMO ([51]). Neither do the measurements of the other variables satisfy the WMO recommendations for accuracy (Table VII) that have been adopted by the WMO Commission for Hydrology. Unfortunately, these are desired accuracies rather than those being achieved at present: indeed the actual measurement errors must be somewhat larger than those stated. Take precipitation as an example: errors can range from 5 to 30% for rain and 10 to 80% for snow depending on gauge type, site, wind speed and other factors.

6 Reduction Of Errors Through WMO Activities

WMO has been working in various ways to combat errors in instrument practice and in the errors which can occur at later stages in data management. One of the main vehicles for this work has been the series of instrument intercomparisons that deal with the measurement of the main hydrological variables.

The initial comparison of liquid precipitation measurements selected one particular reference gauge for the tests undertaken in the 1950s. When the problems of this gauge were recognized, a different reference was chosen for the second intercomparison carried out in the 1970s ([52]). For the third intercomparison, which is of devices for measuring solid precipitation, and which is concluding at the present time, a third interim reference was employed ([53]). For these tests reference instruments were installed at a large number of different sites around the world for comparison with the various national gauges. The results of the different intercomparisons show how national gauges performed against the reference.

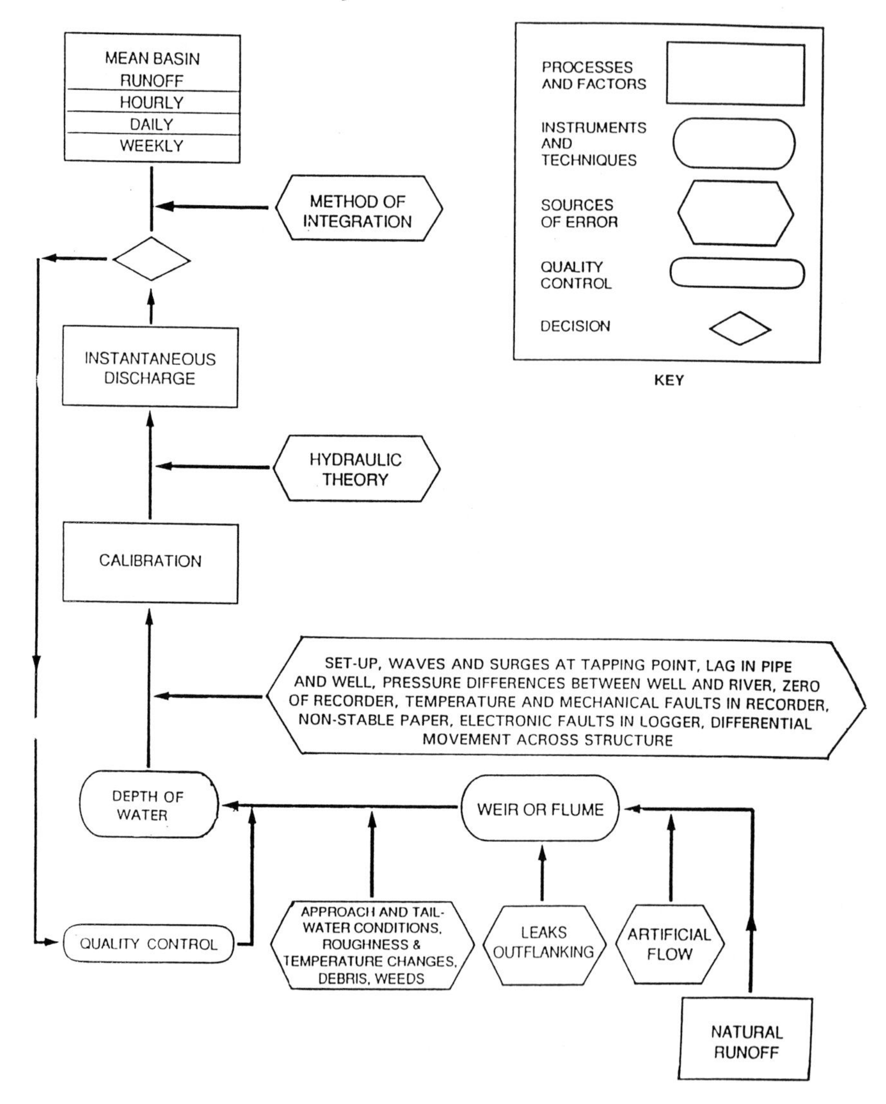

Figure 2. Source of errors in the measurement of flow by weir or flume.

For the comparison of evaporation measurement and estimation which was conducted in the 1960s ([54]), a large amount of data was collected from national tests and by means of a questionnaire. For Phase 1 of the intercomparison of water level recorders and current meters, tests were conducted by the several nations involved of their own instruments, while several other instruments were tested against detailed test specifications which had been previously agreed ([55]). For Phase 2 of this intercomparison, which also included data loggers, suspended sediment samplers and several categories of water level and velocity sensors, the US Geological Survey made available a number of US P61 sediment samplers to participating countries for tests against national samplers ([56]).

Unfortunately, many different interests mitigate against the formulation of the results of these intercomparisons in the manner of a consumer report on the `best buy'. So no clear preferences for devices of particular types are stated in the conclusions. Nevertheless, for the discerning reader, it is usually obvious which of these instruments performed most closely to the specifications of the tests, but it was not found possible to include in the results a statement of the errors of measurement for the variables concerned. However, the tests themselves, through the transfer of technology they promoted, help to raise the standard of performance of the Hydrological Services involved, which in turn assist in error reduction in the long term. The same applies to the transfer of technology facilitated by HOMS ([57]) (the Hydrological Operational Multipurpose System), the WMO technology transfer system which achieved more than 3000 exchanges of components between hydrologists in developed and developing countries between 1981 and 1995. But neither as a result of these comparisons, nor in HOMS does WMO, nor any other UN organization, advocate the use of a particular type of instrument, so that observations might be made in a uniform way world-wide. If this were the case, and all the subsequent data management procedures were also harmonized, then there would be fewer difficulties in ascribing errors to the estimates of water balances. This type of observational homogeneity is what the GEMS Water Quality Programme ([36]) aims to achieve. It is what has been achieved by certain national programmes in water quality, for example: in the United States National Water Information System ([45]).

7 TOTAL QUALITY MANAGEMENT

Perhaps the time is approaching when similar initiatives should be adopted globally so as to embrace the concept of total quality management which is embodied in the ISO 9000 series of standards ([58]). These offer an integrated global system for optimizing the quality effectiveness of a company or organization. By adhering to defined standards, an organization ensures that its product is "fit for the purpose for which it is intended", a hydrometeorological data set being one such product.

This is the approach that has been adopted by New Zealand's National Institute for Water and Atmosphere (NIWA) ([59]) for its environmental data activities, including hydrological data. The reasons given were:

Table VII. Recommended accuracies (uncertainty levels) expressed at the 95 per cent confidence interval for different variables* (from WMO guide to hydrological practices ([46])).

Variable	Recommended accuracy
Precipitation (amount and form)	3-7%
Rainfall intensity	1 mm/h
Snow depth (point)	1 cm below 20 cm or 10% above 20 cm
Water content of snow	2.5-10%
Evaporation (point)	2-5%, 0.5 mm
Wind speed	0.5 m/s
Water level	10-20 mm
Wave height	10%
Water depth	0.1 m, 2%
Width of water surface	0.5%
Velocity of flow	2.5%
Discharge	5%
Suspended sediment concentration	10%
Suspended sediment transport	10%
Bed-load transport	25%
Water temperature	0.1-0.5^{o}
Dissolved oxygen (water temperature is more than 10°C)	3%
Turbidity	5-10%
Colour	5%
pH	0.05-0.1 pH unit
Electrical conductivity	5%
Ice thickness	1.2 cm, 5%
Ice coverage	5% for _ 20 kg/m^3
Soil moisture	1 kg/m^3 _ 20 kg/m^3

* NOTE: When a range of accuracy levels is recommended, the lower value is applicable to measurements under relatively good conditions and the higher value is applicable to measurements in a difficult situation.

1. to eliminate the costs of poor quality control;

2. to ensure that the product meets the user's needs - that the data are confidently usable;

3. to demonstrate to users that the Institute is meeting and anticipating their needs.

The approach incorporates adherence to manuals and procedures and includes certification by an independent body accredited to carry out the inspection work. NIWA implemented new procedures and upgraded instruments and equipment to meet the necessary requirements in order to achieve the standards of quality claimed for them.

8 CONCLUSION

The errors in the majority of assessments of water resources, at virtually every scale, are usually disregarded. The sizes of these errors are unknown for most national assessments, except where total quality management has been implemented, such as in New Zealand.

Although several global systems for hydrological data management have been established, they suffer from a number of problems which may be overcome with the advent of WHYCOS. WHYCOS, when fully implemented, should provide quality assured data with a global coverage.

Without such an approach, future assessments of the water balance are handicapped by the unknown errors that they contain. This is not a desirable position to be in when a world water crisis is looming. In 1997 the United Nations should recognize that reliable data are the key to the sustainable development and management of water resources and endorse the steps being taken to provide them.

ACKNOWLEDGEMENTS

The author would like to express his considerable thanks to his colleagues in the WMO Hydrology and Water Resources Department for their help in the preparation of this paper. He would also like to thank the observers and others who, over many years, have made available the data on which this paper is based.

KEYWORDS

Water balance, global water resources, groundwater resources, hydrological cycle, hydrogeological data, quality control, management, WMO activities, decision-making, anthropogenic effect.

REFERENCES

[1] Ayibotele N.B., The world's water: assessing the resource. International Conference on Water and the Environment, Keynote Papers, 1.4-2.6 (1992).

[2] Falkenmark M. and Lindl G., Water for a Starving World, Westview Press Inc. (1976).

[3] Postel S., The Last Oasis, Worldwatch Environmental Alert Series (1993).

[4] World Conference on Natural Disaster Reduction, Disasters around the World - A Global and Regional View. Information Paper No. 4 (1994).

[5] United Nations, Earth Summit: Agenda 21, The United Nations Programme of Action from Rio (1992).

[6] UNESCO/WMO International Glossary of Hydrology, 2nd Edition, UNESCO/World Meteorological Organization (1992).

[7] International Conference on Water and the Environment, The Dublin Statement and Report of the Conference, 26-31 January, Dublin Ireland (1992).

[8] Rodda J.C., No more 'Water à la Carte', Information Development, 10(2), 71 (1994).

[9] Three centuries of scientific hydrology: Key papers from the celebration of the Tercentenary of Scientific Hydrology, UNESCO/WMO/IAHS (1974).

[10] Perrault P., De l'origine des fontaines, Pierre le Petit, Paris (1674).

[11] Tixeront. L'hydrologie en France au XVIIe siècle, Three centuries of scientific UNESCO/WMO/IAHS , 24-25 (1974).

[12] Dalton J., Experiments and observations to determine whether the quantity of rain and dew is equal to the quantity of water carried off by the rivers and raised by evaporation, with an enquiry into the origin of springs. Mem. Lit. and Phil. Soc. Manchester, V Pt II, 346-372 (1802).

[13] Dooge J.C.I., Water of the Earth, Hydrological Sciences Journal, 29, 149-176 (1984).

[14] Engler A., Einfluss des Waldes auf den Stand der Gewasser, Mitt Schweiz anst für das Forslich Wersuchswesen, 12, 626 (1919).

[15] Rodda J.C., Basin studies: in Facets of Hydrology (Rodda J.C., Ed.), John Wiley, 257-297 (1976).

[16] Robinson M. and Whitehead P.G., A review of experimental and representative basin studies: in Methods of Hydrological Basin Comparison (Robinson M. Ed.), Proc. Fourth Conf. of the European Network of Experimental and Representative Basins, Institute of Hydrology, Report No. 120, 1-12 (1993).

[17] Falkenmark M. and Chapman T., (Eds). Comparative Hydrology, UNESCO (1989).

[18] Roald L., Nordseth K. and Hassel K.A., (Eds). Friends in Hydrology, Proc. 1st International Frend Symposium, International Association of Hydrological Sciences, Pub. No. 187 (1989).

[19] Lvovitch M.I., World Water Balance (General Report). Symposium on the World Water Balance. International Association of Hydrological Sciences, Pub. No. 93, VolII, 401-415 (1970).

[20] Lvovitch M.I., World Water Resources and Their Future (English edition translated by R.L. Nace). American Geophysical Union (1979).

[21] Baumgartner A. and Reichel E., Preliminary results from a new investigation of world's water balance, Symposium on the World Water Balance. International Association of Hydrological Sciences, Pub. No. 93, Vol. III, 580-592 (1970).

[22] Korzun V.I., (Ed. in Chief). World Water Balance and Water Resources of the Earth, USSR National Committee for the IHD Gidromet (1974).

[23] Shiklomanov I.A., The World's Water Resources in: Proc. International Symposium to Commemorate 25 years of the IHP, UNESCO/IHP, 93-126 (1991).

[24] Street-Perrott A., Beran M. and Radcliffe R. (Eds), Variations in the Global Water Budget, Reidel, 518 (1983).
[25] Shuttleworth W.J., Macrohydrology - A new challenge for Processes Hydrology. Journal of Hydrology, 100, 31-56 (1988).
[26] Shuttleworth, W.J. Insight from large-scale observational studies of land/atmosphere interactions. Reviews of Geophysics, 12, 3-30 (1991).
[27] Bolle J.J., Feddes R.A. and Kalma J.D., Exchange processes at the land surface for a range of space and time scales. International Association of Hydrological Sciences, Pub. No. 212, 626 (1993).
[28] ICSU/WMO. Scientific Plan for the Global Energy and Water Cycle Experiment WCRP-40, WMO, TD-No. 376, 84 (1992).
[29] ICSU/WMO. Scientific Plan for the GEWEX Continental Scale International Project (GCIP), WCRP 67 WMO, TD-No. 461, 65 (1992).
[30] Walling D.E. and Webb B.W., Material transport by the world's rivers: evolving perspectives in Water for the Future (Rodda, J.C., and Matalas, N.C. Eds), International Association of Hydrological Sciences, Pub No. 164, 313-329 (1987).
[31] Degens E., Kempe S. and Ittekkot V., Monitoring carbon in world rivers Environment, 26, 29-33 (1984).
[32] Meybeck M., Carbon, Nitrogen and Phosphorus Transport by world rivers, American Journal of Science 282, 401-450 (1982).
[33] Goodison B.E. and Vet R.J., Precipitation data compatibility in North America and the impact of studies on acid deposition: in Acid Deposition, International Association of Hydrological Sciences, Pub No. 179, 47-55 (1989).
[34] Berner E.K. and Berner R.A., The Global Water Cycle-Geochemistry and Environment. Prentice Hall, 397 (1987).
[35] WMO. INFOHYDRO Manual, 2nd Edition. Operational Hydrology Report, World Meteorological Organization (1994) (in press).
[36] WHO. GEMS/Water 1990-2000. The Challenge Ahead. UNEP/WHO/UNESCO/WMO Programme on Global Water Quality Monitoring and Assessment, WHO/PEP 2 (1991).
[37] Rodda J.C., Pieyns S.A., Sehmi N.S. and Matthews G., Towards a world hydrological cycle observing system, Hydrological Sciences Journal, 38, 373-378 (1993).
[38] WMO/UNESCO. Report on water resources assessment, World Meteorological Organization/UNESCO, 64 (1991).
[39] UNDP/WB. Reports of the Sub-Saharan Hydrological Assessment Project (42 regional & national reports) United Nations Development Programme/World Bank (1990-1994).
[40] Toebes, C., and Ouryvaev, V., (Eds.). Representative and experimental basins: an international guide for research and practice, Studies & reports in hydrology 4, UNESCO, 348 (1970).
[41] IAHS. The influence of man on the hydrological regime with special reference to representative and experimental basins, International Association of Hydrological Sciences, Pub. No. 130, 483 (1980).
[42] Swanson R.H., Bernier P.Y. and Woodard P.D., (Eds). Forest Hydrology and Watershed Management, International Association of Hydrological Sciences, Pub. No.167 (1987).
[43] Lang H. and Musy A., (Eds). Hydrology in Mountainous Regions 1 - Hydrological Measurements; the Water Cycle, International Association of Hydrological Sciences, Pub. No. 193 (1990).
[44] Dozier J., Opportunities to improve hydrologic data, Reviews of Geophysics 30, 315-331.
[45] USGS. National Water Summary 1990-91. Hydrologic Events and Stream Water Quality, United States Geological Survey, Water Supply Paper No. 2400 (1993).
[46] WMO. Guide to Hydrological Practices, 5th Edition. World Meteorological Organization (1994).
[47] Ackers P., White W.R., Perkins J.A. and Harrison A.J.M., Weirs and Flumes for Flow Measurement, John Wiley & Sons (1978).
[48] Herchy R.W., Accuracy in: Hydrometry: Principles and Practices (Herschy, R.W., Ed). John Wiley & Sons, 353-397 (1978).
[49] Herschy R.W., Streamflow Measurement, Elsevier, 553 (1985).

[50] ISO. Measurement of liquid flow in open channels, International Organization for Standardization Handbook, 16, 518 (1983).
[51] WMO. Technical Regulations 1988 Vol. III - Hydrology Supplement World Meteorological Organization (1992).
[52] Sevruk B. and Hamon W.R., International Comparison of national gauges with a reference pit gauge, World Meteorological Organization Instruments and Observing Methods Report No. 17 (1984).
[53] Goodison B.E., Elomaa E., Golubev V., Gunther T. and Sevruk B., WMO Solid Precipitation Measurement Intercomparison: preliminary results. World Meteorological Organization Instruments and Observing Methods, Report No. 57, 15-19 (1994).
[54] WMO. Measurement and Estimation of Evaporation and Evapotranspiration WMO Technical Note No. 83, 121 (1966).
[55] Starosolszky O. and Muszkalay L., Intercomparison of principal hydrometric instruments, World Meteorological Organization Commission for Hydrology (CHy), unpublished report (1978).
[56] Wiebe K., Second phase - intercomparison of principal hydrometrical instruments, WMO Technical Reports in Hydrology and Water Resources No. 43 (1994) (in press).
[57] Miller J.B., Making the software available: WMO experience in technology transfer; in Hydroinformatics '94 (Verwey, A., Minns, A.W., Babovic, V., and Maksomovic, C., Eds.), Institute of Hydraulic Engineering, Delft, 437-442 (1994).
[58] ISO. International Organization for Standardization, 9000 News 2, (1993).

[59] Mosley M.P. and McMillan D., Quality Assurance for Hydrological Data, World Meteorological Organization Bulletin 43, 29-35 (1994).
[60] Young G.J., Dooge J.C.I. and Rodda J.C., Global Water Resources Issues. Cambridge University Press, 194 (1994).

System for the Management of Agricultural Pollution : EUREKA-ISMAP Project

Adel ABOUJAOUDE[a], *Jean-Marc USSEGLIO-POLATERA*[a] *and Thierry VANDEVELDE*[b,1]

[a]*Laboratoire d'Hydraulique de France, 6 rue de Lorraine, 38130 Echirolles, France*

[b]*Compagnie Générale des Eaux, 32 place Ronde, 92982 Paris La Défense, France*

Abstract

Industrial and public partners from Italy, France and the United Kingdom, involved at various levels in the problems posed by agricultural pollution (water suppliers, chemical manufacturers, environmental administrations, farmers, research institutes, Information Technology companies), are currently collaborating in the ISMAP project within the EUREKA framework. ISMAP stands for Integrated System and Services for the Management of Agricultural Pollution. Within ISMAP, the development of new approaches, methodologies and tools to improve the management of agricultural pollution is carried out through a close collaboration of specialists in various disciplines (e.g. agronomy, hydrology, biochemistry, socio-economy, information technologies).

Information et Système de Décision pour la Gestion des Polluants Agricoles: le projet EUREKA-ISMAP

Résumé

Différents partenaires industriels et publics venant d'Italie, de France et du Royaume-Uni, impliqués à divers niveaux dans les problèmes posés par la pollution agricole (les fournisseurs en eau, les fabricants de produits chimiques, les administrations environnementales, les agriculteurs, les instituts de recherche, les compagnies de Technologies de l'Information) collaborent régulièrement au projet ISMAP dans le cadre d'EUREKA. ISMAP signifie Système Intégré pour la Gestion de la Pollution Agricole. Le développement de nouvelles approches, de méthodologies et d'outils pour améliorer la gestion de la pollution agricole est mené à terme au sein de l'ISMAP à travers une collaboration étroite entre spécialistes de diverses disciplines (à savoir l'agronomie, l'hydrologie, la biochimie, la socio-économie, les technologies de l'information).

---

[1] To whom correspondence may be addressed

1 INTRODUCTION

In recent decades, thanks to the application of fertilizers and pesticides, farm yield and agricultural product quality have been significantly improved. However, although they are essential to a modern and profitable agriculture, fertilizers and pesticides may contaminate the environment, especially soil and water resources, when accidentally released or intensively used. There is an urgent need to elaborate a reasonable trade-off between water resource protection requirements and agricultural efficiency and profitability.

Due to a variety of agricultural techniques, land uses, applied chemicals and to the extreme variability of geographical, geological and hydrometeorological conditions in time and space, the problem of agricultural pollution control is extremely intricate. A clear assessment and a reliable prediction of contamination risks are still very difficult for decision makers. Furthermore, applicable regulations often lack scientific basis and become controversial.

It is increasingly difficult for drinking water suppliers to meet the current standards. The agrochemical industry is deeply concerned and public opinion and media are increasingly receptive to possible threats. The environmental and economic stakes are high, and the major parties involved in the problem must collaborate.

2 MAJOR CONCERNS AND NEEDS

Reasonable management of agricultural pollution for a sustainable protection of water resources is a complex question.

First, it is necessary to be able to detect chemical residues in surface waters or groundwater. This is currently possible for fertilizers, but there is still a lack of analytical methods for pesticides. Furthermore, when available, analytical multi-residue methods are expensive and difficult to implement, leading to spurious applications and results.

Since these chemicals have been used for years, residues can be present in surface water and groundwater resources currently exploited for drinking. It is already possible to eliminate nutrients through existing treatment procedures. But it is still necessary and urgent to elaborate efficient water treatment procedures for various types of pesticide residues.

In addition to measurement techniques, control of agricultural water pollution on various management scales (plot, farm, small agricultural area, hydrological catchment) requires understanding and decision support tools and methods.

The environment is a delicately balanced whole of a multitude of complex and interacting components. There is no single solution to maintain agricultural pollution at an acceptable level. Various types and spatial distributions of corrective measures must be investigated (e.g. grassed buffer strips, riparian woods, spatial crop distribution, agricultural practices, chemical applications, ...).

Mathematical models can provide powerful support to understand the situation and to forecast possible evolution [2]. However, there are hundreds of models available elaborated by the scientific community, often with competing modeling techniques, limited to specific applications, based upon specific hypotheses, requiring specific data sets.

Consequently, it is very difficult for decision makers to have a clear view in this jungle, and modeling has become very controversial and too poorly exploited for management purposes.

Agricultural pollution concerns extensive geographical areas and data availability is a major preoccupation. The natural environment is extremely heterogeneous, and there are never enough observation results available to characterize this spatial variability. In addition, as mentioned, pesticide detection is very expensive and, whatever the effort, the data available often remains inadequate. Therefore, it is necessary to process these scarce data in an optimal way in order to draw reasonable conclusions. It is also very important to pool all existing knowledge and skill to understand the interacting phenomena, to reliably apply existing tools or to suggest solutions.

The solution seeking process for a contamination problem on a large scale (catchment) is complex. It must integrate sampling strategy, analytical capabilities, understanding of relevant natural/anthropogenic phenomena interactions, predictive testing of remediation decisions, existing know-how or previous experience in similar problems.

This is clearly necessary to elaborate an overall and balanced methodology, adapted to each class of problem, so as to guide the local manager through this intricate maze.

Recent information technology developments make it possible to integrate in computerized user friendly systems various types of complementary tools such as data bases, knowledge bases and predictive models, and it is a major objective of the 90s to develop this kind of tool for environmental applications.

3 ISMAP Project Objectives

To address all the above listed concerns and needs, industrial and public partners from Italy, France and the United Kingdom, involved at various stages in agricultural pollution, are currently collaborating in the ISMAP project within the EUREKA framework [5].

ISMAP stands for Integrated System and services for the Management of Agricultural Pollution. ISMAP involves a water supplier (Compagnie Générale des Eaux), chemical manufacturers (Rhône-Poulenc, Ciba-Geigy, Grande Paroisse), environmental administrations (Agences de l'Eau in France and ENEA in Italy), farming industries (Gruppo Ferruzzi), research institutes and Information Technologies companies (A.I.TEC, LHF, ITCF and CEMAGREF). A.I.TEC, representing Gruppo Ferruzzi, is appointed as project coordinator, Compagnie Générale des Eaux and LHF co-coordinate the French contribution.

This project involves close collaboration of specialists in various disciplines (e.g. agronomy, hydrology, biology, chemistry, socio-economy, information technologies).

ISMAP started in September 1991 by a Definition Phase aimed at a comprehensive state-of-the-art review and at technical specifications of the planned developments. Phase 2 started in October 1993 and is expected to be completed early in 1997 with intermediate operational results. The program of work in Phase 2 is presented below.

4 ISMAP DEVELOPMENT PROGRAM

4.1 Pesticide detection and field water quality measurements

ISMAP is currently identifying, developing and validating multi-residue analytical methods for pesticide detection that could be routinely used with greater efficiency and adequate reliability. A preliminary selection of about twenty pesticides was established for investigation; this selection is based on application quantities, dispersion potential, toxicological and ecotoxicological data and referenced contamination cases.

In addition, ISMAP will address the problem of analysis quality insurance and will try to establish evaluation criteria for laboratories involved in water quality analysis.

ISMAP involves partners with different (even apparently competing) interests in agricultural pollution concerns. Prepared for cooperation by several months of collaborative work, this group intends to contribute to preparing standards and regulations at national and European levels.

4.2 Water treatment procedures

Based on the priority list of pesticides established for analysis, water treatability tests are being developed and validated through laboratory experiments on pilot facilities and in water treatment plants.

Seven treatability tests were identified in the Definition Phase and have already been tested in the laboratory : coagulation-flocculation, biodegradation, stripping, photolyze, adsorption on activated carbon, oxydation with or without hydrogen peroxyde.

A collection of technical documents describing the behaviour of each pesticide and the performance and efficiency of treatment procedures for each pesticide will be established. A knowledge base is being set up in order to support treatment plant operators in selecting and implementing appropriate water treatment procedures.

4.3 Decision-support methodology

The decision-support methodology being developed in ISMAP is twofold :

The first component is a knowledge base aimed at helping the end-user to understand all the interacting aspects of a given problem on a given site: in-depth phenomenological characterization of problems and spatial characterization of application sites. This knowledge base will consist of a collection of information documents, interactively accessible, about the relevant environmental objects and their attributes (parameters), the basic processes (physical, chemical, biological, socio-economic), the principles of interactions (data processing and modeling techniques) and the spatial distribution of georeferenced data.

The second component of the decision-support methodology is a catalogue of procedures, defined on a problem by problem basis, characterizing the overall integrated approach of ISMAP. A preliminary list of typical classes of problems related to agricultural pollution has been established.

Procedures should consist of:

- guidelines and methodologies for data acquisition and monitoring;
- guidelines and methodologies for environmental and/or socio-economical impact assessment;
- guidelines for water treatment;
- advisable sources of information and knowledge;
- catalogue of possible solutions and remedial measures to be investigated.

This decision-support methodology is called ISMAP/ERM (Environmental Reference Methodology). It will be computerized and integrated in the Information and Decision Support System described hereafter.

4.4 Information and Decision-Support System

A computerized framework for information, data and knowledge management was designed. This system consists of the following components:

- information bases for general site-independent data (regulations and standards, characteristics of chemicals, references, information on external data bases, ...);
- information management framework for site applications: historical data bases (hydro-meteorology, hydrology, water quality, socio-economy, ...), forecast data bases (model results), Geographical Information System (thematic maps), general site-dependent information bases (site description, references, images,...);

- data / information processing tool kit including graphical and statistical tools;

- knowledge bases (methodologies, problem-oriented guidelines and educational tools).

A simulation/forecasting system including a library of complementary predictive modeling tools and a framework for model application support is being developed. The modeling systems library will encompass various kinds of models adapted to various processes (water movement, sediment transport, fate of contaminants, socio-economy, ...), to various physical compartments (groundwater, surface waters, overall continental water cycle, ...), with competing but complementary techniques (deterministic or stochastic, mechanistic or conceptual,...).

The computerized framework for information, data and knowledge management and the simulation/forecasting system will be integrated in an overall Information and Decision-Support System (ISMAP/IDSS). The content of the IDSS is presented in more detail in [1] and [4].

An originality of ISMAP is that, in addition to model developments, the emphasis will be placed on the following components of environment modeling:

- elaboration of guidelines for sound applications of models in accordance with relevant classes of operational objectives;

- advisable modeling strategies (appropriate scenarii and spatial descriptions) aimed at rationalizing model exploitation;

- model calibration and validation procedures;

- tentative assessment of results, errors and uncertainties induced by data scarcity and model inadequacies.

The development of appropriate methodologies for vulnerability and risk mapping is under way. The appropriate tools will be available in the Information and Decision Support System, and the methodologies will be integrated in a knowledge base [3].

4.5 Test-site applications

The methodologies and tools developed under ISMAP will be tested on fifteen test-sites located in France and Italy. These sites cover various aspects of the problem (chemicals, agricultural practices, site conditions), various space scales (from farm to watershed), various objectives (water resource protection, research investigations).

A Steering Committee has been constituted on each test site, involving local authorities and farmers in order to define and follow the local operational objectives of ISMAP tests and demonstrations.

5 ISMAP AND DECISION-MAKING

5.1 ISMAP end-users

ISMAP end-users typically are local/regional authorities and national/regional agencies in charge of water resource protection. Recent laws (e.g. "Loi sur l'Eau 1992" in France) prescribe a sustainable management of water resources at regional and local scale in E.C. countries. Appropriate tools, technologies and methodologies to prepare protection plans and to follow them up in the coming years are expected. These legal orientations imply a significant evolution in the capabilities of tools. ISMAP will contribute to bridge the gap between conventional methods and current and future requirements, based on a sound valorization of state-of-the-art knowledge.

It must be clearly stated that, given the complexity of the problem and the diversity of situations, it is not realistic to envisage a comprehensive computerized Decision-support system accessible to all type of users. Expertise will be required, and the end-user here is the user of results and not the direct operator of the tool.

Farmers are also potential end-users of ISMAP results in a different way. Farming models taking into account socio-economic balances and environmental consequences should be used for large farm exploitations (e.g. Po Valley in Italy).

5.2 ISMAP decision-support contribution

ISMAP has no intention of substituting for the decision-maker. The role of the IDSS is to provide a synthetic vision of a given problem and to assist the user in investigating sound elements of solutions to this problem so that a reasonable decision can be made, taking into account the relevant aspects of the problem.

Some possible applications of the Information & Decision Support System are mentioned hereafter:

- Valorization of existing information
    - direct availability and easy retrieval of existing general information
    - easy retrieval of existing local data (environmental system)
    - reasonable spatial interpretation of existing scarce data
    - characterization of timewise evolutions of environmental indicators
    - record of local background and past events (storage of knowledge)
    - education support based on recorded events or educational tools
    - didactic presentation of information

- Support of every day management (follow-up)
    - control board of current situation (contamination level of resources)
    - common platform for dialogue between actors in case of conflict
    - vulnerability and risk assessment and mapping
    - identification of sources of observed pollutions
    - information on possible approaches for problem solving
    - centralized access to supervision and warning systems
    - communication support for the media and public opinion
    - technical support of emergency intervention (accidental pollution)

- Support of predictive management
    - availability of pre-calibrated predictive models
    - forecasting the trend of contamination evolution
    - investigation, intercomparison and tests of remediation measures
    - investigation, intercomparison and tests of land use planning scenarii
    - investigation, intercomparison and test of agricultural practices
    - rationalization of sampling & measurement (cost, localization, methodology)
    - rationalization of future monitoring systems

5.3 Regulations and standards

A major objective of ISMAP has become the elaboration of appropriate and reliable tools and methods in order to contribute to the definition of regulations and standards at national and European levels.

Concerning analytical methods, the French partners are already involved in working groups under AFNOR, the French Standards Institute.

Concerning predictive modeling and risk assessment, the application of models in regulatory setting is current in the U.S. [2]. However, the selection of appropriate modeling tools and especially their conditions of application requires in-depth reflection. ISMAP partners are ready to address this question.

6 CONCLUSION

ISMAP partners have obtained financial support to achieve the programme presented in this paper. Their experience in the relevant disciplines and the promising results of three years of collaborative work are clear evidence of possible success for this ambitious and innovative programme. ISMAP results and progress will be regularly published by the partners.

7 ACKNOWLEDGEMENTS

The developments presented in this paper are carried out in the framework of the EUREKA ISMAP Project (EU 479). The Italian contribution to ISMAP is partially supported by I.M.I., and the French contribution is partially supported by the French Ministry for Research (Phase 1 and Phase 2 partly) under EUREKA budgets.

KEYWORDS

Agriculture, pollution, ISMAP, Eureka, pesticides, water quality, mathematical model, water treatment, decision-making, European Community

REFERENCES

[1] Remotti, D., Gatti, S., Aboujaoudé, A., Usseglio-Polatera, J.M., Brignon, J.M., Mailloux-Jaskulké, E. "Information and Decision-Support System for the Management of Pollution by Fertilizers and Pesticides (ISMAP/IDSS)", *Second International Conference on Water Pollution (Milano)*, Proceedings to be published by Computational Mechanics Publications (1993).

[2] Russell, M.H. and Layton, R.J. "Models and Modelling in a regulatory Setting: Considerations, Applications and Problems", *Symposium Role of Modelling in Regulatory Affairs (Louisville, USA)*, Proceedings in Weed Technology, Vol. **6**, 673-676 (1992).

[3] Soyeux, E., Mailloux-Jaskulké, E., Girard, M.C., Serini, G., Remotti, D., "Methodology for a cartographic approach for the assessment of agricultural non-point source pollution risk - Use of remote sensing and GIS.", *Second International Conference on Water Pollution (Milano)*, Proceedings to be published by Computational Mechanics Publications (1993).

[4] Usseglio-Polatera, J.M. and Gatti, S. "Information and Decision-Support System for the Management of Agricultural Pollution", *IX Symposium Pesticide Chemistry (Piacenza)*, Proceedings to be published (1993).

[5] Vandevelde, Th., Ferry, A., Dutang, M., Usseglio-Polatera, J.M. and Cacciari, A. "Système Intégré pour la maîtrise des pollutions liées à l'activité agricole (ISMAP)", *Water Supply*, Vol. **10**, Florence, 43-50 (1992).

[6] Villeneuve, J.P. and Fortin, J.P. "Evolution de la modélisation à l'INRS-Eau", *Séminaire "Modélisation du comportement des polluants dans les hydrosystèmes"*, Ministère de l'Environnement, Paris (1993).

New Tools for Diagnosing Underground Pollution by Petro-Chemicals: A Creative Concept in Europe

Jean-François BERAUD[a], Paul LECOMTE[b] and Alain PEREZ[c,1]

[a]*BURGEAP, 70 rue Mademoiselle, 75015 Paris, France*
[b]*B.R.G.M., Avenue de Concyr, B.P.6009, 45060 Orleans Cedex 2, France*
[c]*ELF, Tour ELF, 92078 Paris La Defense Cedex 45, France*

Abstract

Petroleum products and chlorinated solvents are the most frequently encountered pollutants in underground site contaminations: they are the most commonly distributed liquids after water. Among the risks they can engender, groundwater pollution, fires and explosions are the most frequently mentioned. In order to avoid these dangers, a good knowledge of the extension and intensity must be obtained; then a correct risk assessment, taking into account the mobility of products in each phase (gaseous, liquid, water...), must be performed, before defining the best adapted remediation method. The RESCOPP EUREKA PROJECT aims at developing a set of technically and economically competitive tools to perform assessment and remediation work. Its methodological approach will encompass all stages of a remediation project, starting with plume detection, to post-remediation assessment. It will rationalize critical steps, such as data collection, goal definition, implementation, site monitoring and clean-up assessment. Among others things, it will help select the most appropriate technique for a given set of geology and pollutants. The project covers three main areas: modeling, investigation and monitoring techniques, all of them developed through laboratory tests, pilot projects, and implemented on real sites. The first step will be to write a handbook on the present state of the art of the available techniques.

Un nouvel ensemble d'outils pour le diagnostic des pollutions souterraines par les produits pétrochimiques: un concept nouveau en Europe

Résumé

Les polluants les plus fréquemment rencontrés dans les contaminations souterraines sont les produits pétroliers et les solvants chlorés: après l'eau, ce sont les liquides les plus communément dispersés dans le milieu naturel. Parmi les risques engendrés, les plus courants sont la pollution des eaux souterraines, les

---

[1] To whom correspondence may be addressed

incendies et les explosions. Afin de prévenir et gérer ces risques, il est nécessaire de connaître précisément la mobilité de chaque phase (gaz, liquide, eau,...) avant de concevoir la méthode de réhabilitation la mieux adaptée. Le projet RESCOP d'EUREKA a pour objet le développement d'un ensemble d'outils compétitifs tant du point de vue économique que technique pour l'évaluation et la réhabilitation d'un site pollué. L'approche méthodologique proposée prend en compte toutes les étapes d'un projet de réhabilitation, à commencer par la détection des alertes jusqu'aux stades de la réhabilitation et de la remise en état des sites. Il a pour objet de rationaliser les étapes critiques telles que: l'acquisition des données, la définition des objectifs, la mise en route, le contrôle et les procédures de réhabilitation.

1 INTRODUCTION

Unlike scattered sources of pollution, originating from application of solids or liquids to the soil surface over large areas (fertilizers, herbicides, pesticides...), localized pollution (which is sometimes also called accidental, but comes often from a leaking pipe, or leaching of a solid deposit) originates from a limited surface, small in relation to the landscape and the extent of underground layers and waters, and clearly-defined within a geographical area.

The nature of pollutants may vary considerably, but those most frequently encountered are the hydrocarbons, and the organo-halogenated compounds (most of them chlorinated solvents). Obviously, it is because they are the most widely distributed liquids apart from water.

In this paper, we only deal with localized pollution by these products, which is the purpose of the Eureka RESCOPP Project.

2 NATURE OF THE PRODUCTS

2.1 Commercial hydrocarbons

These include petroleum products consumed by the general public sector: petrol and domestic fuel. Contaminations induced by these products are very numerous and extremely diverse, constituting almost half of all known contaminations. A petroleum product is not characterized by a chemical composition, but by its physical properties: density, viscosity, boiling point... Each product contains a great number of molecules, different in proportion depending on its crude oil origins, and how it has been refined.
However, by product type, some general characteristics can be accepted such as:

- gasoline contains on the average 55% alkanes and cyclanes, and 45% light aromatics,
- diesel-oil contains on the average 90% aliphatics and 10% aromatics.

The main features of these products, from the point of view of underground contamination, are:

- they are lighter than water,
- they are poorly biodegradable in natural soil conditions,
- gasoline is highly inflammable, and its vapors explosive,
- some components can be dangerous at low concentrations in water, but their threshold of taste and smell detection is very low. That is more especially the case of aromatics which are also more soluble in water than the other components.

2.2 Organo-halogen aliphatic compounds

These products are most commonly used as solvents in the chemical and mechanical industries. Some of them have chlorines (the most frequent), and/or bromines, and/or other components added to their molecular chains. They are:

- heavier than water,
- more fluid than water,
- relatively soluble in water, and volatile,
- strongly resistant to biodegradation in natural soil conditions,
- considered dangerous in water at low concentrations (10 ppb), as they have carcinogenic effects.

The most prevalent of these substances are: trichloroethylene (TCE) 1.1.1., trichloroethane (TCA), tetra chloroethylene (PCE), chloroform, and bromoform.

2.3 Other hydrocarbons

These include a great number of products with very different properties and behavior such as:

- crude oil and heavy fuel, with a high viscosity and a large range of components,
- aromatic solvents, largely used in the paint industry, volatile, inflammable, and relatively soluble in water,
- aviation products (kerosene) intermediate between gasoline and gas-oil,
- residues from coal mining and petroleum industry, encountered for example on old gas production plants, and sometimes used as filling material on old sites. These products (bitumins, tars and creosotes), are complex mixtures of heavy hydrocarbons, more particularly poly-aromatics, and sometimes phenols, which can present a real threat.

3 Sources of Subsurface Infiltration

The study of cases of infiltration of pollutants into the subsurface conclude that some of them are accidental (motorway and rail accidents, rupture of pipelines, etc.). More often, the source of contamination comes from leakages of pipes or tanks and this from service-stations to refineries in petrol treatment and frequently in plants where these products are used. Moreover, the stocking of products on sites unadapted for this activity (unclassified landfill sites, and unauthorized tipping/dumping), can be the source of very large-scale contaminations.

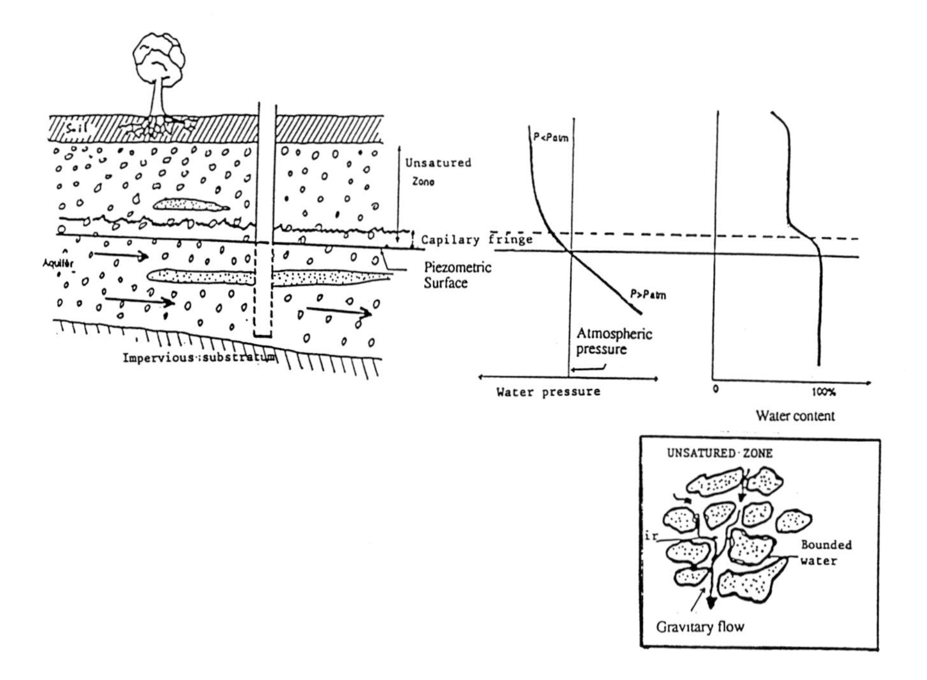

Figure 1. Schematic cross section of an aquifer.

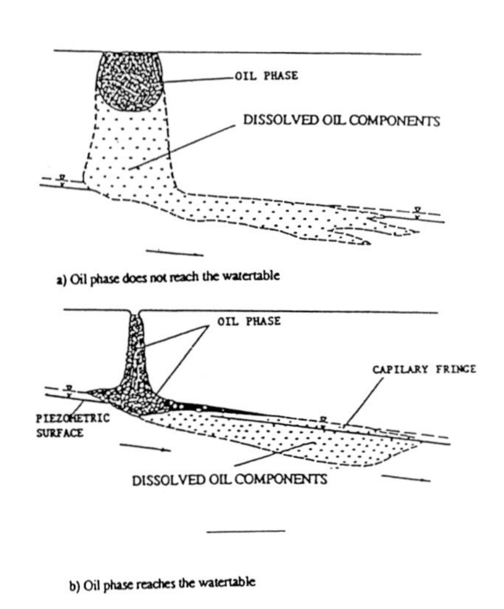

Figure 2. Oil migration patterns.

Although daily infiltration of contaminants can be small, an accumulation over a long period of time can be quite significant. Moreover the subsurface environment is conservative, and flowspeeds of groundwater are very low, so that pollution can be present for a long time in the subsoil.

4 BEHAVIOR OF POLLUTANTS IN THE SUBSURFACE

4.1 The underground environment

Rocks and soils are made up of small fragments and grains separated by pore spaces, creating routes in which fluids and gases can circulate (Fig. 1).

Moving downwards from the soil surface, the following structures can be distinguished:

- the non-saturated zone, where the pores are only partially filled with water, which has a more or less vertical transport characteristic under the force of gravity. Pore spaces which do not contain water permit the circulation of gases.
- the saturated-zone, where all pore spaces are filled with water, and the horizontal flow component is dominant. The boundary at which this pressure is equal to the atmospheric pressure forms the free surface of the aquifer. Above this level is the capillary fringe, with a height ranging from some 12 cm. for coarse sands, to 2.50 m. for silts.

4.2 Pollutant formation

4.2.1 Petroleum products lighter than water

During its vertical migration through the unsaturated zone, the pollutant liquid leaves in its wake soil which is permeated almost to the point of saturation (Fig. 2).

If the spill is big enough, the product as a phase will reach the capillary fringe, then the watertable, and spread over it in a lense form.

4.2.2 Dense chlorinated solvents

The behavior in the unsaturated zone is almost the same as that in the petroleum product zone, with development of vapor (Fig. 3).

A main difference is that, if the product reaches the water-table, it sinks down to the bottom of the aquifer. The most important difficulty is that, contrary to the watertable, the bottom of an aquifer is a very irregular surface. The solvents can then be trapped in holes and be difficult to investigate.

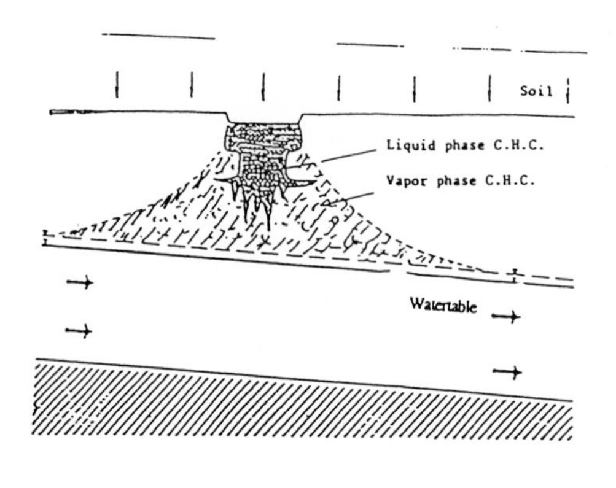

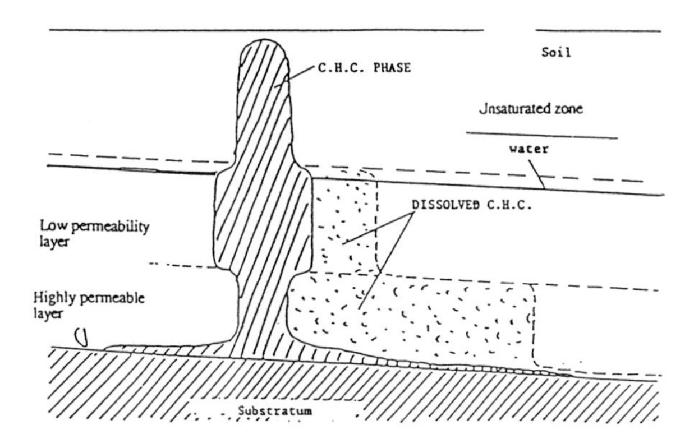

Figure 3. C.H.C (Chlorinated Hydrocarbons) migration patterns.

4.2.3 Evolution of contaminants

For both the saturated and the unsaturated zones, calculation of the transfer time and forecasting of the evolution of substance contents in water, must take into account many phenomena such as:

- convection (directly carried along by water),
- kinematic dispersion due to the tortuous flows between the grains of soil,
- diffusion in the aqueous or gaseous phase,
- biological decomposition, particularly developed in the unsaturated zone, but which may also exist in the saturated zone, etc.

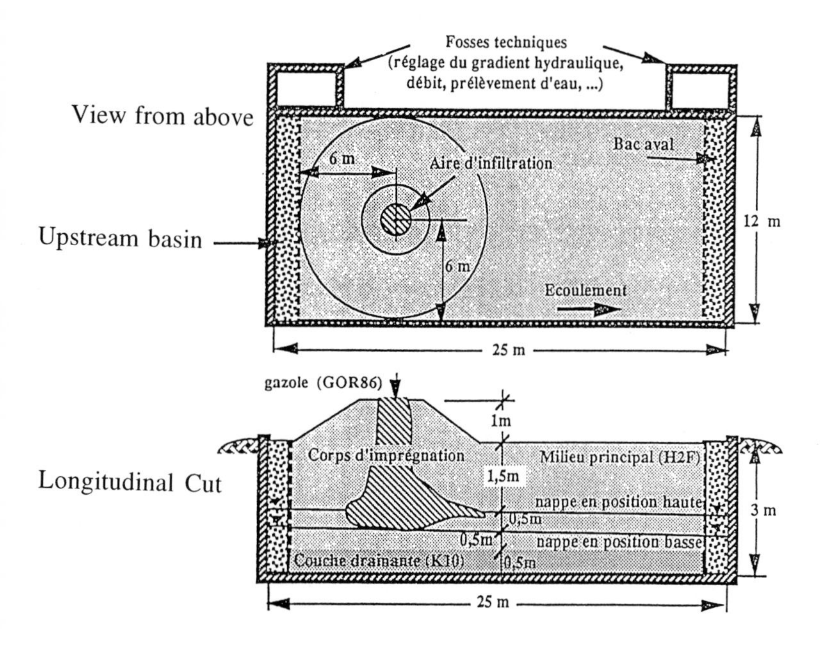

Figure 4. Outline representation of the experimental tool SCERES.

5 Evaluation and Risk Assessment; Remediation

Many deposits of petrol or halogenic hydrocarbons are a significant source of ground water pollution. This does not automatically pose a risk to public health, although on paper, contamination figures may seem high. Certain ground water sources can be condemned after hydrocarbon pollution, as taste detection in water is extremely low (10 micro-grammes per litre), and largely inferior to their toxic threshold.

On the other hand, gasolines can give rise to explosive risk, when a ground water system is contaminated with a floating layer and becomes involved in underground engineering projects, or where a high concentration of hydrocarbon vapors migrate upwards to the level of cellars, foundations, underground car parks, etc.

In relation to public health, halogenic hydrocarbons can contribute a wide range of risks which are being brought more and more to light with increasing environmental legislation and chemical analyses on potable water supplies. These products are very often found on industrial sites overlying large aquifer systems, where in the past, ignorance of pollution and the lack of legislation has led to the problems we see today.

To evaluate with precision the risk associated with this type of contaminant, a thorough knowledge of the pollution, its characteristics and evolution is imperative.

Then many remediation methods (for instance pumping, washing, biodegradation, etc.) can be undertaken.

6 AIMS OF THE RESCOPP EUREKA PROJECT

The RESCOPP EUREKA PROJECT groups Italian and French partners (SNAM, AGIP, ITALGAS, AQUATER, ELF, C.G.E., I.F.P., B.R.G.M., BURGEAP). It aims to develop a set of technically and economically competitive tools to perform assessment and remediation work. Its methodological approach will encompass all stages of a remediation project, starting with plume detection, to post-remediation assessment. It will rationalize critical steps, such as data collection, goal definition, implementation, site monitoring and clean-up assessment. Among others things, it will help in selecting the most appropriate techniques for a given set of geology and pollutants. The project covers three main areas: modeling, investigation and monitoring techniques, all of them developed through laboratory tests, pilot projects as, for instance the SCERES basin, at IFARE (French and German Institute for Environmental Research).

This basin (see Fig. 4), 25 m long, 12 m wide, and 4 m deep, enables us to test the different tools created, on a real scale, in an isotopic and homogeneous material, which cannot be found in nature. Later tests will be implemented on real sites.

A first step will be to write a handbook on the present state of the art of the available techniques.

KEYWORDS

Underground contamination, chlorinated solvent, petroleum product, groundwater pollution, remediation method, clean up assessment, pesticides, European Community, water.

EFFECTS OF TRIBUTARIES ON RIVER WATER - SEDIMENT INTERACTIONS: A MULTI-DIMENSIONAL PRESENTATION

M. PURIC[a], *P. PFENDT*[b,c] *and P. POLIC*[b,c,1]

[a]*Biological Institute, Podgorica, Yugoslavia*
[b]*Faculty of Chemistry, University of Belgrade,*
[c]*Institute of Chemistry, Technology and Metallurgy, Belgrade, Yugoslavia*

ABSTRACT

The river Tara (UNESCO ecological reservation, Montenegro, Yugoslavia) arises from the rivers Verusa and Opasanica. The drainage area encompasses two dominant geological regions - a limestone region and a flysch region. The river Tara is considered to be one of Europe's very few unpolluted rivers. However, there are several pollution sources downstream - beside municipal wastewater, dumps and some minor industrial capacities, it is the Mojkovac lead and zinc mines and metal industry that are regarded as the most severe contaminant sources. The distinction between anthropogenic and naturally occurring heavy metals in such a system could be of high ecological significance. Since heavy metals become incorporated (adsorbed, coprecipitated, etc.) into various sedimentary phases, they are also expected to be remobilized under certain conditions. Similarly, a redox-potential decrease results in reductive dissolution of iron and manganese hydrous oxides. These remobilization processes represent a threat for the aquatic system, even if sorption mechanisms (e.g. onto the bottom mud) have apparently purified the aqueous phase.

Effets des Affluents sur les Interactions Eaux-Sédiments d'un Système Fluviale : une Représentation Multidimensionnelle

RÉSUMÉ

Le fleuve Tara (réserve écologique de l'UNESCO, Monténégro, Yougoslavie) a pour principaux affluents le Verusa et l'Opasanica. La zone de drainance recouvre deux formations géologiques principales - l'une à dominance carbonatée, l'autre constituée de flysch. Le fleuve Tara est parmi l'un des rares fleuves non pollué d'Europe. Toutefois, il existe plusieurs sources de pollutions anthropiques situées en aval du fleuve. En plus des rejets d'eaux usées, des décharges publiques d'origine urbaine et quelques pollutions mineures d'origine industrielle, la principale source de pollution est la mine plomb-zinc de Mojkovac et l'industrie métallurgique associée. Pouvoir distinguer les métaux lourds d'origine anthropique et d'origine naturelle d'un tel système peut être d'un très grand intérêt

---

[1] To whom correspondence may be addressed

pour la compréhension des systèmes écologiques. Puisque les métaux lourds sont fixés (par absorption, par coprécipitation, etc....) sur différentes phases minérales, ils peuvent également être remobilisés sous certaines conditions. Ainsi, une réduction du potentiel redox a pour conséquence une mise en solution des hydroxydes de fer et de manganèse. Ces processus de remobilisation représentent une menace pour les systèmes aquatiques, même si des mécanismes d'absorption (i.e. sur les argiles) ont apparemment purifié la phase aqueuse.

1 EXPERIMENTAL

In order to obtain a clear insight into the nature of river water/sediment interactions, most relevant river water parameters (e.g. t, pH, COD, BOD, O_2, Corg, major cation and anion, Cr, Co, Cu, Zn, Cd, Pb content), as well as the contents of heavy metals in two sediment phases (2M $MgCl_2$ and 2M HCl-extractable) were determined at 26 locations (Fig. 1), in the river Tara and in its main tributaries (Puric, 1993).

2 RESULTS AND DISCUSSION

The data were processed by factor analysis, a multivariate method that enables the extraction of factors which predominantly determine the system's variability (Flury, Riedwyl, 1983). The distribution of samples (Fig. 2) clearly indicates the influence of the Rudnica river (sample 18), draining the Mojkovac mine area, as well as the purification process downstream - samples 19, 20, 21 - the last sample, practically indistinguishable from the main population, indicates the termination of the self-purification process, approximatively 10 km downstream from the Rudnica confluence. The corresponding distribution of parameters (Fig. 3) shows the nature of the contamination process - most points on the left-hand side of Fig. 3 represent heavy metal values - dissolved, as well as sedimentary. Factor 2 is obviously geochemically determined (hardness).

In order to detect more details in the interaction mechanisms, the uncontaminated area, upstream from the Rudnica mouth, was investigated separately. Factor analysis of sediment and water sample characteristics clearly shows the clustering of the following groups of samples (Fig. 4):

- Samples of tributaries originating from the flysch region, as well as Tara River samples in the immediate vicinity of the tributaries' mouths, where the flysch region influence still dominates (4,5,6,7);
- Tributaries originating from the limestone region, and, again, a similar Tara River sample downstream (8,9,10);
- A main group in the centre, representing Tara River samples - in fact, a result of thorough mixing and averaging of tributary water (3,11,13,15,17).

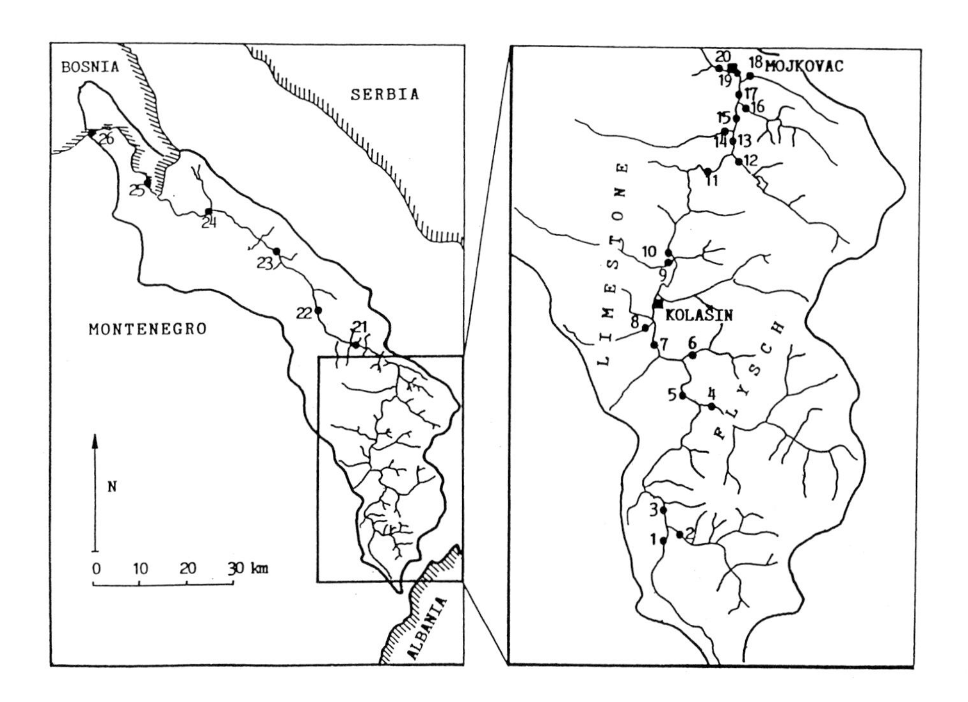

Figure 1. Tara drainage basin with sampling locations.

The corresponding distribution of parameters reveals the existence of two factors (Fig. 5), primarily determining the total uncontaminated model. The first factor ('Factor 1', horizontal axis) represents the observed geochemical contrast between the flysch and the limestone regions, reflected in water and sediment characteristics. Significantly correlated with the 'carbonate pole' (HCO^{3-}, CO_3^{2-}, Ca, etc.) is the Cd-content of sediments, 'CdM' and 'CdH' (suffix 'W' for dissolved, suffix 'M' for $MgCl_2$-extracted and suffix 'H' for HCL-extracted); Cd has already been shown to become associated with carbonates. The 'flysch pole' is represented by other heavy metals, indicating their sorption onto 'non-carbonate' substrates (iron and manganese oxides, silicates, etc.).

The second factor ('Factor 2', vertical axis), found to contribute significantly to the model variability, represents the mobility of metals. The poles identified within this factor are the 'temperature pole' ('t'), and the 'oxygen pole' ('O_2', Fig. 5), revealing much more than the simple fact that higher temperature results in lower O_2 content.

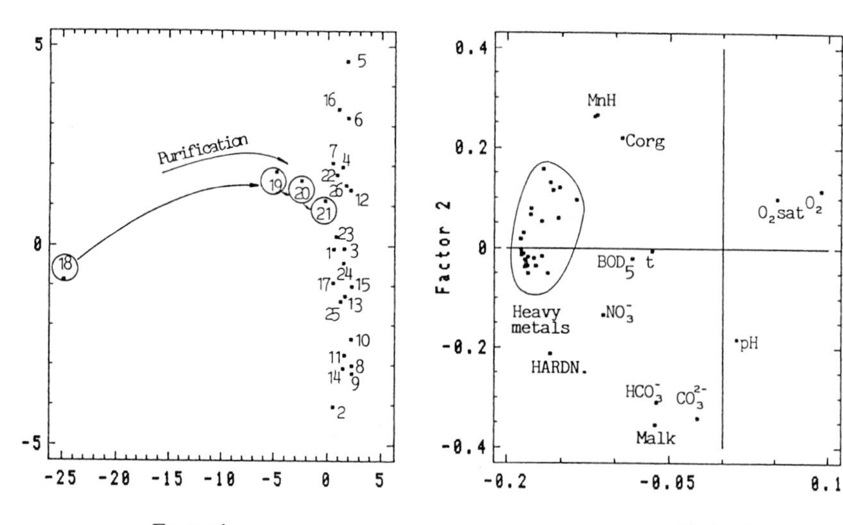

Factor 1
Figure 2. Sample distribution (N=26).

Factor 1
Figure 3. Parameter distribution (N=26).

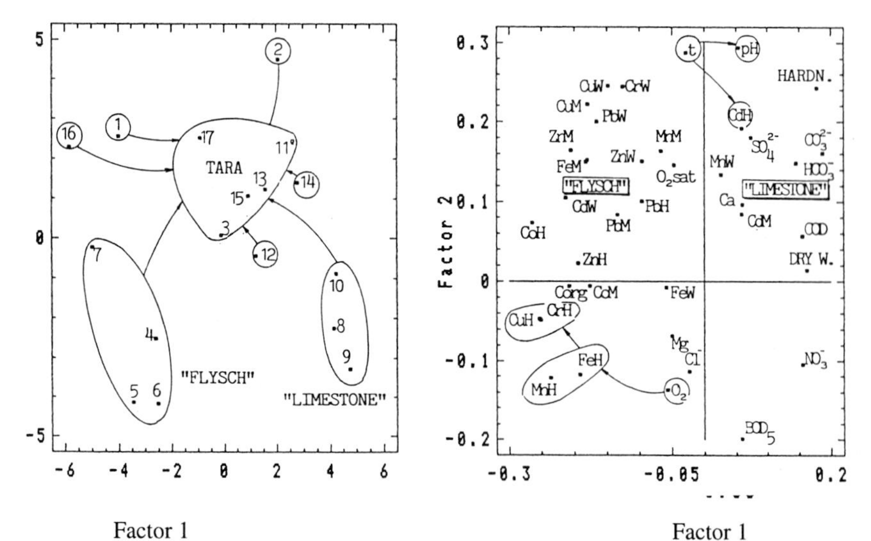

Factor 1
Figure 4. Sample distribution (N=17).

Factor 1
Figure 5. Parameter distribution (N=17).

Figures 2-5. Factor analysis of total (N=26), and uncontaminated drainage area (N=17).

Namely, in correlation with the temperature we also find the pH (as a result of the CO_2-carbonate equilibrium), and, consequently, the less mobile Cd-phases ('CdH', precipitated as carbonates). On the other hand, the less mobile iron and manganese phases ('FeH', 'MnH' - precipitated oxides), together with the metals (especially 'CrH', 'CuH'), probably associated with these substrates, are located close to the 'oxygen pole', indicating the well-known redox-potential dependence of iron and manganese chemistry.

The applied multivariate model enables a fundamental ecogeochemical characterization of the drainage area, as well as the differentiation of various sediment-water interactions and heavy metal associations.

KEYWORDS

Water pollution, heavy metals, sediments, anthropogenic effects, factor analysis.

REFERENCES

[1] Deurer R., Foerstner U., Schmoll G. *Geochim. Cosmochim. Acta* **42** (4), 425-427 (1978).
[2] Finney B.P., Huh C.A. *Environ. Sci. Technol.,* **23** (3), 294-303 (1989).
[3] Flury B., Riedwyl H. *Angewandte multivariate Statistik.* Gustav Fischer Verlag, Stuttgart, New York (1983).
[4] Jacobs L.A., von Gunten H.R., Keil R., Kuslys M. *Geochim. Cosmochim. Acta*, **52** (11), 2693-2706 (1988).
[5] Moore J.N., Ficklin W.H., Johns C. *Environ. Sci. Technol.,* **22** (4), 432-437 (1988).
[6] Puric M. "Sediment/water interactions as a factor of metal chemo-dynamics in the Tara River", PhD Dissertation, Belgrade (1993).
[7] Shea D. *Environ. Sci. Technol.,* **22** (11), 1256-1261 (1988).
[8] Wersin P., Charlet L., Karthein H., Stumm W. *Geochim. Cosmochim. Acta*, **53** (11), 2787-2796 (1989).

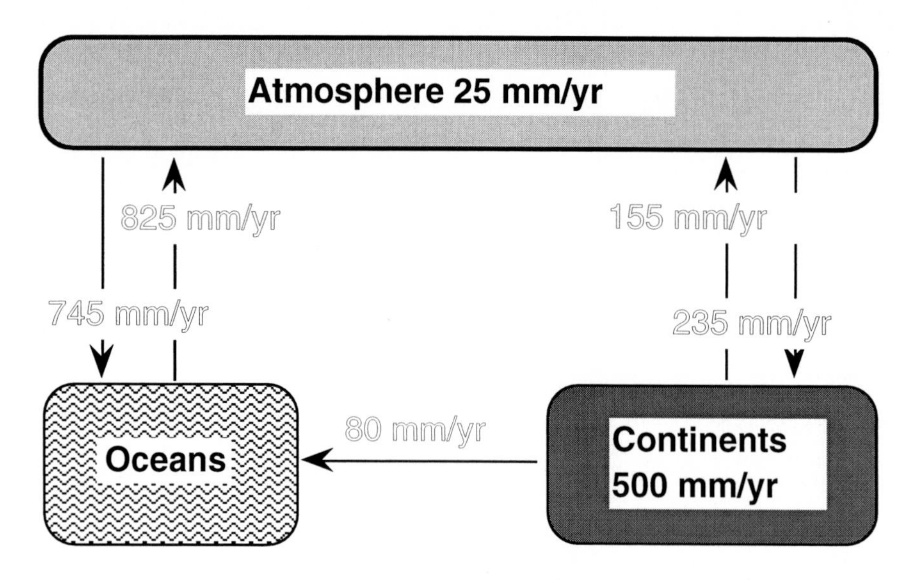

Schematic hydrological cycle of water on a world-wide scale
Water exchanged between the main reservoirs are given in mm/yr

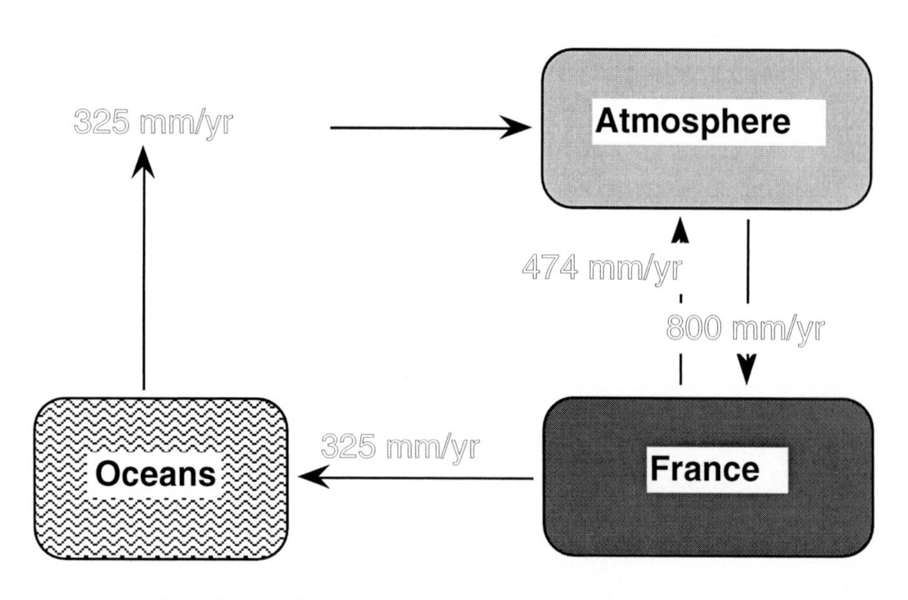

Water balance for France - Quantity of water exchanged between the main reservoirs are given in mm/yr.

After J.P. Laborde "Le milieu hydrologique et l'acquisition des données" édition revue par L. Demassieux, C, Cachet, B. Renaud.

ENVIRONMENTAL ASSESSMENT AND MANAGEMENT IN MOLDAVIA

Olga VEDINA

Institute of Plant Physiology, Moldavian Academy of Sciences, Padurii 22, Kishinev 277002, Moldavia

ABSTRACT

Moldavia is a densely populated agrarian republic. Agricultural lands occupy 68,2 % of all arable land. Consequences of the intensive development of agriculture are most dangerous for the environment in Moldavia. To determine the potential impact of the agriculture on the environment and human health, the method known as Environmental Impact Assessment (EIA) has been applied [1]. In some countries EIA is a legal requirement. In Moldavia it is enforced indirectly by the general planning and activities of the Ministries of Health and Agriculture, Department for Environmental Protection and Natural Resources. The research carried out has shown that agricultural activities and environment are strongly interrelated. The intensification of agricultural production leads to a conflict between agricultural and environmental objectives, but beside negative changes, positive ones occur in the environment as well. In order to harmonize agricultural and environmental objectives, integrated policies should be carried out based on increasing the positive role of agriculture, pollution prevention and control, and encouragement of the adaptive system of agriculture.

Etude d'Impact et Gestion de l'Environnement en Moldavie

RÉSUMÉ

La Moldavie est une république agraire, à forte densité de population, les terres agricoles occupent 68,2% de la surface totale des sols utilisables. Le développement intensif de l'agriculture dans le pays est une cause principale de dangers pour l'environnement. Afin de déterminer les effets potentiels de l'agriculture sur l'environnement et sur la santé humaine, la méthode connue sous le nom d'Étude d'Impact sur l'Environnement (EIA) a été utilisée [1]. Dans certains pays, les études d'impact EIA sont imposées par la législation. En Moldavie, ces études sont imposées indirectement dans le cadre du plan et des activités des Ministères de la Santé et de l'Agriculture, du département de la Protection de l'Environnement et des Ressources Naturelles. La présente étude démontre indubitablement, que les activités agricoles et l'environnement sont très fortement liés. L'intensification de la production agricole entraîne un conflit entre les objectifs agricoles et environnementaux. Afin d'harmoniser ces objectifs, une politique intégrée basée sur le rôle positif de l'agriculture, la prévention de la pollution et son contrôle devrait être mise en place ainsi que l'incitation à utiliser des techniques agricoles adaptatives.

1 INTRODUCTION

At present in Moldavia a rather complicated ecological situation has been created. Due to the lack of integration of agricultural and environmental policies significant problems have arisen :

- The negative impact of pesticide and fertilizer residues, heavy metals and other contaminants in soils and food products on human health.
- Contamination of ground and surface waters, eutrophication of surface waters by nitrates and phosphates leading to local health risk and lowering the quality of aquatic resources.
- Agricultural pollution problems associated with the growth of intensive animal husbandry.
- Salinization, compaction and erosion of soils causing losses in soil productivity and landscape amenity.

2 SOIL ASSESSMENT AND MAJOR GROUPS OF POLLUTANTS

Soils constitute the greatest wealth of Moldavia. 74 % of the country is covered by very fertile chernozem soils with a high content of humus. During the last 20 years the soil fertility significantly declined mainly because of the acceleration of the soil degradation process. The most important reasons for this are: agriculture technology violations; imperfect territorial organization; increasing erosion and slide formation.

About 2 million ha, or 82 % of agricultural lands in the republic, are located on slopes. 40,000 ha are subject to ravine and slide formation. According to the data of the Moldavia Institute of Soil Science and Agrochemistry, the area of eroded soils increases annually by 0,86 %.

Annually, as a result of erosion processes, an average of 18 million tons of fertile soils are washed away [2]. The main cause of the increase in soil erosion is excessive development of slopes without the implementation of antierosion measures. Clearly, the problem of soil erosion in the republic presents a serious danger for the environment.

The contamination of biosphere components in Moldavia occurs because of the intensification of agricultural production. The main sources of pollution in agricultural activities are the following: mineral fertilizers, wastes of livestock production units and toxic chemicals.

3 FERTILIZER USE AND ENVIRONMENTAL IMPACT

Over the past fifteen years an excessive use of fertilizers has had a very negative environmental impact on soils, leaving substantial amounts of chemical residues. According to the data of the Central Statistics Department of the former USSR, in 1977 in Moldavia the mineral fertilizer consumption was approximately 1 million tons.

During recent years the use of mineral fertilizers in the country decreased considerably (Table 1) [3]. This decrease is chiefly explained by economic difficulties. These have had an impact on enhancing environmental quality. Improper transportation and storage, irrational utilization of mineral fertilizers, violating scientifically determined rates and terms of their application caused negative environmental impacts. It has been estimated that only 50 % of fertilizers are utilised by agricultural crops, the remaining portion being lost from the soil system without benefit for crops. Losses occur through leaching, run-off and volatilization. It should be noted that main nutritive substances (N, P, K) behave in quite different ways under use in soils. Nitrogen fertilizers constitute the gravest threat for environmental contamination in Moldavia. When fertilizers are applied, nitrogen penetrates into the soil in the form of ammonium, amide or nitrate.

The available moisture leads to the movement of the nitrate nitrogen that has not been utilised by the plant into lower soil layers which causes a danger of water body pollution. According to the research carried out in Moldavia on chernozemic soils, in those locations where fertilizers have been applied for a long time a significant increase of nitrate content is determined at the depth of 9-10 m, with 7 to 17 % of nitrogen being leached from the fertilizers. A part of leached nitrate nitrogen can return into agricultural rotation. It is indisputable that nitrates at a depth of over 3 m are lost for agriculture. The data received convincingly prove that in Moldavia nitrogen fertilizers are one of the main pollution sources of surface and ground waters.

Phosphorus fertilizers are relatively less harmful for the environment than nitrogen ones. Only a minor proportion of phosphorus fertilizers is leached from the soils into drainage water since it is immobile in the soils of Moldavia. However, the soil cover is strongly dissected and most lands are subject to erosion.

Therefore phosphorus absorbed on the surface of soil particles may contaminate water bodies through soil erosion. The proportion of the agricultural pollution of waters caused by phosphorus does not exceed 10-15 %. Phosphorus utilization efficiency from fertilizers is not high, reaching 20 %. The excess accumulation of phosphorus in soils can lead to its pollution, plant toxicosis, and yield reduction. Potassium fertilizers do not have a harmful impact on the environment. Potassium is held sufficiently strongly by chernozemic soils, it migrates slowly and only small amounts reach drainage waters.

However, together with potassium fertilizers high amounts of chlorine enter the soil. Chlorine is more mobile and contaminates soils and ground waters through rapid leaching. Hence, excessive application of fertilizers creates a number of environmental problems.

Chief among these problems is the contribution of phosphorus and nitrogen fertilizers to surface water eutrophication, due to which water quality deteriorates. In addition, the durable application of high doses of mineral fertilizers results in the accumulation of contaminants in soils, particularly heavy metals, which is an undesirable phenomenon requiring research and control. According to the classification of soils by the level of heavy metal contamination some soils on the territory of Moldavia can be identified as weakly polluted ones [4].

Table 1. Mineral fertilizer use in Moldavia (thousand tons)

Fertilizer	1977	1991	1992
Mineral	1000	204	135
Nitrogen		87	65
Phosphorus		81	46
Potassium		36	24

4 THE IMPACT OF INTENSIVE LIVESTOCK HUSBANDRY ON THE ENVIRONMENT

The pollution caused by stock breeding is also an important problem affecting the environment in Moldavia. A very complicated ecological situation is created due to the accumulation of animal husbandry unit wastes (8,22 million t) and the unsatisfactory condition of accumulators. More than half of the accumulators are not concrete and therefore a possibility of ground water pollution is not excluded.

Most of the livestock systems lack devices for waste effluent purification. Animal residues are stored in accumulators without any treatment and farmers refuse to use them for fertilization and irrigation. Sufficiently high levels of nitrogen, of nitrates and ammonia, chlorine, sodium, sulphur, some heavy metals as well as pathogens were discovered in manure. There are several different ways in which intensive livestock production units can pollute the environment. The presence of such units creates local problems through noise and smell, and a modification of the visual landscape. Specific problems arise from the storage and spreading of effluents, through run-off or leaching of manure components into the soil and also through the emission of acidic gases into the air.

In addition, manure pollution can affect health, local ecosystems and ecosystems in general. Most of the effects on health concern the deterioration of the potable water quality through the increased concentration of nitrates which may be converted into cancerogenic compounds. Furthermore, water resources can be polluted by pathogens from manure.

This urgent problem could be solved through the following approaches: decrease of the manure mineral residues through its decrease in feeds; encouragement of development of manure treatment and marketing arrangements; introduction of regulations to control production, spreading and removal of manure.

Table 2. Pesticide use in Moldavia

	1985	1989	1990	1991	1992
Quantity of pesticide use, t	58591	25842	21451	19644	15167
Area of pesticide use, thousand ha		4149,7	3222,4	2123,2	2702,9
Average pesticide use, kg of active matter per ha		7,3	5,3	5,6	4,6

5 Environmental Assessment of Pesticide Use and Managing Risk

In Moldavia the level of pesticide use has decreased sharply since 1984. Despite this the quantities of pesticides applied in agriculture production are very high (Table 2). Pesticides may persist in the soil for a long time. From the soil they enter plants, ground and surface waters. The soil type has a strong influence on pesticide mobility. In Moldavia where hard, organically rich soils are typical, pesticides persist a long time. For example, though the application of DDT has been prohibited since 1972 in the republic, the presence of its residues at higher than the permissible levels was discovered in the chernozems of the south and north parts of Moldavia. Taking into consideration insufficient humidity and the high intensity of erosion processes, pesticide migration is supposed to occur mainly with clay particles through erosion.

In addition, a high level of pesticide application without mitigation measures leads to the deterioration of agricultural product quality and, considering the geographical position of the country, to the pollution of water bodies of the Black Sea basin.

This problem can be solved through the following approaches :

- Accomplishment of erosion preventive measures.
- Decrease in pesticide application due to the development of an integrated protection system based on resistant cultures.
- Synthesis of unstable and rapidly decomposing products.
- Standardization of residual amounts of harmful substances in soils and their content control.
- Interchange of pesticides and agriculture crop rotation.
- Reduction of pesticide treatment of cropped lands.

6 Conclusion

Environmental Impact Assessment plays a major role in the efforts for achieving rational solutions in the conflict between social goals of agricultural development on the one hand, and preservation of the quality of life of the environment on the other hand.

Acknowledgements

This research was supported by the Research Support Scheme of the Central European University: grant N 124/93.

Keywords

Nitrates, phosphorous, fertilizers, heavy metal, pesticide, agriculture, polluted water, human health, environmental impact assessment, ground water pollution, Moldavia, erosion.

References

[1] "Environmental and Health Impact Assessment of Development Projects. A Handbook for Practitioners". *WHO, CEMP by Elsevier Applied Science.* London and New York (1992).

[2] "Raport privind calitatea factorilor de mediu si a activitatii Inspectoratului Ecologic Republica de Stat 1992". *Chisinau* (1993).

[3] Sinkevich, Z. Strijova, G. "Chemical contamination of Moldavian soils". *Kishinev, Mold. NIINTI* (1980).

[4] Vedina, O. Liach, T. "Effect of long-term utilisation of fertilizers on heavy metal content in soils". *J. Moldavian Agriculture.* N7, p 16 (1984).

Time and Space Forecasting of Nitrate and Salinity Contents in the Aquifers of Campo de Dalias (South East Spain)

R. JIMENEZ-ESPINOSA[c], *J.-J. ROYER*[b,1], *L. MOLINA-SANCHEZ*[a] *and F. NAVARRETE*[a]

[a]*Dpt. Hidrogeologia y Quimica Analitica. University of Almeria. La Canadã 04120, Almeria, Spain.*

[b]*CNRS - Centre de Recherches Pétrographiques et Géochimiques. B.P. 20, 54501 Vandoeuvre-Les-Nancy Cedex, France*

[c]*Dpt. Geologia, University of Jaén. Escuela Politécnica Superior, 23071 Jaén, Spain*

Abstract

The Campo de Dalias is situated in the eastern part of Andalusia (southern Spain), and west of the Bay of Almeria, occupying nearly 330 km^2*. This is economically the most important agricultural area in the province of Almeria, where intense greenhouse activities are widespread.*

Water for agriculture is supplied principally from many wells drilled in the aquifers of the region. These aquifers are composed of a geometrical complex succession of Triassic limestone and dolomites, Miocene conglomerates and calcareous sandstone, Pliocene calcarenites and gravel, and Quaternary sands and silts.

Three hydrogeological units can be distinguished at the Campo de Dalias: Aguadulce, Balanegra and Balerma-Las Marinas. The mean annual renewal for all the aquifers is around 50 Hm^3*, while approximately 100 Hm*^{3} *are pumped out. Two main important problems occur in this area: processes of marine intrusion, due to high pumping activity, and nitrate pollution caused by intensive agriculture.*

Constant checking has been carried out in these aquifers through time, in order to study the evolution of pollution. The aims of this work are to locate the most polluted areas and to characterize the spatial and temporal evolution of salinity and nitrates from 1988 to 1990, during spring and autumn. 110 wells spread along the Campo de Dalias were considered, distributed as follows: 44 wells belonging to the Aguadulce aquifer, 22 to the Balanegra one, and 44 to the Balerma-Las Marinas one.

---

[1] To whom correspondence may be addressed

Estimation de la Salinité et des Nitrates dans l'Espace et le Temps des Aquifères de Campo de Dalias (Sud-Est de l'Espagne)

RÉSUMÉ

La région de Campo de Dalias située entre l'Est de l'Andalousie (Sud de l'Espagne), et l'Ouest de la baie d'Almeria occupe près de 330 km^2. Du point de vue économique, il s'agit de la plus importante région agricole de la province d'Almeria pour son activité de cultures en serres intensives fortement dispersées. L'eau utilisée en agriculture provient essentiellement de nombreux puits forés dans les aquifères de la région. Ils se composent d'une succession géométrique complexe de calcaires et dolomites du Trias, de conglomérats et de grès calcaires du Miocène, de calcarénites et de graviers du Pliocène, et enfin de sables et de limons du Quaternaire. Trois unités hydrogéologiques sont identifiées à Campo de Dalias: Aguadulce, Balanegra et Balerma-Las Marinas. La recharge moyenne annuelle de ces aquifères est d'environ 50 Hm^3, alors que le pompage est d'environ 100 Hm^3. Les principaux problèmes rencontrés sont: les intrusions d'eau marine qui s'expliquent par une activité de pompage importante, et la pollution par les nitrates d'origine agricole. Un contrôle régulier et permanent au cours du temps de ces aquifères est effectué afin d'étudier l'évolution de la pollution. L'objet de ces travaux est de localiser les zones les plus polluées et de caractériser l'évolution spatiale et temporelle de la salinité et des nitrates, de 1988 à 1990 en particulier au printemps et en automne. Environ 110 puits dispersés dans la région de Campo de Dalias ont été pris en compte: 44 puits appartenant à la couche aquifère de Aguadulce, 22 à celle de Balanegra et 44 à celle de Balerma-Las Marinas.

1 INTRODUCTION

The Campo de Dalias is situated in the eastern part of Andalusia (southern Spain), and west of the Bay of Almeria, occupying nearly 330 km^2. The relief is relatively smooth and the area lies between the Sierra de Gador, an important carbonic mountain alignment, and the Mediterranean sea (Fig. 1). This region has a semi-arid Mediterranean climate: dry and warm summers, and springs and autumns with localised and violent rains, concentrated mainly in the Sierra de Gador mountain. This is the only way to renew the aquifer layers in the area. This zone is important economically because of its farming activities, being one of the most important agricultural areas in the Almeria province, and even in Spain. The climatic conditions (many hours of sunshine during the year) allow the cultivation of numerous products at times when the market has not been well supplied, therefore intensive greenhouse activities are wide spread. This increase in farming implies a parallel rise in the need for water. Water for agriculture is supplied principally from several wells drilled in the aquifers of the region, the mean yearly renewal for these aquifers being around 50 Hm^3, while approximately 100 Hm^3 are pumped.
Geologically, the outcrops of the area can be grouped in two broad categories: pre-orogenic and post-orogenic materials (Fig. 2). In the former, called Alpujarride materials, two distinct groups of units appear: the Gador and the Felix. In the Gador units, from bottom to top, a kind of phyllite and quartzite can be distinguished, upon which rests a carbonate series about 1000 m thick.

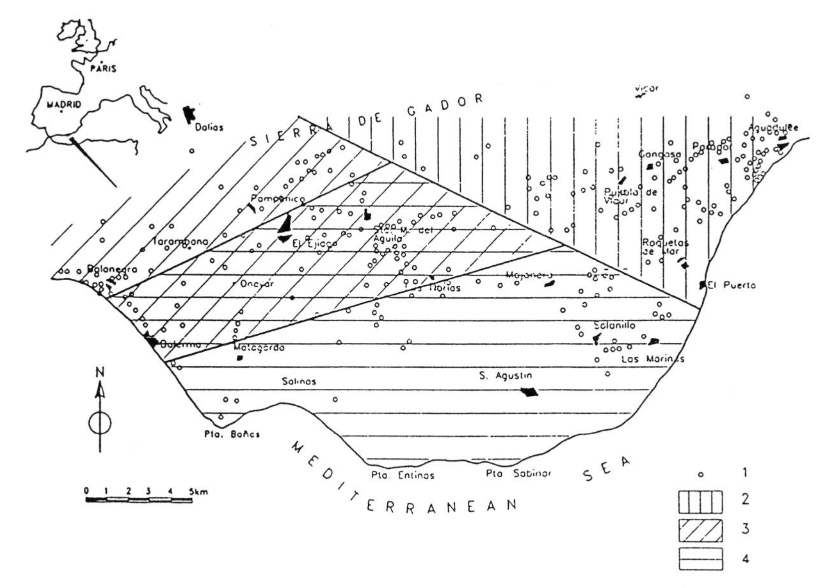

Figure 1. Geographical situation, monitoring network and different aquifer units 1.- wells; 2.- Aguadulce unit; 3.- Balanegra unit; 4.- Balerma-Las Marinas unit. (from Pulido et al., 1993b).

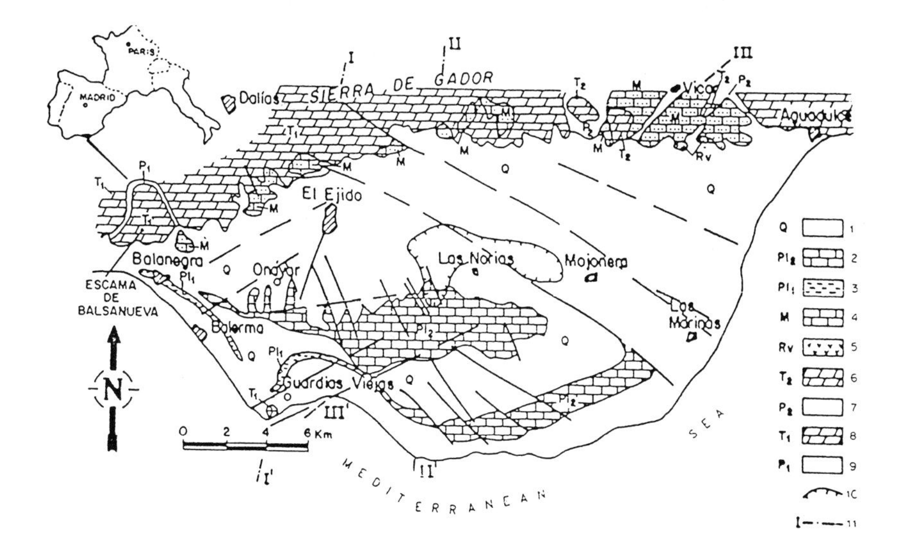

Figure 2. Schematic geological map of Campo de Dalias. 1: Quaternary materials; 2: Pliocene calcarenites; 3: Pliocene marls; 4: Miocene calcarenites; 5: conglomerates and volcanic rocks; 6: Felix carbonates; 7: Felix metapelites; 8: Gador carbonates; 9: Gador metapelites; 10: endoreic zone; 11: representative cross-section. (from Pulido et al., 1993a).

In the Felix units, there are also two members; the lower one comprised of phyllite, schist and quartzite; and the upper carbonate member, composed mainly of dolomite. The age of these units is approximately Permo-Werfenian to Middle-Late Triassic (Pulido et al., 1993). No deposits are known between the Late Triassic and the Miocene. The Late Miocene materials outcrop in the northern margin of the Campo de Dalias, over limestone and dolomite and also over phyllite and quartzite of both units. The Pliocene is represented by a lowest conglomerate upon which a marl marine formation was deposited in varying thicknesses of possibly more than 700 m. Calcarenites, reaching over 100 m, are at the top of this formation. Quaternary materials are widely represented, with diverse facies (Fourniguet, 1975): from alluvial fans to older marine Quaternary deposits.

The most important tectonic phase was the one that gave rise to the movement of the Felix units thrusting over the Gador ones. The deformations that contributed mainly to the current morphology were neotectonic, involving stages of fracturing, tilting and uplifting. After the Late Miocene, the swell of Gador fractured in the dominant strike N 70 E. Normal faults with great thrusts produced a series of compartments within the basin; a horst and a graben appeared, occupying the central sector of the Campo de Dalias (Pulido et al., 1989).

In relation with the hydrogeological features of this area, three hydrogeological units can be distinguished at the Campo de Dalias (IGME, 1982; Dominguez et al., 1988; Pulido, et al., 1989) involving a complex structure to be studied (see Fig. 1):

1.1 Balerma-Las Marinas Unit

This aquifer has the greatest surface area of the Campo de Dalias, with nearly 225 km^2 of outcrop. This is essentially composed of Pliocene calcarenites, whose thickness decreases from north to south, while the proportion of terrigenous components increases. Balerma-Las Marinas presents a great variety of transmissivity (IGME, 1982), showing values from 1800 m^2/day to 120 m^2/day.

Most farming activity is located in this aquifer unit. The whole area is below sea level and the temporal piezometric evolution indicates a general tendency to rise or stabilise (Thauvin, 1986). This situation is reflected in the poor natural quality of the water (Pulido et al., 1989). Pumping in this unit reaches 15 Hm^3/year and the total mean influx is estimated at 18-24 Hm^3/year.

1.2 Balanegra Unit

This water reservoir occupies the western half of the Campo and is partly under the Balerma-Las Marinas aquifer, being separated by Pliocene marls that act as a confining top. The aquifer material is mainly made of dolomites from the Gador sheet, and local Miocene calcarenites. The unit is bordered on the west and north by the lowermost metapelites of the unit, and on the south by the sea, although the sea is not in contact with the aquifer along the coast, but only with a slice of schist ("Escama de Balsa Nueva" slice of Balsa Nueva, Fig. 2).

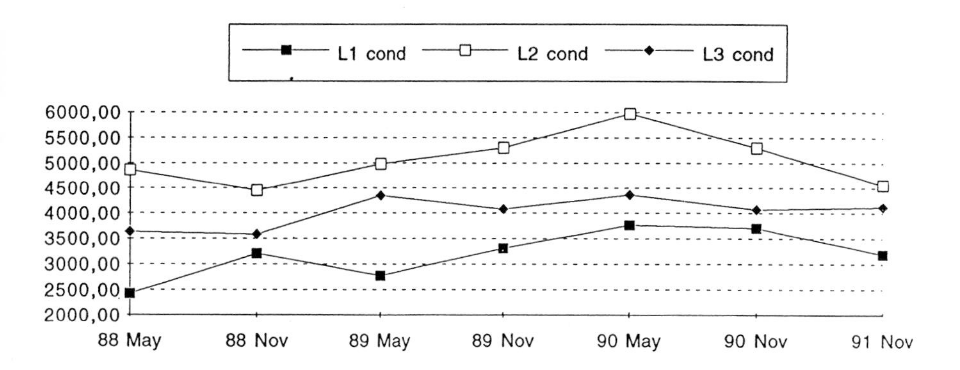

Figure 3. Time evolution of conductivity of the different layers in Aguadulce aquifer.

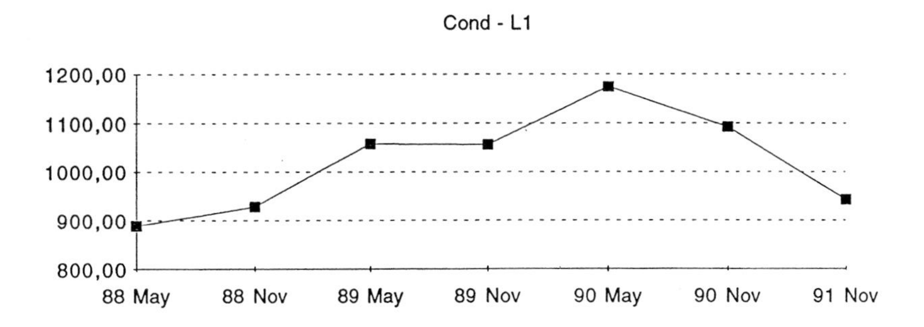

Figure 4. Time evolution of conductivity of the different layers in Balerma aquifer.

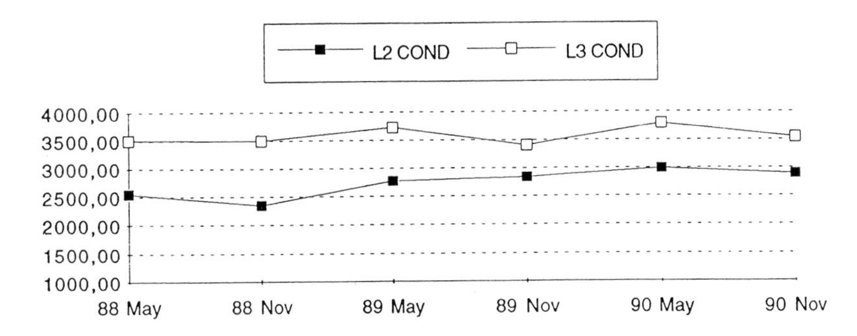

Figure 5. Time evolution of conductivity of the different layers in Balerma-Las Marinas aquifer.

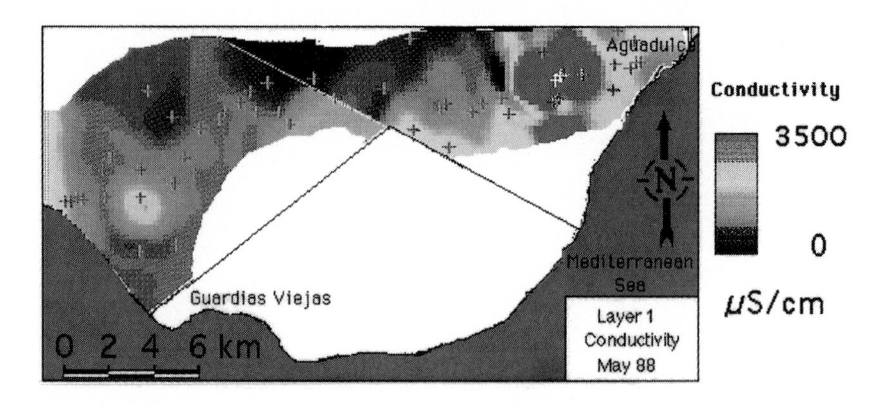

Figure 6. Map of conductivity in Layer 1, May 1988. (For color figure, see page 297)

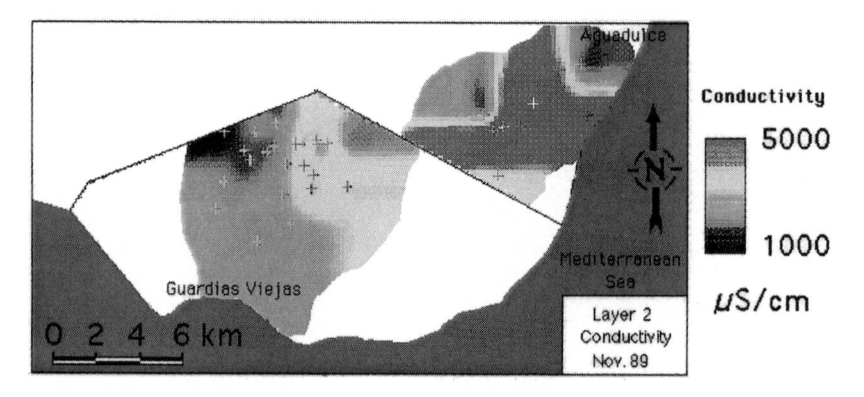

Figure 7. Map of conductivity in Layer 2, November 1989. (For color figure, see page 297)

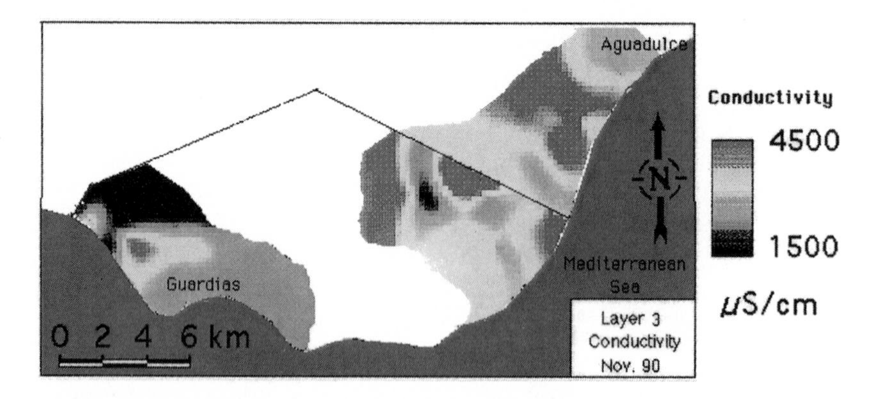

Figure 8. Map of conductivity in Layer 3, November 1990. (For color figure, see page 297)

Hydraulic parameters give values between 15,000 and 22,000 m^2/day (IGME, 1982). Below sea level piezometric values are found, and the rate of continuous decrease is about 0.75-1 m/year. In May 1989 there were some conoids with piezometric levels of more than 10 m below sea level (Pulido et al., 1993). Whereas the average influx is around 15 Hm^3/year, the amount of pumping water is double.

1.3 Aguadulce Unit

This is the hydrogeological unit which has the greatest tectonic complexity. In addition to the presence of the majority of permeable formations which occur in the Campo de Dalias, various aquifers layers appear, separated by other levels of low permeability, giving a true multilayer aquifer. Nevertheless, laterally some formations could be wedged out, allowing more hydraulic connection. Such lithological variety is reflected in the values of hydraulic parameters: 5000 to 7000 m^2/day in Gador dolomites; around 6000 m^2/day in Miocene calcarenites; or 200 m^2/day in Quaternary materials. As in the previous unit, the piezometric levels are decreasing gradually; most of the piezometric surface is below sea level, reflecting an over exploitation. In fact, while pumping averages near 50 Hm^3/year, the influx is only around 15 Hm^3/year.

The critical lack of balance between pumping and influx in the aquifers of Campo de Dalias, in addition to the intensive farming activities in this area, have produced serious consequences in the physico-chemical characteristics of this water. For this reason, in 1986 a network was set up to check water quality by daily, weekly and seasonal surveys (Pulido et al., 1989). In this work we have used data from the seasonal control network, taking into account mainly the evolution of conductivity (a measure of salinization) and nitrates (indicating agricultural pollution).

2 EXPERIMENTAL DATA

All wells, in the different aquifers of Campo de Dalias, pump mainly in three layers. These are layer *L1* composed of limestone's and dolomites from Alpujarride Triassic sheets; layer *L2* made up of Pliocene calcarenites; and layer *L3* of Quaternary detritic lithologies.

These layers can belong to each of the three different hydrogeological units. Specifically, the Aguadulce aquifer pumps in L1, L2 and L3; the Balanegra one in L1; and the Balerma-Las Marinas unit in L2 and L3. The study was carried out analysing layer by layer in each aquifer. This is a more realistic point of view than a treatment of the global aquifers, where all lithologies are mixed. Therefore, a spatial revision of these three layers through time has been made.

The hydrogeochemical study of this area was carried out using two seasonal surveys during the years 1988 to 1990, specifically, a spring survey (May) and an autumn one (November). In this way, 110 wells spread out along Campo de Dalias were considered. All the 110 points are distributed as follows: 44 wells belonging to the Aguadulce aquifer, 22 to the Balanegra one, and 44 to the Balerma-Las Marinas unit (Fig. 1).

In this work there are two important objectives to attain, related to the anthropic influence: the analysis of the behaviour of water salinization, and the study of nitrate evolution in the ground water. A good way to study the problem of water salinization could be the use of conductivity as a marker of marine pollution.

3 POLLUTION ANALYSIS OF CAMPO DE DALIAS

3.1 Evolution of Salinity

***Layer L1*:** This layer is in relation to the Aguadulce (AG) and Balanegra (B) aquifers. This is the deepest layer, representing Triassic limestone and dolomite (see Fig. 3 and 4). In general, L1 presents high levels of pollution in relation to fresh water, but lower than the other two layers. Moreover, values in AG are always higher than in B. This fact could be due to: (i) a *different behaviour of the aquifer*: in the area of B there are some natural borders that protect it from marine intrusion. As we could see in the above hydrogeology introduction, these borders are some impermeable metasediments ("Balsa Nueva slice"). (ii) a *different amount of pumping*: usually higher in AG. This layer is partially connected between the two aquifers. In fact, a hydraulic connection between them in the northern part of the area, along a fault, can be found. If we compare AG and B aquifers, in general, value ratios in both aquifers are as follows:

$$\frac{AG}{B} = \frac{3}{1} \quad \text{or} \quad \frac{2}{1}$$

A constant rise through time together with spatial extension of the saline zone in both aquifers of these areas indicate that the marine invasion is increasing. In the Aguadulce aquifer the main polluted area is restricted to known "Sector III" (Pulido et al., 1989) in 1988, and progressively this area is growing in the direction of Aguadulce village and the sea. Moreover, a new polluted area appears in the north of Aguadulce, which in 1990 was linked to the other one (see for example Fig. 5, May 1988).
In part of Balanegra in 1988 the entire zone has values less than 2000 µS/cm, where two high value areas are evidenced. One of these is around the Balanegra village and the other SE of this village, on the small road to Guardias Viejas village. In these two zones, salinity is increasing in time, mainly in the Balanegra area; in the other part a decrease was produced in 1990, due to a well whose value decreased that year (Fig. 6).

***Layer L2*:** This layer is made up of calcarenites from the Pliocene and is represented by AG and BM aquifers. This layer presents important values of marine pollution and, in relation to AG, this layer has the highest values of conductivity, indicating salinization (see Fig. 3 and 5). Usually concentrations are greater in AG than BM in ratios like:

$$\frac{AG}{BM} = \frac{1.5}{1} \quad \text{or} \quad \frac{2}{1}$$

The range of values in L2 in relation to L1 (aquifer of AG) is higher. Conductivity varies from approximately 6000 to 4500 µS/cm. In general, the range of variation in relation to BM is less important than the AG one. The behaviour in both aquifers in relation to L2 is quite similar and can be considered parallel, always taking into account the higher values in AG.
In this layer the number of wells is greater in BM than in AG, and once again, the Aguadulce aquifer is more polluted than the Balerma-Las Marinas one.

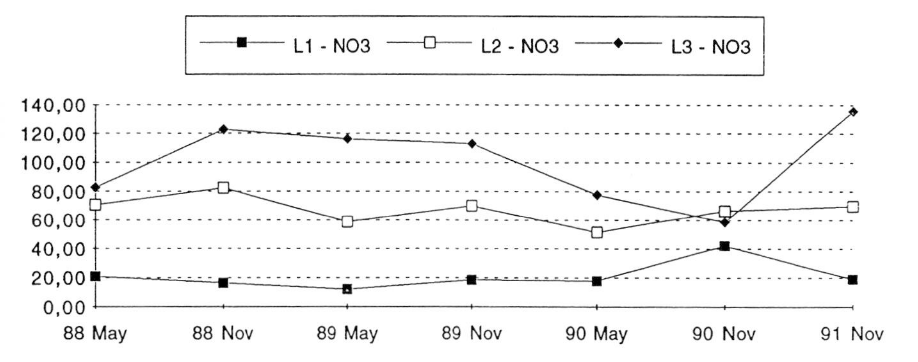

Figure 9. Time evolution of nitrates of the different layers in Aguadulce aquifer.

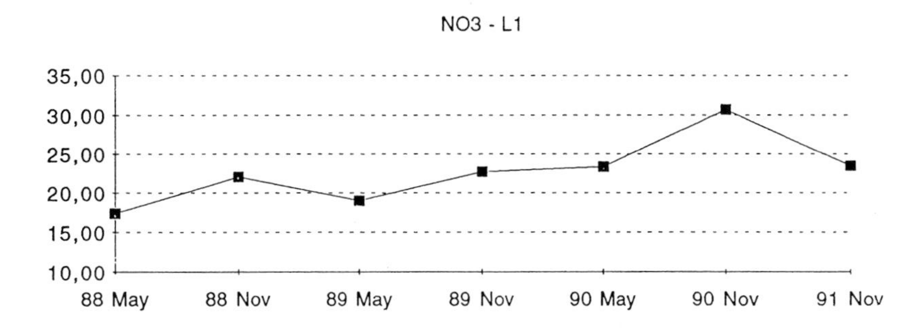

Figure 10. Time evolution of nitrates of the different layers in Balanegra aquifer.

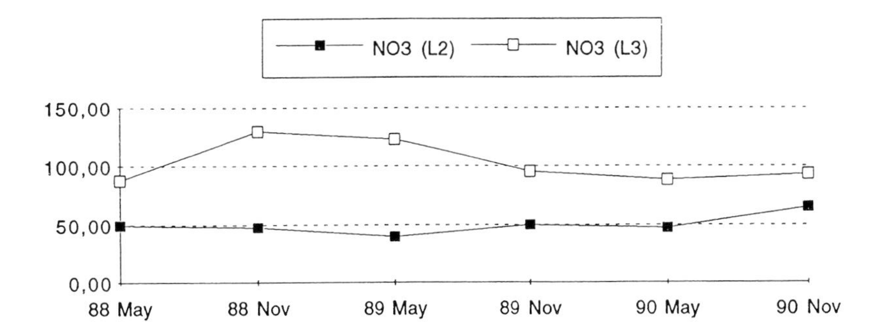

Figure 11. Time evolution of nitrates of the different layers in Balerma-Las Marinas aquifer.

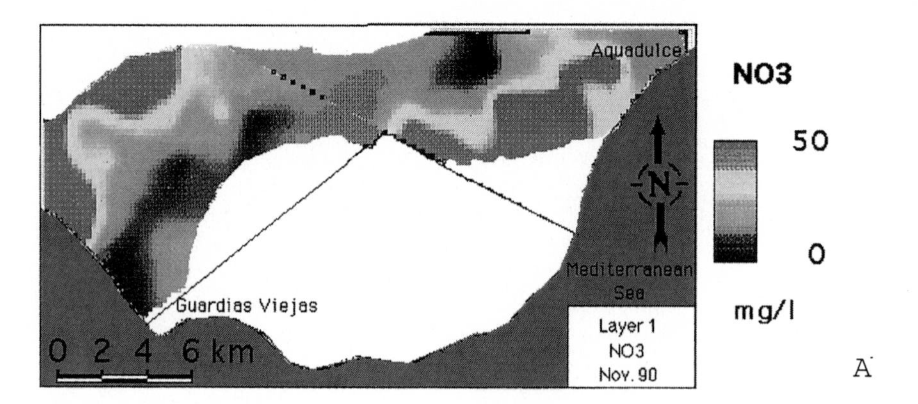

Figure 12. Map of nitrates in Layer 1, November 1990. (For color figure, see page 298)

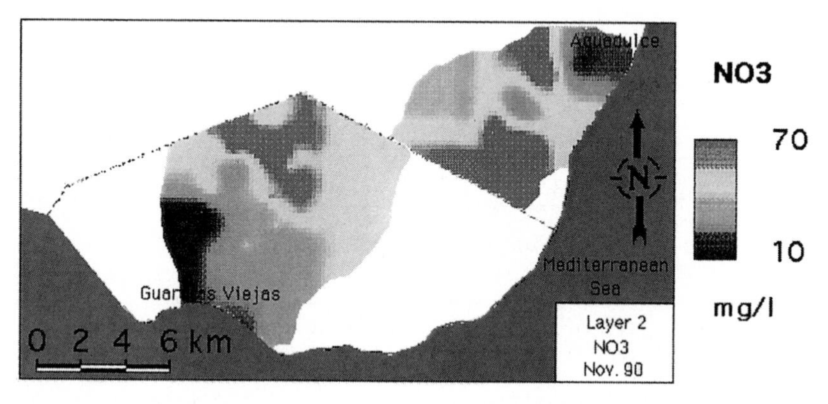

Figure 13. Map of nitrates in Layer 2, November 1990. (For color figure, see page 298)

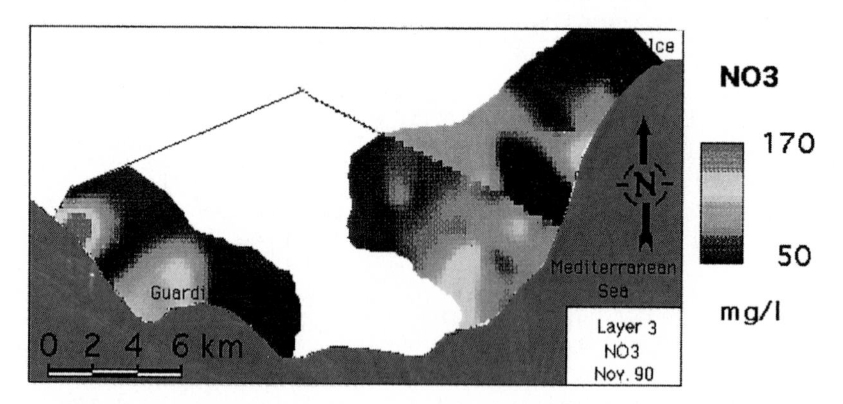

Fig. 14. Map of nitrates in Layer 3, November 1990. (For color figure, see page 298)

In the AG reservoir there is a concentration of high values (~ 5000 μS/cm) in the zone of *Sector III* (Southwest of Aguadulce village) through the different months with a dangerous area that is spreading towards the SE (sea) and in the direction of western Aguadulce village. A constant increase in the spreading of the polluted area is seen. Seasonal variations (Fig. 7) are not clear. The BM zone is less polluted, with values less than 4000 μS/cm. and a clear and constant increase of this area through time (Fig. 7). A more polluted zone appears in May of 1988, located next to the *Sector 111* area. This could indicate some type of connection between both aquifers. In the western part, a zone with smaller values (1000 μS/cm) appears. This disposition could suggest a separation between the two parts, the most and the least polluted. This separation could denote the presence of a main fault in this area.

Layer L3: This layer is in relation to AG and BM aquifers and corresponds to the more superficial material, the detritic Quaternary. In L3 both aquifers present similar values of conductivity (see Fig. 3 and 6), although values of AG tend to be superior to BM ones. Values of these variables in BM in layer L3 are higher than in layer L2, although not very different, because a connection between these two layers might exist.

In the AG aquifer, values of L2 are bigger than those of L3, but we have already commented on the fact that values of L2 in AG are in the most pumped area of the Campo de Dalias. In order to do a spatial study of conductivity through time, an analysis of the interpolation of this variable was made. Experimental data for this layer are concentrated in several areas and not spread over the Campo de Dalias. Specifically, in the BM aquifer these points appear in the western part of the Campo (zone of Balerma) and in the area of Mojonera and Las Marinas villages. In AG aquifer wells are located mainly along the coastal border (Fig. 1).
The most polluted areas in the Aguadulce reservoir are located around Roquetas village and this area increases through time northwards and to the coastal line. However, special care must be taken because of the few points that exist (Fig. 8). In the zone of Mojonera and Las Marinas of the BM aquifer, the area of highest values appears in 1988 to be restricted around Mojonera zone, and in 1989 a new polluted area arises in Las Marinas. In general, an increase in conductivity values through time is seen in this aquifer (Fig. 8).
In the zone of Balerma, well 295 presents a high value in conductivity that produces a rich zone in the area of Guardias Viejas village. In successive years a clear reduction in the value of this point produces a diminution of the anomalous area and even its disappearance. However, another high value appears in 1989 in the area of Balerma village, giving a zone of elevated values.

3.2 Evolution of nitrates

Layer Ll: This layer is the least polluted for this variable in the region. Average values in the Aguadulce and Balanegra aquifers are similar, around 20 mg/l; although a trend to increase this average value is shown in both aquifers, reaching 30 mg/l in B and 40 mg/l in AG. The behaviour in the two aquifers is parallel, and in this case, the definition of a single layer for this lithology is fully justified (Fig. 9 and 10). In general, we can consider that this layer Ll is not very affected by nitrate pollution, but the trend seems to be towards an increase in $N0_3$ concentrations. Therefore, special care must be taken to preserve this layer from nitrification. The spatial analysis through time indicates the appearance, in the aquifer of Aguadulce, of an area of higher nitrate concentrations (> 35 mg/l) in the zone of Las Hortichuelas, related to wells 961, 962 and, further east, well 1137, the high point appeared in 1988 and later reduced in concentration. In 1989 a new polluted area appeared, due mainly to well 1038, in Aguadulce village. In 1990 these two areas converged, becoming, in November, the most widespread polluted area (Fig. 12). In the Balanegra aquifer the most concentrated area is located in the Balanegra village zone and north of the aquifer. These areas are maintained through time, and in 1990 they increased considerably (Fig. 12).

Layer L2: This layer, comprised of the aquifers of AG and BM, presents high average values of $N0_3$, generally over 40 mg/l in both aquifers and higher in AG. Its trend is slightly different in AG and BM (Fig. 9 and 11).
AG presents a continuous decrease, varying between around 50 to 80 mg/l, and with a characteristic seasonal variation, higher values in autumn and lesser ones in spring. This could be due to a concentration during summer, when there is no rain and the heat is extreme. Later, rain can dilute concentrations and reduce these values. On the other hand, the BM aquifer has a rather constant evolution through time, with values around 50 mg/l, and a slight decrease in May 1989 and a rise in November 1990. Seasonal variations are not evident in this aquifer. Analysing spatial and temporal evolution (Fig. 13), a seasonal variation is observed on the maps, with higher values in spring and lesser ones in autumn. In the Aguadulce aquifer, practically all wells have values over 50 mg/l, and all the area of this reservoir is affected by nitrate pollution. Moreover, there are no satisfactory explanations for the results. In the Balerma-Las Marinas more experimental data exist, and it is possible to establish a polluted area (> 50 mg/l), located in the east of El Ejido village. Towards the western part of this village, values are very low (< 20 mg/l). This indicates the presence of a fault that separates high values from low, similar to the way conductivity acts (Fig. 13).

Layer L3: Obviously, the nitrate values in the most superficial layer are the highest. AG and BM are represented in this layer, and they present very close concentrations in this ion. A similar trend is seen in this layer: a general increase from May 1988 (~90 mg/l) to May 1989 (~120 mg/l), and later, a constant decrease is found in the layer until November 1990 (~60 mg/l in AG, ~90 mg/l in BM). No clear seasonal variations in L3 are observed for AG and BM (Fig. 9 and 11). The spatial and temporal study shows that all values are over 70 mg/l and a large part is over 100 mg/l. However, with the years a reduction of the values in the whole layer is clear, due mainly to filtering through lower layers that increase their values through time. The different polluted areas are located in positions similar to the conductivity ones (Fig. 14).

4 DISCUSSION

The study carried out allowed us to characterise hydrogeochemically the Campo de Dalias aquifers, on the basis of concentrations of nitrates and conductivity. This physical parameter is quite representative of the mineralization of water and a good indicator of some processes affecting aquifers in this zone. In the study three different layers, from which practically all the wells absorb water, were considered: Layer 1, corresponding to the deepest layer and consisting of Triassic carbonate materials; Layer 2, that represents Pliocene clastic carbonate materials; and Layer 3, made up of Quaternary lithologies.

In these layers, the analysis and evolution of salinization and pollution by nitrates were the main goal of this work. The hydrogeochemical study of this area was carried out using two seasonal surveys during the years 1988 to 1990, specifically, a spring survey and an autumn one. In relation to salinization, layer 1 presents parallel trends of pollution in the Aguadulce and Balanegra aquifers, but with a high level of pollution in Aguadulce, restricted to the area of *Sector III,* a zone highly overexploited, that has produced an important level of marine intrusion.

Moreover, the Balanegra reservoir has a natural barrier to oppose marine intrusion, since the pollution is less than the other one. In the Balanegra aquifer the area most affected by salinization is located in and around the Balanegra village and in the zone of Guardias Viejas village. Layer 2, that involves Pliocene calcarenites and AG and BM aquifers, appears more polluted in relation to the Aguadulce aquifer, and again in the zone of *Sector III.*

On the other hand, the Balerma-Las Marinas aquifer is less polluted than the previous one, but the area where values of conductivity are highest is located near *Sector III,* indicating some kind of connection between aquifers. In the western part of BM, a zone with smaller values appears. This disposition could suggest a separation between the two parts, the most and the least polluted, and might indicate a main fault. Layer 3, in relation to AG and BM and linked to Quaternary lithologies presents similar values of salinization in both aquifers, but slightly higher in the AG reservoir.

Most polluted areas in the Aguadulce aquifer are located around Roquetas village, increasing towards the north and the coastal line with time. However, special care must be taken because of the few points that exist. In the zone of Mojonera and Las Marinas of the BM aquifer, the area of highest values appears restricted around the Mojonera zone in 1988, and in 1989, a new polluted area arises in Las Marinas. In general, an increase in conductivity values through time is evidenced in this aquifer. In the zone of Balerma, one well presents a high value in conductivity that produces a rich zone in the area of Guardias Viejas village. In successive years a reduction in the value of this point is clear and produces a diminution of the anomalous area and even its disappearance. However, another high value appears in 1989 in the area of Balerma village, giving a zone of high values.

In relation to nitrates, the behaviour of L1 in the two aquifers is parallel, and in this case, the definition of a single layer for this lithology is fully justified. In general, we can consider that this Ll layer is not greatly affected by nitrate pollution, but the trend seems to be towards an increase in $N0_3$ concentrations. Therefore, special care must be taken to preserve this layer from nitrification. Analysing the spatial and temporal evolution of L2, a seasonal variation is observed on the maps, with higher values in spring and lower ones in autumn.

In the Aguadulce aquifer, practically all wells have values greater than 50 mg/l, and the whole area of this reservoir is affected by nitrate pollution. Moreover, there are few points to explain the results satisfactorily. In the Balerma-Las Marinas, more experimental data exist, and it is possible to establish a polluted area (> 50 mg/l), located in the east of El Ejido village. Towards the western part of this village, values are very low (< 20 mg/l).

This indicates the presence of a fault that separates high values from low, similar to the action of conductivity. In relation to L3, no clear seasonal variations are observed for AG and BM. The spatial and temporal study shows that all values are over 70 mg/l and a large part over 100 mg/l. However, with the years a reduction of the values in the entire layer becomes clear, due mainly to filtering through the lower layers, that increase their values through time. The different polluted areas are located in similar positions to the conductivity ones.

5. CONCLUSION

In order to avoid an increase in the deterioration of water quality some recommendations are given: (i) in relation to salinization, it is recommended to concentrate pumping far from the coast, in order to reduce the risk of marine intrusion; (ii) it is very important to make a water mass balance for establishing the evolution of the water inside the aquifer system; (iii) moreover, a rational policy of extraction is strongly advised; (iv) the problem of nitrate pollution increases with time. The contribution of this ion comes from the surface, and tends to filter down to the deepest water reservoirs. In this sense, it is important to protect the less polluted layers from nitrates. Further advice is to extend the use of fertilizers throughout the year, and not to concentrate it in several months.

KEYWORDS

Water quality, nitrate, salinity, ground water, pollution, remediation, agriculture, Spain, GOCAD.

REFERENCES

[1] Dominguez, P., Franqueza, P., and Gonzalez, A. "Sintesis hidrogeologica del Campo de Dalias y su entorno", *TIAC'88.* Almunecar, Granada. **11**:69-144 (1988).

[2] Foumiguet, J. "Néotectonique et Quaternaire marin sur le littoral de la Sierra Nevada", *Thèse Univ. Orléans,* Andalousie (Espagne). 234 p. (1975).

[3] "Estudio hidrogeologico del Campo de Dalias", *IGME.* **12 vol.** (unpublished) (1982).

[4] Mallet, J.L. "Discrete smooth interpolation", *ACM Transaction on Graphics.* **8-n2**:121-144 (1989).

[5] Pulido, A., Martinez-Vidal, J. L., Navarrete, F., Benavente, J, Molina, L., Gonzalez -Murcia, V., Macias, A., and Padilla, A. "Caracterizacion hidrogeoquimica del Campo de Dalias (Almeria)", *IARA.* Junta de Andalucia. Sevilla. 256 p. (1989).

[6] Pulido, A., Navarrete, F., Molina, L., Vallejos, A., Martin-Rosales, W. and Martinez-Vidal, J. L. In: "Some Spanish Karstic Aquifers", *Pulido Ed.* Univ. Granada. 143-158 (1993a).

[7] Pulido, A., Molina, L., Navarrete, F., and Martinez-Vidal, J. L. In: "Some Spanish Karstic Aquifers", *Pulido Ed.* Univ. Granada. 183-194 (1993b).

[8] Thauvin, J. P. "Riesgos de intrusion marina en el Campo de Dalias (Almeria)", *Tecniterrae.* **40**:19-26 (1986).

INCIDENCE OF FLUOROSIS IN INDIA WITH SPECIAL REFERENCE TO ANDHRA PRADESH

D.K. Samarasimha REDDY[a] *and K.L. Narasimha RAO*[b,1]

[a]*Minister for Panchayat Raj and Rural Development, Government of Andhra Pradesh, Hyderabad - 500 001, A.P., India*
[b]*Professor of Geology, S. V. University, Tirupati-517 502, A.P, India*

ABSTRACT

After reminding us that safe drinking water is as essential for mankind as any other nutritional component, the authors devote themselves to excess fluorides in ground water - a global phenomenon existing mostly in developing countries. It is assuming alarming proportions in India. A survey has been conducted in the districts of Andhra Pradesh; different maps and statistics included in the paper show fluoride concentrates in that region.
The disastrous effects of fluoride excess in the human body are reflected as a disease called "Fluorosis". There is no treatment nor cure for this disease; the only means to prevent it is to use defluoridated water. Various techniques more or less advanced and more or less costly are experimented in India. The Indian government tries to improve sanitary conditions and health awareness among local populations.

Incidence du Fluor sur la Santé en Inde, en particulier dans l'État de Andhra Pradesh

RÉSUMÉ

Après un rappel de l'importance pour l'humanité d'une eau potable saine au même titre que toute autre composante nutritive, les auteurs s'attachent à analyser les excès de fluor dans l'eau du sol - phénomène existant essentiellement dans les pays en développement. Ce phénomène atteint des proportions alarmantes en Inde. Cette étude a été effectuée dans la région d'Andhra Pradesh. Des cartes et des statistiques contenues dans ce texte dévoilent la concentration en fluor dans cette région.
Ce papier décrit les effets désastreux de l'excès de fluor dans le corps humain, qui se manifeste par une maladie appelée fluorose. Il n'existe pas de traitement ni de guérison possible en cas de maladie. Seule reste la prévention par l'utilisation d'une eau défluorée. On expérimente de nombreuses techniques de défluoration en Inde, plus ou moins avancées, plus ou moins coûteuses. Le gouvernement indien essaie par divers moyens d'améliorer les conditions sanitaires et de faire prendre conscience aux populations locales des méthodes de prévention pour la santé.

---

[1] To whom correspondence may be addressed

1 INTRODUCTION

Historically, civilisation began and centered around regions of abundant water supply. People knew very little about the quality of water and its impact on human health. With the passage of time, however, it has been felt that 'potable/safe' drinking water is as essential as any other nutritional component for societal health. Increasing population pressure, evolving industrial-chemical society, and advances in science and technology have contributed inter alia to contamination of drinking water with bacteria and virus, containing excess iron, fluorides, sulphates, carbonates and bicarbonates. The presence of these constituents in excess make sources unsafe for drinking purposes and in fact, are harmful, when consumed consistently.

One such constituent is the presence of excess fluorides in ground water. This is almost a global phenomenon, prevalent in USA, Japan, Soviet Union, Afghanistan, some African countries, some countries in Europe and also in many parts of developing countries, Algeria, Argentina, China, India, Kenya, Morocco, Senegal, Turkey and Thailand. In particular the fluoride problem is assuming alarming proportions in many states in India (Fig. 1) - Andhra Pradesh, Bihar, Delhi, Gujarat, Haryana, Himachal Pradesh, Karnataka, Madhya Pradesh, Maharashtra, Tamil Nadu, Uttar Pradesh and small pockets in other states (Susheela, 1987).

2 FLUORIDE IN THE ENVIRONMENT

2.1 Fluoride in biological tissues

Fluoride is a component of human and animal tissue deposited mostly on bones and teeth. While fluoride in limited quantities is desirable for the healthy growth of bones and teeth in human beings, it has toxic effects if present in excess quantities in waters and foods taken by them. This toxicity is reflected as a disease called 'Fluorosis'. Thus dental disorders are caused by ingestion of fluorides from drinking water. It has been found that while about 1.0 mg/l of fluoride is good for health, 1.5 mg/l would cause yellowing of teeth and their gradual decay. A quantity of 3.0 mg/l would lead to skeletal fluorosis resulting in the weakening of bones, especially at the joints.

In a bid to treat fluorosis in its early stages and prevent the irreparable deformities, doctors are trying to diagnose the early symptoms of the disease. One of the systems in the body which is affected very early is the gastro-intestinal system.

Severe damage is caused to the mucous lining, absorptive surfaces and mucous production in the gastro-intestinal tract resulting in nausea, loss of appetite, vomiting, gas formation, constipation and intermittent diarrhoea.

Other manifestations due to ingestion of fluoride ions are extreme weakness due to damage to RBC'S leading to low haemoglobin levels as well as muscle destruction. Fluoride poisoning can cause infertility in some cases, heart problems in others due to calcification of blood vessels. Children under twelve are more prone to fluorosis.

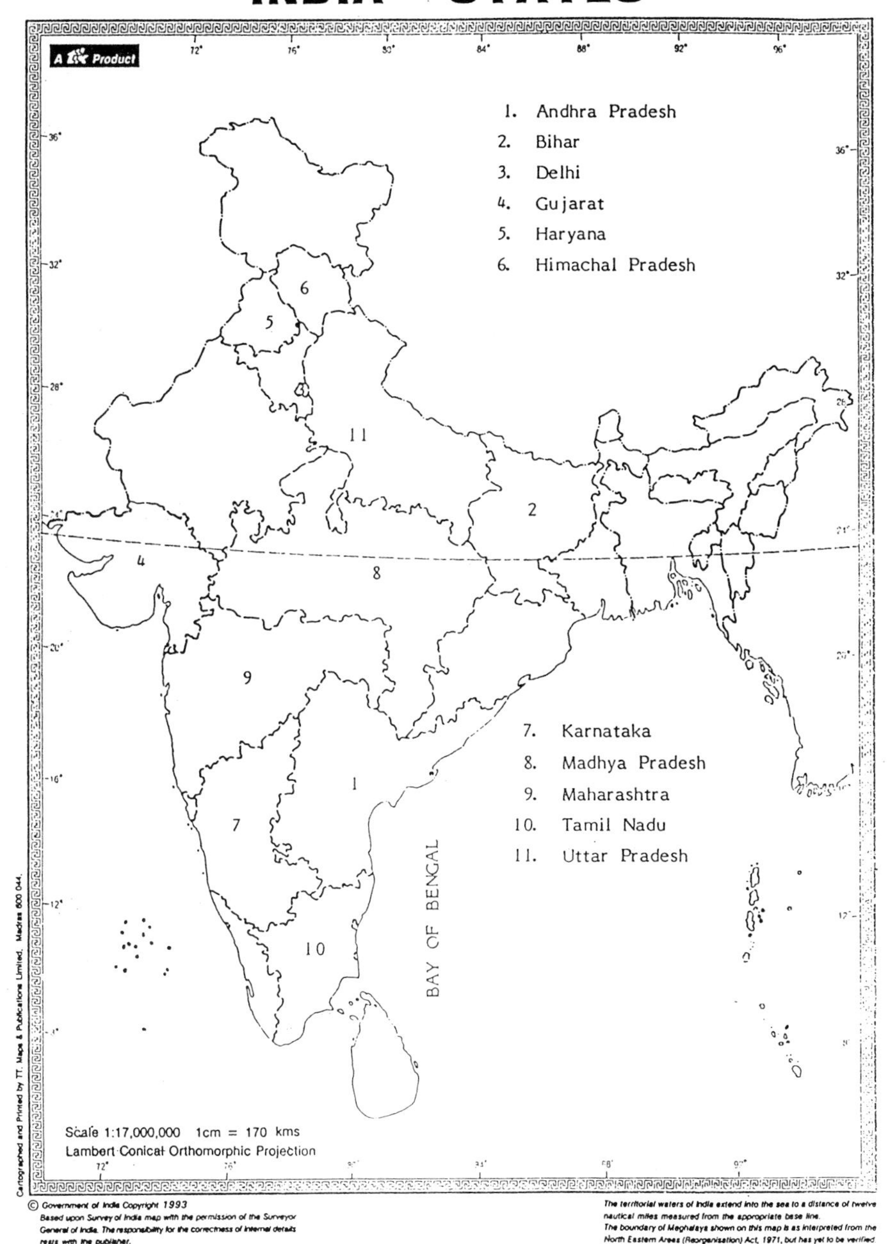

Figure 1. Geographical location map of India.

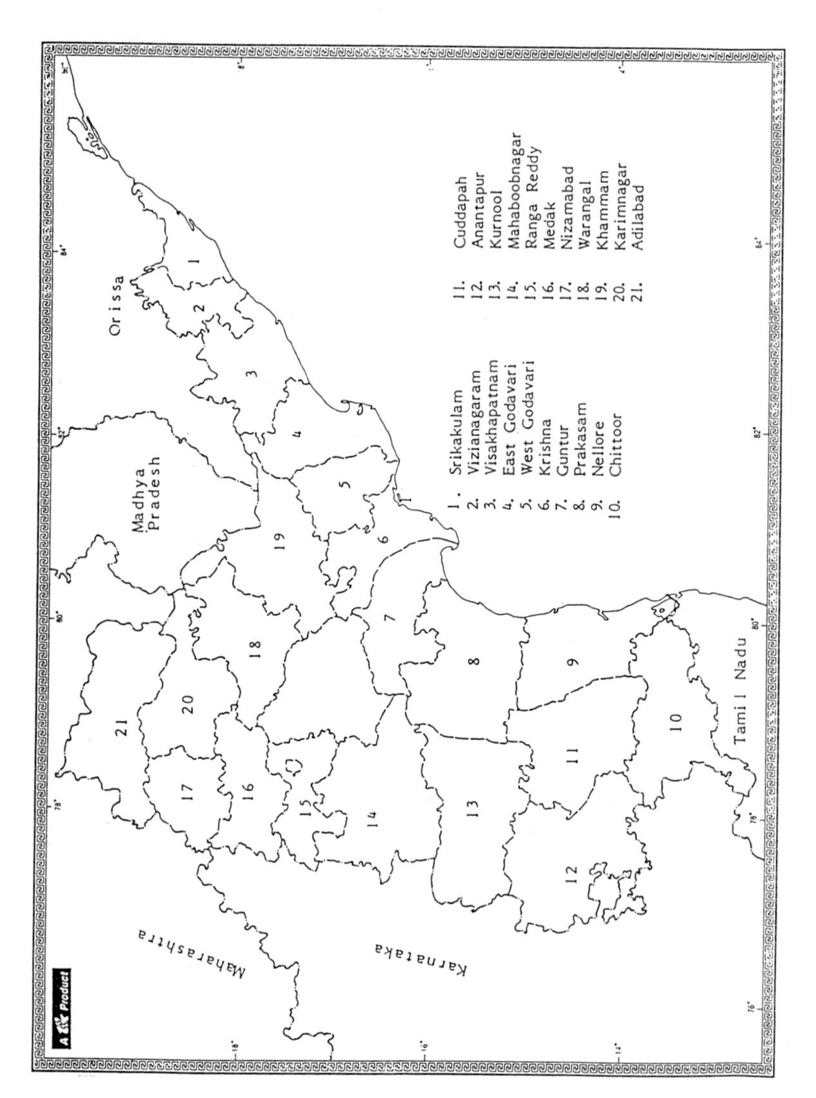

Figure 2. Map of the districts of the state of Andhra Pradesh.

2.2 Fluoride in soils

Leachable fluoride absorbed in the soil and clays are the main sources of fluoride in ground water. The fluoride content of ground water in India varies considerably. Apart from drinking water, the other sources of fluoride intake are food, cosmetics, drugs and industrial fumes (WHO, 1984). Among factors which control the concentration of fluorides are the climate of the area such as tropicality of regions with abnormal temperatures, scanty rainfall and rocky terrain, and the presence of the accessory minerals, fluorite and/or apatite in the rock-mineral assemblage through which ground water circulates. Regional and tectonic factors also play some part in affecting the fluoride concentration of ground water.

There is no treatment or cure for this disease, but it can be prevented by using defluoridated water. Roholm (1937), Largent (1954), Forbes et al (1978) and others have noted that humans removed from excessive fluoride exposures continue to excrete markedly elevated fluoride concentrations in the urine for long periods of time, indirect evidence of the mobilization of fluoride from the skeleton. So fluoride from the skeleton can be mobilized and excreted primarily in the urine by using defluoridated water.

In the present investigation, a survey was conducted to pinpoint the districts of Andhra Pradesh (Fig. 2), affected by "excess fluoride". These districts are Prakasam, Anantapur, Guntur, Krishna, Mahaboobnagar, Cuddapah, Nalgonda and Ranga Reddy. The data is presented in Table 1.

A total of 6586 water samples from each village in 21 districts of Andhra Pradesh was taken and analysed for fluoride content only (Table 1). The lowest fluoride concentration (1.0 mg/l) is recorded in Nellore and Cuddapah districts, and the highest (18.0 mg/l) in Medak district. The average fluoride concentration in Andhra Pradesh is 2.49 mg/l; minimum average fluoride concentration (1.94 mg/l) is in Srikakulam district and maximum (3.38 mg/l) in Chittoor district.

Only 13 villages (0.20%) of Andhra Pradesh have fluoride content in water within the permissible limit (1.5 mg/l); 5490 villages (83.36%) have fluoride content between 1.5 and 3.0 mg/l and 1083 villages (16.4%) beyond 3.0 mg/l. The maximum number of villages i.e., 1385 (25.23%) affected by dental fluorosis is in Anantapur district and the minimum i.e., 12 (0.21%) in West Godavari district. Similarly the maximum number of villages i.e., 184 (16.99%) affected by skeletal fluorosis is in Prakasam district and the minimum i.e., 2 (0.18%) in West Godavari district.

By and large, the number of villages in Andhra Pradesh with fluoride concentration in drinking waters within the permissible limit seems negligible. Most villages have been affected by fluorosis, either dental or skeletal, or sometimes both together and this is shown in phi diagram (Fig. 3).

Table I. Descriptive Statistics of fluoride content

Sl. No.	Name of the District	N° of villages	Min.	Max.	Ave.	S.D.	1.5 mg/l	1.5-30 mg/l	3.0 mg/l
1	Srikakulam	124	1.51	12.32	1.94	1.51	0	120	4
2	Vizianagaram	101	1.52	13.70	2.80	2.67	5	85	11
3	Visakhapatnam	37	1.60	11.50	2.43	1.74	0	29	8
4	East Godavari	24	1.60	12.00	3.27	3.32	0	20	4
5	West Godavari	14	1.52	4.90	2.08	0.88	0	12	2
6	Krishna	393	1.53	8.00	2.53	0.97	0	304	89
7	Guntur	247	1.55	9.60	2.74	1.23	0	168	79
8	Prakasam	715	1.60	14.00	2.72	1.25	0	531	184
9	Nellore	198	1.00	15.00	2.35	1.38	1	161	36
10	Chittoor	66	1.54	15.00	3.38	3.18	0	47	19
11	Cuddapah	215	1.00	15.00	2.61	1.91	5	177	33
12	Ananapur	1412	1.52	4.45	1.99	0.35	0	1385	27
13	Kurnool	409	1.05	3.50	2.65	1.02	1	293	115
14	Mahaboobnagar	319	1.60	10.10	2.18	0.75	0	295	24
15	Ranga Reddy	253	1.52	4.40	2.21	0.61	0	223	30
16	Medak	327	1.51	18.00	2.17	1.38	0	298	29
17	Nizamabad	214	1.60	3.80	2.21	0.83	0	190	24
18	Warangal	730	1.50	11.00	2.45	0.85	0	550	180
19	Khammam	239	1.54	4.90	2.28	0.68	0	197	42
20	Karimnagar	414	1.40	17.00	2.59	1.45	1	317	96
21	Adilabad	135	1.60	4.50	2.66	0.76	0	88	47
		6586			2.49		13	5490	1083

3 DISCUSSION

The Ministry of Rural Development, the Government of India, Public Health and Panchayat Raj Departments of State Governments, voluntary organisations like UNICEF have recognised fluorosis as a serious problem and have been working towards the provision of safe drinking water.
Different technologies world-wide have been studied to treat water for the removal of excess fluorides.

These techniques involve use of lime and alum, bone ash, activated alumina, electrolysis etc. The concept of a lime and alum treatment was developed (NEERI, 1987 - 'The Hindu' dated April 17, 1994) and had been modified and implemented to put up defluoridation plants suitable for hand pump and community models (Venkateswara Rao and Mahajan, 1988). The concept is known as 'fill and draw'.

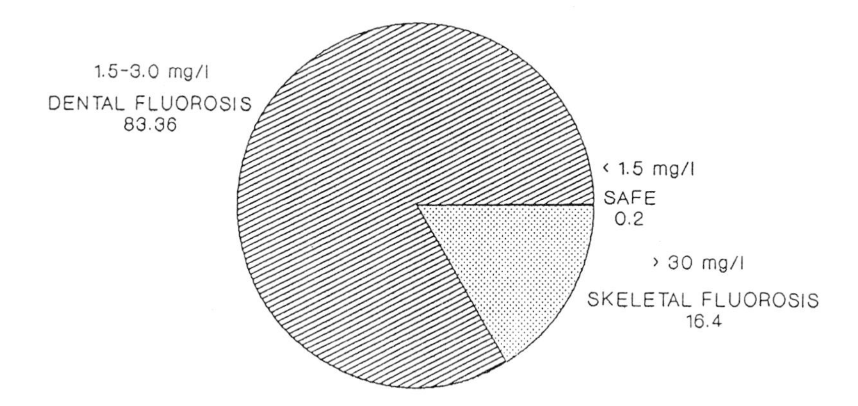

Figure 3. Statistics of fluoride content in Andhra Pradesh

The 'Fill and draw' technique, though viable, has field problems, e.g. it has to be attended by an operator for each batch of treatment.

The cost of treatment per litre is high due to input cost of lime and alum. Techniques other than activated alumina are in infant stages of study in India. The Government of India is preparing a master plan to provide safe drinking water to all villages in the country by the turn of this century at an estimated cost of Rs.450/- crores.

The Government of Andhra Pradesh is sparing no pains to eradicate this problem by providing safe drinking water with an annual investment of Rs.100/- crores on rural water supply through programmes such as the minimum needs programme (Rs.40/- crores), augmented rural water supply scheme (Rs.40/- crores) and under various drought relief schemes (Rs.20/- crores). In addition under the integrated fluorosis control project, Anantapur district has been chosen for UNICEF aid of about Rs . 14/- crores . So far 104 'fill and draw' type defluoridation plants have been set up in the country and about 377 hand pump defluoridation plants have been installed. Other defluoridation methods Nalgonda Technique, Prasanthi Technique, Activated Alumina and the Anion Exchange Technique are also being tried in Andhra Pradesh.

The Nalgonda Technique is most suitable and economical for large communities though it is time consuming, demanding the involvement of a skilled engineering staff and chemists for effective implementation. The Anion Exchange Technique, though reliable and acceptable in theory, has to await the positive results of this technique being experimented at present.

4 Conclusion

An integrated approach is sought to prevent the spread of fluorosis and other allied ailments by creating health awareness, providing better sanitary conditions and good hygiene, making people take part in activities related to the problem on hand and to evolving ways and means for improving living conditions. What can not be cured/eradicated can be contained and prevented by an integrated approach in the wake of various schemes and plans contemplated to call a halt to the spread of fluorosis.

Keywords

Water quality, F, India, fluoride, toxicity, soil contamination, pollution.

References

[1] Susheela A. K., Fluoride in India, The magnitude and severity of problem. Sci. Dev and Environ: 147-157 (1987).

[2] Venkateswara Rao K. and Mahajan C. L., Defluoridation of drinking water in developing countries in the "Hazardous and Industrial Wastes" (Manmohan Varma and James H. Johnson Jr., Eds.), Hazardous Materials Control Research Institute, Silver Spring, M.D., USA 1988, 55-64 (1988).

[3] WHO, Fluoride and Fluorides: Environmental Health Criteria 36, WHO, Geneva (1984).

Contamination of Lake Water by Pesticides via Atmospheric Transport

Pierre MASCLET

Laboratoire d'Études des Systèmes Atmosphériques Multiphasiques, Université de Savoie, Campus Scientifique, 73376 Le Bourget du Lac, France

ABSTRACT

High altitude lakes cannot be contaminated by a direct introduction of xenobiotic chemical compounds such as pesticides. Nevertheless they can be contaminated by these compounds after meso scale atmospheric transport. Laboratory experiments were carried out to determine the life time of pesticides in the atmosphere. Simulation of atmospheric degradation by UV and/or atmospheric oxidants of pesticides absorbed on terrigenic aerosols (kaolin) or anthropic aerosols (fly ash) shows that some pesticides such as atrazine, simazine or trifluraline have sufficient life times to be transported over several hundred miles from the source areas to the receptor sites and then contaminate the water after wet or dry deposition, while other pesticides such as diuron, carbaryl or fenitrothion are rapidly destroyed during atmospheric transport and cannot reach remote areas. Simultaneously, field measurements were carried out in the Savoy region in France to see if these pesticides were present in the troposphere and in the Alpine lake water. Significant amounts of atrazine, simazine, pendimethalin, and trifluralin were found both in the air and in the water, while no trace of isoproturon, diuron and fenitrothion were detected in the atmosphere in accordance with their expected life times. The presence of carbaryl, however, in lake water is strange since this compound has a short life time. These preliminary results will be used to keep track of the water in parallel with ecotoxicological tests and biochemical assays.

Contamination des Eaux de Lac par les Pesticides Transportés par Voie Atmosphérique

RÉSUMÉ

Les lacs en haute altitude ne peuvent pas être contaminés par une introduction directe de composés chimiques xenobiotiques tels que les pesticides. Néanmoins, ils peuvent être contaminés par ces composés après transport par voie atmosphérique à des échelles intermédiaires. Des expériences en laboratoire ont été conduites afin de déterminer le temps de séjour des pesticides dans l'atmosphère. La simulation de la dégradation atmosphérique par les U.V. et/ou par les agents oxydants des pesticides absorbés sur des aérosols terrigènes (kaolin) ou des aérosols anthropiques (cendres volantes), montre que certains pesticides tels que

l'atrazine, la simazine, ou la trifluraline ont une durée de vie suffisante pour être transportés sur plusieurs centaines de miles à partir des zones d'utilisation jusqu'aux sites de réception où ils contaminent l'eau par déposition sèche ou humide. Par contre, d'autres pesticides, tels que le diuron, le carbaryl ou le fenitrothion sont rapidement détruits lors de leur séjour dans l'atmosphère et ne peuvent atteindre des zones éloignées.
On a procédé simultanément à des mesures sur le terrain pour ce qui est des eaux des lacs alpins en Savoie (France) afin de voir si ces pesticides étaient présents dans la troposphère et dans les eaux des lacs alpins. On a trouvé des quantités significatives d'atrazine, de simazine, de pendimethalin et de trifluralin dans l'air et dans l'eau, alors que l'on n'a pas trouvé de trace d'isoproturion, de diuron ni de tenitrothion dans l'atmosphère, ce qui était conforme à leur durée de vie. Cependant la présence de carbaryl dans les eaux des lacs est surprenante puisque ce composé chimique a une durée de vie courte. Ces résultats préliminaires seront utilisés pour étudier la qualité de l'eau en parallèle avec les essais écotoxicologiques et biochimiques.

1 INTRODUCTION

Because of technological progress modern societies produce greater and greater amounts of exogenous or xenobiotic compounds, often referred to as anthropic chemical compounds, since they are produced directly by human activity. The major activities are industry, transport and modern agricultural practices.

The first and second of these activities are accompanied by fossil fuel combustion, such as coal, fuel oil, heavy oil, petrol and diesel fuel combustion. All these forms of combustion emit into the atmosphere chemical compounds, amongst which a family of organic compounds particularly dangerous for human health because they are often mutagenic and carcinogenic: PAHs.

Some of these compounds are famous, such as BaP rightly considered to be the most carcinogenic member of this family. These compounds are often resistant to chemical and photochemical attacks, since they are unreactive with atmospheric oxidants. They are hydrophobic and therefore poorly eliminated by rain and snow events. Consequently, they can be transported over long distances from the source zones - such as urban or industrial areas - towards the most remote regions of the Earth, thus contributing markedly to the background pollution.

Furthermore, modern agricultural activity requires the use of fertilizers and insecticides, designed to increase yields, and used in large quantities. Their main purpose is to eliminate agricultural parasites.

In the same category, the upkeep of the road and rail networks requires the use of herbicides, and the building industry uses fungicides. All these compounds, known collectively as pesticides, also represent a danger for human health and for the environment, since many of them are toxic. Some of these pesticides are rapidly destroyed in the atmosphere by photochemical processes or are sufficiently soluble to be found in the soils and in the surface water close to where they are used.

Some, however, are unreactive and less soluble and, like PAHs, can be transported over long distances and contaminate the atmosphere and all the media of our ecosystem.

The problem is therefore the same for these two families of compounds. In biogeochemistry, they are often used as excellent tracers, since by the monitoring of their atmospheric concentrations, it is possible to determine the nature and the location of the main pollution sources.

Furthermore the identification and the quantification of these compounds in the various "nonpolluted" media of the globe can be used to evaluate the total toxic capacity of the atmosphere. Studying the composition of the air, that of high-altitude or polar snow, sediments or marine water leads to the same result: knowledge of the global diffusion of chemical compounds transported to a large extent in the atmosphere, from populated areas to remote areas.

Three phases are involved in the atmospheric transport:

- gaseous form, close to that in which the compounds are emitted,
- particulate form (aerosols); the hydrophobic compounds have a high affinity for carbonaceous or mineral particules, and they are especially adsorbed on small particles (< 3μm),
- water droplets of clouds or snow, when these compounds are sufficiently soluble. This atmospheric transport occurs in three or four days. In the aerosol form, PAHs and pesticides are easily transported over long distances because of the fine aerosol association and the low reactivity of these compounds in the particulate form. The information related to the presence of these compounds in atmospheric particles is practically unaltered throughout their transport and even after their wet or dry deposition towards the soil or ice, for example.

In this study, one campaign in polluted area (Paris) and three campaigns in non polluted areas were performed:

- marine zone: the Mediterranean Sea,
- mountainous zone: lakes in the French Alps,
- polar zone: Greenland

Air (atmospheric particles), surface water in the mountain lakes and surface snow of the Arctic ice sheet were sampled over the last four years (1990-93).

An account of two families of toxic or carcinogenic compounds, PAHs and pesticides, has been established.

This study presents a quantitative account and a data set of these compounds at the beginning of the 1990s, in these zones considered to be non-polluted. It can be used to follow the subsequent evolution of the overall toxicity of the atmosphere and the media, contaminated in the next years or in the next decade.

2 Experimental Section

2.1 Samples and sampling sites

14 samples of atmospheric aerosols were taken in the west of Corsica, on the Mediterranean Sea coast during the spring of 1990. 13 Particulate PAHs: Naphthalene; Anthracene, Phenanthrene, Fluoranthene, Pyrene, Benzo a Anthracene, Chrysene, Benzo a and e Pyrenes, Benzo b and k Fluoranthenes, Indeno Pyrene, Benzo ghi Perylene and Coronene were identified and measured in these atmospheric samples.
8 atmospheric aerosol samples and 29 surface snow samples were collected in the ambient air and on the neve of the ice sheet (the upper 3.5 m) at the Summit station in central Greenland, during the summer of 1992.
6 atmospheric aerosol samples and 8 surface water or fresh snow samples were collected in Savoy and Lake Carro situated in the central French Alps (2600m altitude) during the summer of 1993.
14 pesticides, often used in Europe, were sought: Atrazine, Simazine, Terbuthylazine, Diuron, Trifluralin, Fenithrothion, Carbaryl, Alachlor, Pendimethalin, Isoproturon, Carbofuran, Malathion, Isoxaben and Fluzilazin.

2.2 Analytical method

The PAHs and the pesticides are extracted from the various solid or liquid matrices (particulate matter, water or snow samples) by liquid-liquid or solid-liquid extraction, in a Soxhlet, by means of mixtures of polar and nonpolar organic solvents: dichloromethane/cyclohexane for the PAHs and hexane/acetone/benzene for the pesticides. After purification these extracts are analyzed by gas phase chromatography with electron capture detection (pesticides) or by inversed phase HPLC, coupled with UV and fluorimetric detectors (pesticides and PAHs).

As has been shown elsewhere, by careful choice detection conditions, the sensitivity and the selecticity are optimized and analysis of these compounds at very low concentrations (of the order of 220 $pg.m^{-3}$ of air) is possible.

3 Results and Discussion

3.1 Polycyclic Aromatic Hydrocarbons

Relevant data on PAHs are presented in Table 1. The range of concentrations and the average concentration are given for the various sites sampled and the different periods.

PAHs are present in all the atmospheric aerosols: urban, rural, marine and polar aerosols. The order of magnitude of the concentrations varies considerably: four orders of magnitude between the total mean concentration (TMC) of the urban aerosol and of the polar aerosol. In urban dust the TMC is close to 100 $ng.m^{-3}$ (Paris), with a very pronounced seasonal variation (factor of 10) between summer and winter periods.

Table 1. PAH concentrations in polluted and non polluted areas

	PAHs (ngm^{-3})	Paris 1987	Mediterranean Sea 1990	Ivory Coast savanna fires	Ivory Coast back-country	Alps rural area 1992
TMC		summer 12 winter 100	North sector 4 West sector 0.5	50	0.3	0.8
range		5-150	0.4 - 5.5	5 - 90	0.2 - 0.5	0.3 - 2

	PAHs	Greenland atmosphere 1991 ($pg.m^{-3}$)	Greenland snow 1992 ($pg.kg^{-1}$)	Greenland deep ice 1993 ($pg.kg^{-1}$)
TMC		20	1350	800
range		5-120	500 -1950	250 - 1450

In warm season the TMC is low (10 $ng.m^{-3}$), while in winter the PAH concentrations are high (> 100 $ng.m^{-3}$), partly because emissions are higher during the cold season (domestic heating) partly because of the greater photochemical degradation of reactive compounds (such as BaP) in the intense sunshine period and finally partly due to the displacement of the phase distribution; the gas fraction increases and the particulate phase decreases with the higher temperature during the warm period.

Furthermore, there is a daily cycle, less pronounced than the seasonal cycle, which leads to high concentrations in the morning and to low concentrations in the middle of the afternoon. This behaviour is the same whatever the urban site studied, (Los Angeles, Paris, Stockholm, Tokyo, etc....) in agreement with the similarity of the sources from one town to another, (vehicle traffic, energy production from fossil fuels and waste incineration) and the similarity of the way of life. The various forms of anthropic combustion produce mainly submicronic aerosols with heavy PAHs - the most carcinogenic compounds.
Biomass combustion, particularly frequent in the tropical zones and in the boreal regions, is also a major source of particulate PAHs, because of domestic practices (savanna fires and charcoal fabrication for cooking) in hot regions such as Africa and deforestation all over the world - especially in tropical regions and Canada or Siberia.

The PAH levels vary considerably from one study to another, because of the non-standardization of the measurements due to difficulties in sampling in biomass burning areas. Nevertheless the general assessment of the TMC is 50 $ng.m^{-3}$. (5-90 $ng.m^{-3}$) (South Africa and Ivory Coast savanna fires). Also the biomass burning PAH flux is comparable to that of anthropic sources. The biomass burning emissions are characterized by PAH with generally four aromatic rings. These compounds are less toxic than the heavy PAHs found in urban aerosols.

In *rural areas*, PAHs are present at much lower concentrations: TMC is 100 to 300 times less than in urban areas (0.3 to 2 $ng.m^{-3}$). The concentrations depend considerably on the meteorological conditions and the contribution of air masses. In *Western Europe*, minima are observed during rain episodes and when the humid air masses come from the ocean, and maxima are observed when the air masses come from the North or the North-East sector. In rural atmosphere all PAHs are present, showing that the various pollution sources are mixed. Nevertheless, PAHs characteristic of petrol-engined vehicles (Benzo ghi Perylene) and fuel oil combustion (Fluoranthene and Pyrene) remain predominant in the atmospheric aerosols.

The background level of tropical atmospheres is about the same as that in rural European zones: TMC value is of the order of 200 $pg.m^{-3}$ (*Ivory Coast*). The air quality is the same in *European rural regions* and in remote tropical regions, but in the latter regions PAHs come mainly from biomass burning.

In a marine atmosphere the concentrations depend mainly on the sampling site, the proximity of the sources and the prevailing wind. In the *Mediterranean Sea*, concentrations are similar to those found in continental media (0.4-5.5.$ng.m^{-3}$). The many urban and industrial sites around the *Mediterranean Sea* explain these relatively high levels. On the other hand, the atmospheric levels in the Pacific ocean are 10 to 70 times lower (about 50 $pg.m^3$). Although low, these concentrations are not negligible and prove that particulate PAH can be transported over very long distances. During transport, the aerosols are diluted by atmospheric dynamics. Rain or snow events scavenge the air masses and their aerosol content.

These two phenomena result in a marked drop in the concentrations. Also it is logical to find very low concentrations in receptor zones remote from the pollution sources.

Particulate PAHs are also found in polar media (*Greenland*). The concentrations are very low - at the limit of detection. The TMC is 20 $pg.m^{-3}$. We observe important seasonal variations, with maxima at the end of winter and minima in summer.

The presence of significant amounts of PAHs in these regions shows that the whole globe is contaminated by these potentially carcinogenic compounds. The PAH profile is different from that observed in urban atmosphere or in biomass burning aerosols. The polar PAH profile is a mixture of both; however the latter predominate, showing the importance of the biomass burning source in the Arctic regions.

In *Arctic media*, after their transport, the aerosols are deposited on the earth by snowfall, covering the surface of the polar sheet, then accumulating in the ice. In samples taken from surface snow the TMC value is 1350. $pg.kg^{-1}$ of snow and in a deep ice core collected at the same site (75m deep), PAHs were detected in all strata of the ice; the TMC is around a value of 800 $pg.kg^{-1}$ of ice.

This result proves that these compounds have been emitted, transported and accumulated in the environment at least since the beginning of the industrial era (200 years, corresponding to the 75 m sampled).

Table 2. Pesticide concentrations in the Alps (1993).

Pesticides	Alps atmosphere aerosols ($ng.m^{-3}$)	Alps atmosphere gas ($ng.m^{-3}$)	Alps snow or lake water ($\mu mole.l^{-1}$)
TCM	4	8	25
range	2.3-10	2.5 - 14	12- 55

3.2 Pesticides

Table 2 gives the results for pesticides collected in the air, water, snow and ice of the alpine region. The presence of pesticides in the air and in natural media which can only be contaminated atmospherically - such as mountain lakes - has been clearly observed, even if the number of results is more limited, since only one campaign was carried out in the Alps. Of the dozen new pesticides, recently on sale, six pesticides are found systematically: Atrazine, Simazine, Terbuthylazine, Carbofuran, Diuron and Trifluralin.

These compounds are observed in all media of the environment: atmospheric gases, aerosols, rain, surface water and snow (other authors also identify them in the soil) according to simulation studies performed in the laboratory. This simulation shows that these compounds are resistant to photochemical degradation, except Diuron. For the five other compounds, no atmospheric degradation can occur, meso scale transport can be observed and contamination of water and soils, far from the sources, is possible. The presence of Diuron in media remote from pollution sources is unexplained.

Mean values for the sum of these six compounds are:

- 4 $ng.m^{-3}$ in aerosols;
- 8 $ng.m^{-3}$in the gas phase;
- 25 $\mu mole.l^{-1}$ in lake-water and surface snow.

As is shown in Table 2 there is a marked preference for aqueous media, in agreement with the high solubility of these products in water, in contrast to what is observed for the PAHs.

It seems likely therefore that clouds are the main vectors of these compounds and that the rain or snow events are the principal agents of deposition on the surface of the earth. Their persistence in media, *a priori* clean, such as the high-altitude lakes, which are not contaminated by run-off, shows that these compounds do not degrade in the atmosphere and in the water. The various compartments of the environment do not therefore allow the elimination of these often toxic compounds. Nevertheless, it does not appear possible to find these compounds very far from their source zones, because of their high solubility. In continental zones rain events are too frequent to allow long-distance transport. In the absence of data concerning very remote regions it is not possible to say more.

However, earlier studies on pesticides, no longer used, because of their high toxicity, such as Lindane (y HCH), Toxaphene and Chlordane, show that these compounds can also reach the polar regions. They are also found in the Great Lakes of America, where they persist. The concentrations are high, of the order of some 10 μmole.l^{-1}.

One should therefore remain prudent about the possibilities of general contamination of the ecosystem by these new compounds which can persist in the environment because of their resistance to degradation. One can nevertheless think that the load in the atmosphere must remain limited. It is not perhaps the same for uncultivated soils, such as the prairies, and the water of rivers and lakes in remote regions.

4 CONCLUSION

The study concerning the atmospheric transport of xenobiotic compounds, produced by human activity and often dangerous for human health, shows that these compounds, usually considered to be present in trace amounts in the atmosphere, can accumulate in the environment close to or very far from the sources of production and use. Mapping their geographical distribution is a long task, since it necessitates many campaigns and long periods of analysis. Nonetheless, this work provides a set of data which can eventually be used to follow the future evolution of these compounds in the environment. It remains also to put a figure to the impact of these compounds on health. At present there are no epidemiological studies showing direct correlations between morbidity and the levels of these compounds in the environment. However, more attention should be given to them in estimating global atmospheric pollution and health effects.

KEYWORDS

Contamination, pesticide, water, atmospheric, pollution, ecosystems, lake contamination, PAH, aquatic system.

REFERENCES

[1] Atlas E., Long range transport of organic compounds. *in* Knap A. H. (ed.) *NATO ASI Series, Dordrecht, The Netherlands, KLUWER, Acad. Publ.,* 105-135 (1990).

[2] Barrie L. A., Gregor D., Hargrave B., Lake R., Muir D., Shearer R., Tracey B. and Bidleman T., Arctic contaminants: sources, occurence and pathways. *Sci. Total. Environ.*, **122**,1-74 (1992).

[3] Masclet P., Mouvier G. and Nikolaou K., Relative decay index and sources of PAH. *Atmos. Environ.*, **20**, 439 (1996).

[4] Muir D. C. G., Grift N. P., Ford C. A., Reiger A. W., Hendzel M. R. and Lockhart W. L., Evidence for long range transport of Toxaphene to remote arctic and subarctic waters from monitoring of fish tissues. *in* "Long range transport of pesticides" Kurtz D. A. (ed), *Lewis Publ.*, 329-347 (1990).

[5] Pelizzetti E., Maurino V., Minero C., Carlin V., Pramauro E., Zerbinatii O. and Tosato M., Photocatalytic degradation of Atrazine and others s-Triazine Herbicides. *Environ. Sci. Technol.*, **24**, 1559-1565, (1990).

[6] Voldner E. C. and Schroeder W. H., Long range transport and deposition of Toxaphene. *in* "Long range transport of pesticides" Kurtz D. A. (ed), *Lewis Publ.*, 223-233 (1990).

A New Methodology Concept in Aquatic Ecotoxicology: the Integrated Approach from Laboratory to Outdoor Experiments

Gérard BLAKE

Laboratoire de Biologie et Biochimie Appliquées, École Supérieure d'Ingénieurs de Chambéry, Université de Savoie, 73376 Le Bourget du Lac, France

ABSTRACT

The newly emerging field of ecosystem health requires much data on disturbed functions and structures of these systems. New concepts of data collecting in aquatic systems, for ecotoxicological studies, are expanding and we present a multiscale approach based on laboratory (monitoring and plurispecific experiments) and outdoor systems. Microcosm and mesocosm studies make it possible to monitor the fate of pollutants and the effects on the structure and the main functions of the ecosystems. For aquatic ecosystems, this approach seems necessary in order to get a complementary set of information for medium and/or long term evaluations. This methodological concept will be developed through examples dealing with pesticides. In this paper, the effects of the contaminants on energy transfers in aquatic ecosystems (photosynthesis, respiration) will be analyzed for the possibility of monitoring complex ecosystems and for risk assessment.

Un Nouveau Concept Méthodologique en Écotoxicologie Aquatique: Une Approche Intégrée du Laboratoire aux Expérimentations Extérieures

RÉSUMÉ

Le récent domaine émergent de la santé des écosystèmes exige de nombreuses données sur les fonctions et les structures perturbées de ces écosystèmes. De nouveaux concepts de collecte des données des systèmes aquatiques pour les études écotoxicologiques sont en expansion et nous présentons une approche multi-échelle reposant sur les contributions des expériences de laboratoire (mono et plurispécifiques) et celles en plein-air. Des études en microcosmes et mésocosmes permettent de surveiller le devenir des polluants et leurs effets sur les structures et les fonctions principales des écosystèmes. Pour les systèmes aquatiques, cette approche semble nécessaire pour acquérir un ensemble d'informations complémentaires pour diverses évaluations. Ce concept méthodologique sera développé sur l'exemple des pesticides. On analysera les effets des contaminants sur les transferts d'énergie dans les écosystèmes aquatiques (photosynthèse, respiration) afin d'évaluer les risques ainsi que les possibilités de suivi de ces écosystèmes complexes.

1 INTRODUCTION

The ecological catastrophes of these past years (Bopal, pollution of the Rhine at Basel,...) combined with the spectacular effects of acute toxicity resulting in a high mortality rate for fish, clearly point to the need for better knowledge of the results and effects produced by chemical compounds in the environment.

Ecotoxicology, at the cross-roads of biological and chemical knowledge, is in full development because compulsory requirements and increased growth of new products go hand in hand with a concern for the protection of ecosystems. Recent debates, centered on acid rain or on the origin of eutrophication of aquatic ecosystems, have only served to bring out the limits of our capacity to make medium or long range forecast for xenobiotics introduced, voluntarily or not, into nature ([1]).

It is clearly still very difficult to cross the boundary between the results obtained from biological tests (bioassays) and scientific data from the field ([2]), and coherent scientific methods must be applied before we can attempt to generalize behavior, set up rejection thresholds and, a fortiori, establish estimated ecotoxicological risks, based essentially on risk analysis (cf. EPA, 1992 and Fig. 1).

The main difficulty is the need to predict medium and long term effects. The procedure for estimating ecological risk must, above all, make use of recent advances in methods for detecting and quantifying ecotoxicological effects of xenobiotics and complex sewages.

Recent methodological progress and new measures applicable to various steps from the laboratory to the field, permit better knowledge of toxic action sites and better selection of endpoints and perturbation bio-indicators. Research objectives focus on correlating these ecotoxicity results obtained at different structural and functional levels of natural environments in order to propose coherent regulation procedures.

Bio-tests and bio-indicators must not be considered separately but should be integrated in a single procedure; bio-tests study the effects and action mechanisms of contaminants in laboratory controlled conditions, while bio-indicators reveal the effects in ecosystems and in mesocosms (these latter are comprehensive units placed in outdoor conditions so as to recreate the "natural" functioning of a pond or a lake, for example) ([3], [4]).

In the case of chemical stress, work published over a period of many years reveals the increasing rarity of the most sensitive species. Functionally, perturbations are closely linked to the trophic dynamics of the system. The disappearance of such groups from an ecosystem means that the balance of production and decomposition of organic matter is perturbed for a very long time, and sometimes permanently, thereby affecting the evolution of all species present in the ecosystem.

Such problems encouraged recent measures for parallel follow-up of the perturbations in the composition and functioning of the disturbed ecosystem.

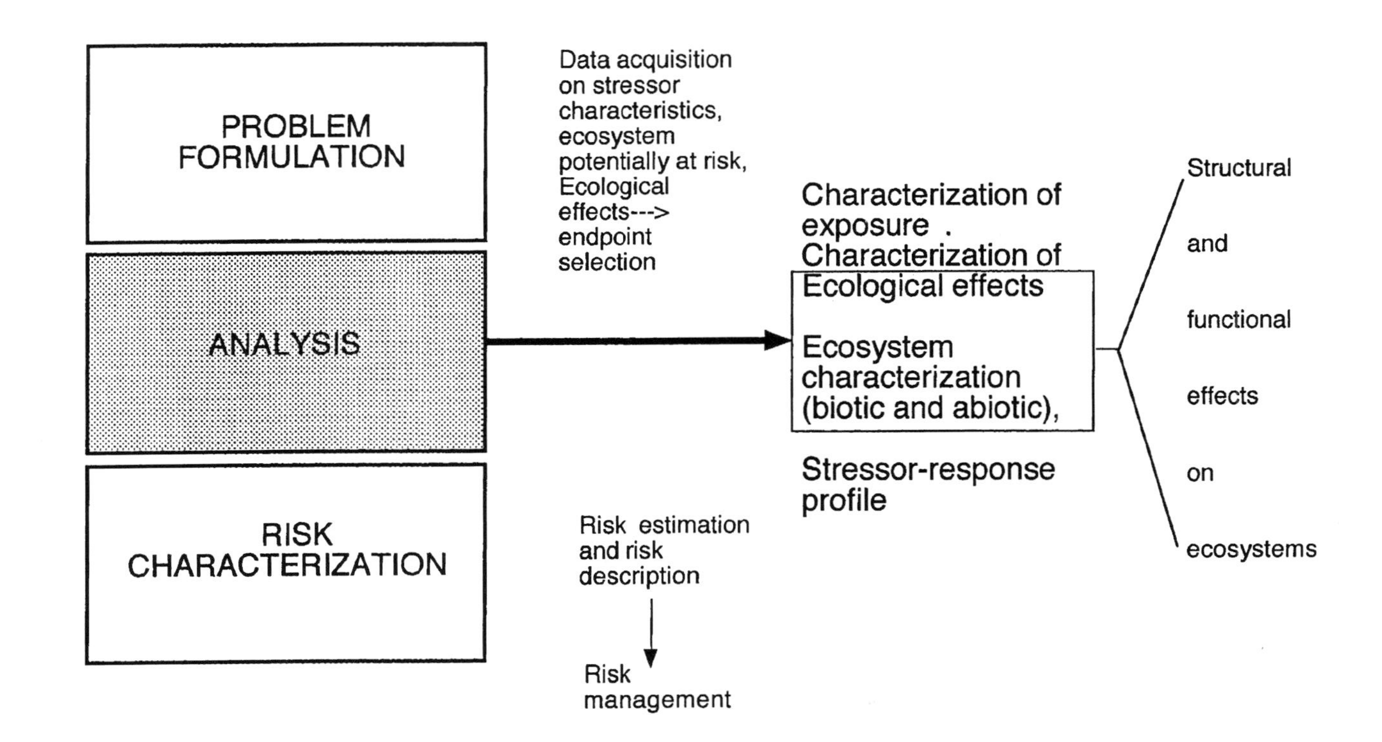

Figure 1. FRAMEWORK for Ecological Risk Assessment (from EPA, 1992)

Microcosms have been used in ecotoxicology for some twenty years now. It is impossible in the scope of this chapter to completely analyze the theory and all the application aspects required by such an approach. We shall, therefore, only present those points that seem basic to us.

Table I recalls the main work and the more important steps in this methodology. Most of these experiments were undertaken with microcosms reproducing processes of competition between species and prey/predator relations; these are particularly important in ecology.

Table 1. Examples of simplified ecosystems (laboratory or outdoor micro/mesocosms)

Waterways:			
Artificial river	Effects of nutrient elements	3m.	1964
Artificial waterway	Effects of pesticides	6m.	1964
Outdoor waterway	Effects of metals et al.	90m.	1977
Canal	Effects of currents and perturb.	100m.	1985
Stagnant water:			
Laboratory (microcosms):			
Aquarium	Primary production	4 l.	1962-63
Cylinders	Fate of Zn, Cr, Cd, Pb...	14 l.	1980
Aquariums	1 year, general evolution	170 l.	1973
Outdoor: (mesocosms):			
Pond enclosures	Fate of arsenic	700 l.	1980
Lake enclosures	Pesticides	1000 l.	1982
Lake enclosures	Primary production	2000 l.	1970
Enclosures	Pesticides	125 000 l.	1985
Limnocorres	Nutritive elements ...	18 000 000 l.	1973...

In addition to mesocosms, planktonic or benthic enclosures can be used. Such techniques are more demanding from the standpoint of data acquisition follow-up and are not exempt from methodological artefacts (wall effects, ecological evolution...).

Many studies presently carried out in natural environments show us that long range evolution remains extremely difficult to predict by mathematical models. The most reliable results come from methods combining:

- precise knowledge of the mechanisms unleashed by toxics on key organisms and the key functions of the ecosystem, and this can be implemented by normalized laboratory tests and complementary experimentation on a sufficiently broad scale to reveal toxic action mechanisms.

Experiments derived from the MERL (Ecotoxicology Marine Ecosystem Research Laboratory: more than 60 experiments). ([4], [5])

- reconstituting a structure and a systemic functioning sufficiently complex to validate the results of biotests and to reveal valid information on an ecological scale. In our opinion, this approach can only be supplied by a laboratory or outdoor microcosm scale.

Justifying the use of large reconstituted systems is particularly worthwhile for ecosystem divisions that require a minimum size to be considered valid:

- the case of sediments that, in insufficient amounts, might show reactivities more linked to wall effects (oxido-reduction artefacts) than to their own specific mechanisms (decomposing organic matter, water-sediment exchanges, bioturbation...);

- the case of biological organisms that require a space (or a time) that cannot be easily simulated in limited laboratory conditions: long-lived organisms (invertebrates and fish, vascular plants, ...). These organisms, by their ecological role, necessarily affect the future of ecosystems.

The variety of physical factors over a period of years lead to responses from the community of organisms that no sophisticated laboratory arrangement could reproduce; these conditions can only be reproduced by transposing ecological systems in a real situation. Such variations of the medium are also the source of different evolutionary possibilities of the contaminant agent, and we were able to verify that the specific physicochemical characteristics of the chemical compound reveal themselves in these conditions:

- revelation of the transient nature of a solvent that will be very quickly eliminated from an aquatic ecosystem into which it was introduced and verification that it is but slightly harmful to biological communities;

- on the other hand, the general distribution of a highly hydrosoluble compound with repercussions on those compartments with important exchange capacities, etc...

2 SOME ADVANTAGES AND DISADVANTAGES OF OUTDOOR MICROCOSMS AND MESOCOSMS

Various parameters, controlled or not, can take place in micro and mesocosm experiments and can limit the validity of the results. In large reservoirs, it was possible to follow their evolution for two to three years parallel to that of the ecosystems they were supposed to represent, and, in this case, the results are very favorable. In the case of microcosms, the ecological drift of the system can be reproduced much more often and must be watched.

In the case of laboratory microcosms with a high degree of environmental control, there must be the same procedure as that followed in mathematical modeling, i.e. one must intervene by varying, during the validation phase, the physicochemical factors and thereby follow the effects of this variation on the microcosm ([6]). In this way, the microcosm undergoes a sensitivity analysis; at the same time its range of application is also validated, as can be done for a mathematical model.

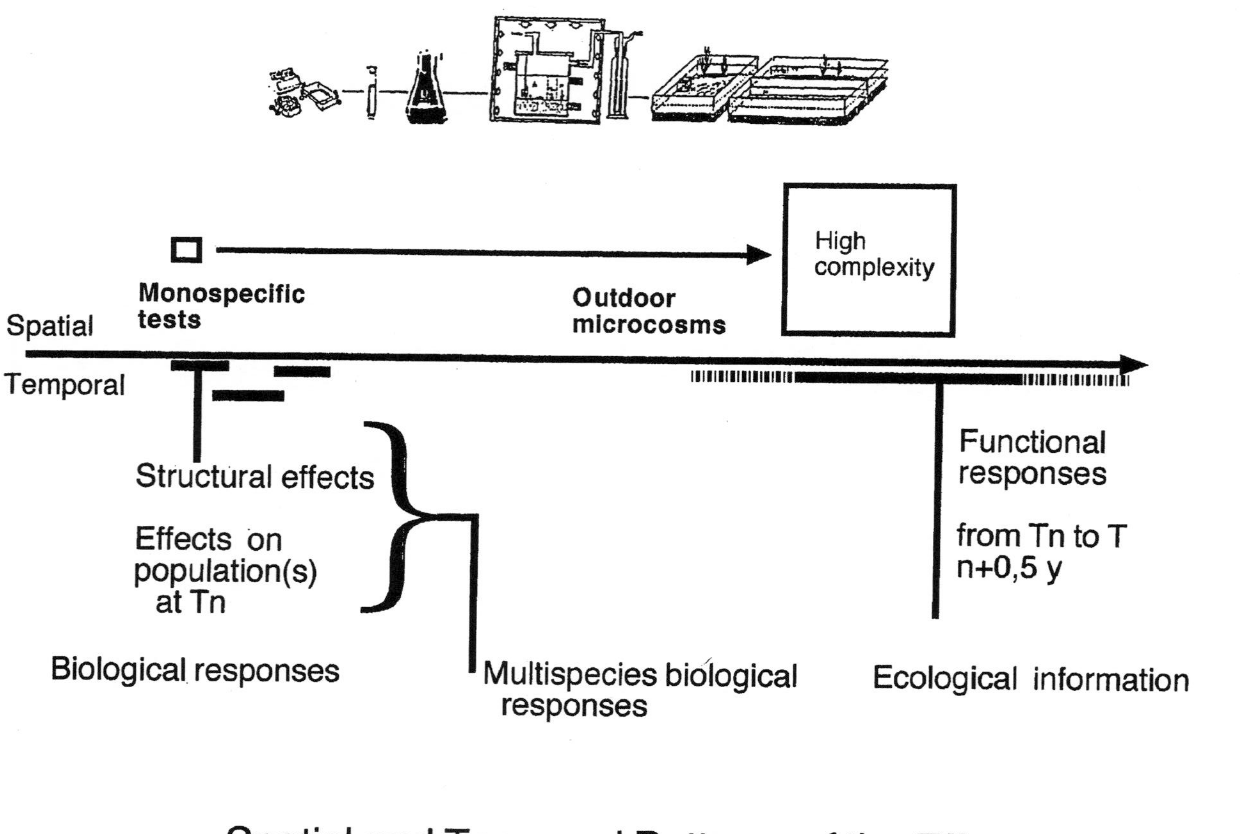
High complexity
Monospecific tests
Outdoor microcosms
Spatial
Temporal
Structural effects
Effects on population(s) at Tn
Functional responses
from Tn to T n+0,5 y
Biological responses
Multispecies biological responses
Ecological information
Spatial and Temporal Patterns of the Effects:
Biological and Ecological Significance

Figure 2

What are the *application limits* of outdoor microcosms and mesocosms?

Too large systems are difficult to handle, expensive and hard to duplicate. As for mathematical models, the increased validity gives rise to extreme complexity, making it difficult to exploit the results.
The use of mesocosms has definite advantages:

- in a natural ecosystem:
    - possibility of using radioactive tracers,
    - better control of sampling and of ecosystem representativity,
    - better control of experimental conditions,
    - evaluation possibility...
- in laboratory experiments (cultures, monospecific breeding):
    - mingling of several selected species with visualization of their relationships,
    - populations whose densities are more representative of the medium (plankton, invertebrates...),
    - more faithful to the natural behavior of organisms,
    - experiments over several seasons or even longer,
    - complexity close to that of a natural system,
    - truer distribution and chemical forms of the pollutant introduced ...

A certain number of *constraints* must be respected, and physical models can have their disadvantages as well:

- The biological populations of mesocosms show more stability in their evolution, since the accumulation of organic matter leads to a greater variety of heterotrophic populations.
- Mesocosms require a very ponderous follow-up procedure of their biological populations.

In fact, it is by an integrated approach, combining monospecific tests, laboratory and outdoor microcosms, that one can best integrate the different levels of biological and ecological information that can result in a more realistic approach of the multispatial and multitemporal dimension of ecological risk evaluation (Fig. 2).

3 CONCLUSION

As we know, biological tests constitute a partial method for learning about the real consequences of xenobiotic impact on aquatic ecosystems. Current procedures involve searching for and following, through systems of various and complex dimensions and structures, the effects of a whole range of contamination by a toxic, primarily via a key process (energy exchanges: respiration and photosynthetic activity).

This approach makes it possible to clearly determine the link between the physiological perturbations undergone by cellular cultures or organisms, the effects noted by simple and reproducible tests, the evidence of structural and functional disorders present in mesocosms situated in conditions very similar to those of the natural environment. The different observation scale also highlights certain effects that are specific to relationships between populations or trophic levels. These results clearly show the need to tackle simultaneously the structural and functional aspects, since they are very dependent on each other, just as the fate of the toxic and its effects over time are equally interdependent ([7], [8]).

Clearly the procedure implied by these methodologies is relatively cumbersome and costly, but it is entirely complementary to the laboratory approach with its monospecific tests. The complementarity of these procedures enables ecotoxicology to contribute to the improved protection of aquatic ecosystems by making it easier to limit pollutions and work out quantities better adapted to needs.

ACKNOWLEDGEMENTS

The author would like to express his thanks to Mrs Dubois for her help in the translation and in the preparation of the English version of this paper.

KEYWORDS

Ecosystems, pesticide, contamination, aquatic system, water, pollution, ecotoxicology.

REFERENCES

[1] Ford J., The effects of chemical stress on aquatic species. Composition and community structure In: Ecotoxicology: Problems and Approaches by Levin S.A., Harwell M.A., Kelly J.R. and Kimball K.D. (eds) (1989).

[2] Franco P.J., Giddings J.M., Herbes S.E., Hook L.A., Newbold J.D., Roy W.K., Southworeth G.R. and Stewart A.J., Effects of chronic exposure to coal-derived oil on freshwater ecosystems. I. Microcosms. Environ. Toxicol. Chem., 3:447-463 (1984).

[3] Giesy J.R. (ed.), Microcosms in Ecological Research. Conf-781101, U.S. Department of Energy, Washington, D.C. (1980).

[4] Giddings J.M., Microcosms for assessment of chemical effects on the prperties of aquatic ecosystems. pp. 45-94 in J.Saxena (ed.) Hazard Assessment of Chemicals-Current Developmeznt. Vol. II Academic Press, New-York (1983).

[5] Levin S.A., Harwell M.A., Kelly J.R. and Kimball K.D., (eds), Ecotoxicology: Problems and Approaches, Springer-Verlag, 547p. (1989).

[6] Touart L.W., Aquatic Mesocosm Tests to support Pesticide Registrations. EPA. 540/09-88-035. US Environmental Protection Agency, Washington, D.C. (1988).

[7] Weinstein D.A. and Birk E.M., Terrestrial Ecosystem structure and chemical stress.in: Ecology:problems and approaches, Levin S.A. et al. (eds), 181-209 (1990).

[8] Persoone G. and Janssen C. R., Field validation of predictions based on laboratory toxicity tests.in. Freshwater field tests for hazard assessment of chemicals (I. Hill *et al* eds.).CRC Press, 379-398 (1994).

STOCHASTIC IMAGING OF ENVIRONMENTAL DATA

Jean-Jacques ROYER[a]***, and Arben SHTUKA***[a,b,1]

[a]*CNRS Centre de Recherches Pétrographiques et Géochimiques, B.P. 20, 54501 Vandoeuvre-Les-Nancy Cedex, France*

[b]*LIAD - ENSG, B.P. 40, Rue du Doyen Marcel Roubault, 54501 Vandoeuvre-Lès-Nancy, France.*

ABSTRACT

Interpretation of environmental data is currently confronted with the problem of estimating the spatial variation of a parameter from a limited number of sample points irregularly distributed in space. The challenge is to extract the relevant information for a given problem from the individual observations at control points. For example, in risk analysis (water resource monitoring, overflow forecasting, pollution monitoring), the emphasis is on detecting the maximum value or the anomalies of the appropriate parameter.

Within this framework, classical numerical mapping techniques such as spline interpolation, multivariate regression or kriging, are of little use, because they provide an estimation of the mean local value while the expected distribution of the parameter would be more relevant.

In this paper, a stochastic simulation technique based on indicator functions is presented. It provides at each unknown point an estimation of the conditional probability function which can be further used to produce a "Stochastic image" or a realistic simulation map of the unknown parameter. This technique is also used to build various non-parametric estimators of an expected value (such as minimum, maximum, median, inter quartile range) which can be used in risk analysis. A case study in environmental monitoring is presented to illustrate the procedure.

Traitement d'Images par Méthodes Stochastiques des Données Environnementales

RÉSUMÉ

L'interprétation des données environnementales se trouve confrontée habituellement au problème de l'estimation de la variation spatiale d'un paramètre à partir d'un nombre limité de points d'échantillonnage irrégulièrement distribués dans l'espace. La difficulté est alors d'extraire des informations pertinentes compte tenu du problème posé à partir des observations élémentaires effectuées sur des points de contrôle.

---

[1] To whom correspondence may be addressed

Par exemple, en analyse des risques (gestion des ressources en eau, prévisions des inondations, gestion de la pollution), il est important de pouvoir estimer la valeur maximum, ou de détecter des anomalies du paramètre mesuré.

Dans ce cadre, les techniques classiques de cartographie automatique telles que les méthodes splines, les méthodes de régression multivariables ou d'interpolation tel que le krigeage, s'avèrent peu utiles, car elles fournissent une estimation locale de la valeur moyenne du paramètre étudié, alors que la loi conditionnelle du paramètre serait bien plus utile.

Nous présentons dans cet exposé une technique de simulation stochastique basée sur les fonctions indicatrices. Cette méthode fournit en chaque point de l'espace une estimation de la loi de probabilité conditionnelle du paramètre étudié. Cette estimation peut ensuite être utilisée pour construire une "image stochastique" ou bien une carte de simulation réaliste d'un paramètre inconnu. Cette technique est également utilisée pour construire différents estimateurs non-paramétriques de la variable étudiée (telle que sa valeur minimum, maximum, médiane, variation inter-quartile) qui peuvent être utilisés dans une analyse des risques.

Une étude de cas concernant la gestion de polluant est présentée afin d'illustrer la méthodologie proposée.

1 INTRODUCTION

Compared to other types of geodata, environmental data have specific features. They are defined on a spatial support (surface, volume, time, ...); they present a high spatial variability implying non robustness in estimators; they are multivariate with several components measured at the same location; their sampling is highly uncertain, and, finally, but not least, they are most of the time irregularly sampled in space.

Such features make the interpretation of environmental data difficult because one is confronted with the problem of estimating the spatial variation of a parameter from a limited number of sample points irregularly distributed in space. The challenge is then to extract the relevant information for a given problem from the individual observations at control points.

For example, in risk analysis (water resource monitoring, overflow forecasting, pollution monitoring), the emphasis is on detecting the maximum value or the anomalies of the appropriate parameter.

The objective is to produce reliable maps from irregular sampling points and to forecast maximum (or minimum) values of the given parameters, including a confident range of the estimated values.

Within this framework, classical numerical mapping techniques such as spline interpolation, multivariate regression or kriging, are of little use, because they provide an estimation of the mean local value while the expected distribution of the parameter would be more relevant.

Stochastic simulations of random fields have been suggested as possible non linear techniques to produce a "Stochastic image" or a realistic simulation map of the unknown parameters. Such a simulation method is of fundamental importance to a number of other technological applications (risk analysis, environmental risk assessment) including hydrogeology (fracturation or ground water flow modeling) and the petroleum industry (reservoir modeling).

This work presents a stochastic simulation method for continuous random variables using an indicator coding technique. It is based on a DSI type of algorithm under a set of constraints which guarantees that the cumulative distribution function (**cdf**), the variogram or the drift of the estimates are respectively the same as those of the initial data. The technique can be applied to estimate a set of risk assessment maps in environmental problems.

2 Overview of Geostatistic or Stochastic Simulation Techniques

The stochastic simulation technique has received increasing attention in geostatistics since the early theoretical work of Matheron (1968). Because of its practical importance, particular interest arose during the last decades in methods based on Indicator Functions (Journel, 1989), especially in risk analysis. Much of the work on this topic has been concerned with two kinds of problems:

- the simulation of a Random Field imposing a model of spatial variability using the covariance or variogram function;
- the simulation imposing a statistical property of the estimated parameters such as the distribution function (Monte Carlo methods).

Most of the time, these two constraints are processed independently according to the following steps:

- estimation of the mean Random Fields Z_m using a classical estimation procedure such as least squares, spline, ARMA or kriging methods;
- Monte-Carlo simulation of the random component Z_r imposing the distribution (Gaussian anamorphosis) function and the variogram (turning band methods, LU decomposition);
- the resulting simulation field Z_s is then the sum of Z_m and Z_r ;

Such a two-step procedure does not guarantee that the simulated field Z_s has the same cumulative distribution function (**cdf**) as the initial sampled value Z_i, except when Z(x) has a Gaussian distribution function. To tackle this problem a considerable amount of work has been done in the last decade (Sequential Gaussian Methods, Orthogonal Polynomials, Indicator kriging ...) [Deutch and Journel, 1992].

Other works treat such simulation problems using the Indicator kriging. The initial RV Z(x) is coded by a vector of indicators [Journel, 1986], then using a Monte-Carlo procedure the Z_s is simulated at each unknown point together with the estimation of the indicator function.

Such a procedure implies that the **cdf** of the simulated values Z_s is the same as the **cdf** of the experimental Z_i. However, it does not guarantee the variogram constraints (i.e. the variogram computed on Z_s is the same as that computed on Z_i). This last constraint is generally assumed by a postponed treatment using simulated annealing (Deutch and Journel, 1992). This approach involves a considerable amount of computer time and memory.

3 Some Basic Geostatistical Concepts

Etymologically, the term *geostatistics* designates the statistical study of the Earth. The first to use this term was G. Matheron (1962), and his definition was: "*Geostatistics is the application of the formalism of random functions to the reconnaissance and estimation of natural phenomena*". Several basic notions will be briefly presented below. The reader will find an exhaustive text book presentation in Matheron (1968), Journel and Huijbregts (1978), Deutch and Journel (1992).

3.1 Regionalized Variables

A natural phenomenon can often be characterized by the distribution in space of one or more variables, called *regionalized variables*. The distribution of concentration in space, the level of pesticides, the porosity or the permeability of a soil can be considered as regionalized variables. The regionalized variables have two apparently contradictory characteristics:

- a local, random, erratic behaviour which resembles that of a random variable;
- a general (or average) structured aspect on a larger scale which can be represented as a function;

Geostatistical theory is based on the observation that the variability of regionalized variables has a particular structure. The values z(x) and z(x+h) of a given regionalized variable Z(x) at points x and x+h are auto-correlated. This auto-correlation depends both on the vector h and on location x.

The direct study of the mathematical function z(x) is excluded because its spatial variability is usually extremely erratic, with all kinds of discontinuities and anisotropies. The random and structured aspects of the regionalized variable are expressed by the probabilistic language of *random functions*.

A random function Z(x) can be seen as a set of random variables $Z(x_i)$ defined at each point x_i of the domain D: $Z(x) = \{ Z(x_i), \forall x_i \in D \}$. The x_i represent the space coordinates { x, y, z } of a given point i.

3.2 Estimation of a Distribution

The cumulative frequency distribution (**cdf**) of a RV, denoted $F(z_c)$, is defined as the proportion of values below the thresholds z_c. By convention, the cumulative frequency below the minimum value is zero and the frequency below the maximum value is one:

$$F(z_{min}) = 0$$
$$F(z_{max}) = 1$$

The local expression of the cumulative distribution function will be written below as F(x;z). The relation between the global and the local cumulative distribution function is:

$$F(z) = \frac{1}{||D||} \int_{x \in D} F(x;\, z)\, dx \qquad (1)$$

where D is the domain of definition of Z(x).

3.3 The Variogram

The variability of a given regionalized variable is characterized by the variogram function $2\gamma(x,h)$, which is defined as the expectation of the random variable $[Z(x) - Z(x+h)]^2$, i.e.,

$$2\gamma(x,\, h) = E\,\{\, [Z(x) - Z(x+h)]^2 \} \qquad (2)$$

In general, the variogram $2\gamma(x,h)$ is a function of both the point x and the vector h, but in geostatistical applications it is assumed that the variogram depends only on the distance h between data point x and x+h and not on location x (*intrinsic hypothesis*). It is then possible to estimate the variogram $2\gamma(x,h)$ from sample points as the arithmetic mean of the squared differences between two experimental measures $[z(x_i), z(x_i+h)]$ at any two points separated by the lag h.

$$2\gamma^*(x,\, h) = \frac{1}{N(h)} \sum_{i=1}^{N(h)} [z(x_i)-z(x_i+h)]^2 \qquad (3)$$

where N(h) is the number of experimental pairs $[z(x_i)-z(x_i+h)]$ of data separated by the vector h .

3.4 Indicator Function

The indicator of a random function Z(x) is the binary function $I(x;\, z_c)$ defined by:

$$I(x;\, z_c) = 1 \text{ if } Z(x) \leq z_c \text{ else } 0$$

where z_c is a threshold value defined in $[z_{min}, z_{max}]$. The stationary mean of the binary indicator is the cumulative distribution function of the random function Z(x) itself; indeed:

$$E\{I(x;z)\} = 1 \cdot Prob\{Z(x) \le z\} + 0 \cdot Prob\{Z(x) > z\} = Prob\{Z(x) \le z\} = F(z)$$

The local estimation of the cumulative distribution function is very important in many applications. The indicator map would separate at high thresholds the distribution of the maximum values in the studied area, while at low thresholds, it would show the distribution of the minimum values.

4 STOCHASTIC SIMULATION USING DSI

4.1 Goals and Constraints

Let $D = \{U \cap K\}$ be a domain on which the RV Z(x) has been sampled at location $\{x_i \in K\}$. The purpose is to estimate the value of Z(x) at an unknown location x in U. Figure 1 shows a real case study in which the values are known on a regular grid 50x50. The RV is sampled by 140 data reported in Figure 2. Classical interpolation techniques such as kriging are applied to reconstruct the initial image (Fig. 3). It is obvious that the estimator used in Figure 3 is inefficient to reconstruct the initial image because the result is too smooth, especially for the maximum and the minimum values. Moreover, kriging provides an estimated value plus a standard deviation for the estimate but not a probability distribution of the errors (or of the risk).

The proposed technique consists in estimating the local **cdf** of Z(x) at each unknown location using the indicator function approach. However, this involves some theoretical difficulty because general estimator techniques (kriging, cokriging) cannot be applied directly to estimate I(x,z) because they do not assume the following properties:

- *Order inequality constraints*: for a set of thresholds $\{z_i;\ i=1,n\}$ the different indicators must verify the cdf properties: $I(x,z) \in [0,1]$, $I(x,z) \le I(x,z')$ for $z < z'$;
- *Global **cdf** constraint*: The mean of the estimated indicators I(x,z) must be equal to the cdf of Z: F(z), according to relation (1);

These different constraints have been implemented in the DSI algorithm (Royer and Shtuka, 1994; Shtuka, 1994).

4.2 Indicator based simulation using DSI

The DSI (Discrete Smooth Interpolation) method was developed by Mallet (1989, 1992) to model complex 3D geological surfaces (reverse faults, salt domes,...) and to interpolate physical properties (velocity, porosity, grades, ...). This method is a new interpolation method based on the optimization of an objective function (such as the roughness of the surface) taking into account precise and imprecise data (hard or soft data) using several local linear constraints. Examples on modeling the geometry of 3D complex objects both in Earth Sciences and Medecine can also be found in Royer et al., 1995 (CODATA Vol. I, Chambery, 1994). To take into account the constraints defined in § 4.1, two types of constraints were added to the classical DSI approach.

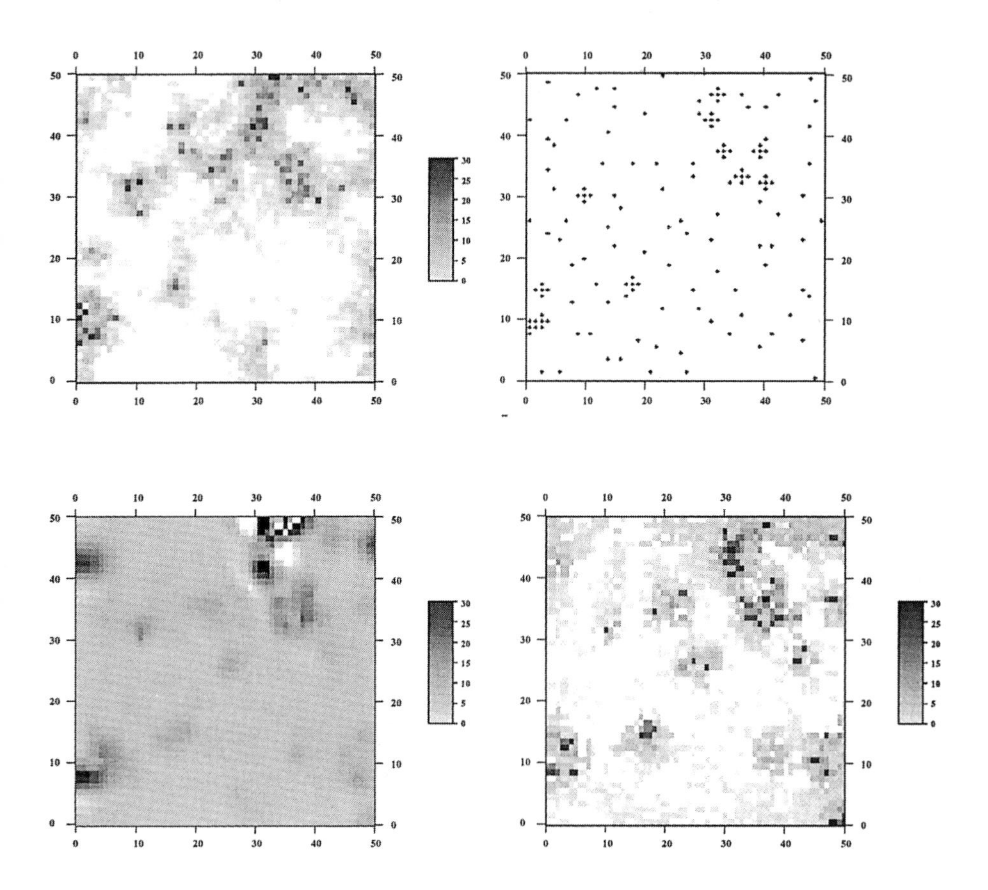

Figure 1. (Top left) Reference image used for testing the Stochastic Simulation proposed technique.

Figure 2. (Top right) Sample data set containing 140 points extracted from the reference image (Fig. 1).

Figure 3. (Bottom left) Interpolated image from the dataset (Fig. 2) using DSI.

Figure 4. (Bottom right) Simulated image using the sequential indicator method (see text).

Inequality constraints for Indicators; Suppose that the Indicators have been defined using n distinct real values in the interval $[z_{min}, , z_{max}]$:

$$\{z_{min}, < z_1 < z_2 \ldots z_i < z_{i+1} < \ldots < z_n < z_{max} \}$$

Noting $\varphi^i(x)$ the Indicator function $I(x, z_i)$, a classical DSI notation, then, the vector function $\varphi(x) = [\varphi^1(x), \varphi^2(x), \ldots \varphi^i(x) \ldots, \varphi^n(x)]^t$ can be estimated using the DSI accounting of the global **cdf** constraints (see below). In order to respect the *Inequality constraints for the Indicators* a post-processing modified slightly the $\varphi^i(x)$ at each iteration according to the following code included in the main loop of the DSI iteration schema (see Mallet, 1992) used for updating $\varphi(x)$:

```
for ( i =1; i ≤ n; i ++ ) {
    if ( φ^i(x) < 0 ) φ^i(x) = 0
    if ( φ^i(x) > 1 ) φ^i(x) = 1
    if ( (i > 1) && (φ^i(x) < φ^(i-1)(x)) ) φ^i(x) = φ^(i-1)(x)
}
```

This modification is similar to a numerical optimization searching for the minimum of a function with inequality constraints.

Global **cdf** *constraint*: The global **cdf** constraints, given by integral relation (1), can be written as follows:

$$\frac{1}{N} \sum_{k \in \Omega_D} \varphi^i(k) \quad = \quad F(z_i) \qquad \forall\, i = (1, ..., n)$$

This equation includes n constraints and can easily be formulated as a set of DSI constraints with the general form :

$$A_i \,.\, \varphi = \sum_{\nu=1}^{n} \sum_{k \in \Omega_D} A^{\nu}_i\,(k) \,.\, \varphi^{\nu}(k) \; = \; b_i \qquad \forall\, i = (1, ..., n)$$

where $A^{\nu}_i\,(k) = I/N$ if $\nu = i$ or equal to 0 otherwise, and $b_i = F(z_i)$. This type of constraint is taken into account by the DSI local algorithm by defining:

$$\gamma^{\nu}_i\,(\alpha) = \omega^2_i \,.\, (\frac{1}{N})^2 \text{ if } \nu = i \text{ or equal to 0, otherwise}$$

$$\Gamma^{\nu}_i\,(\alpha) = \omega^2_i \,.\, (\frac{1}{N})^2 \,.\, \{\, (\frac{1}{N}) \sum_{\beta \neq \alpha}^{n} \varphi^{\nu}(\beta) - F(z_i)\} \text{ or equal to 0, otherwise}$$

These constraints have been included in the DSI algorithm.

5 Indicator Simulation

Two types of simulation techniques were investigated: the *simple* and the *sequential simulation.*

5.1 Simple simulation

The *simple simulation* is performed using the Indicators interpolated by DSI at all grid points. The implemented algorithm consists of two steps:

- DSI interpolation of the indicators $I(x;z)$ at any grid point including the indicator constraints in such a way that $I(x;z) \equiv F(x;z)$;
- Computation of a simulated value Z at grid point using a Monte-Carlo method (*i.e.* generation of a uniform random number $u \in [0;1]$, followed by a simulation of Z using $z(x) = F^{-1}(x; u)$);

This technique is very fast and provides a set of realizations $Z_S(x)$ with the following properties:

- $Z_S(x)$ honors the observed data being equal to $Z(x)$ at each location $x_i \in K$ where data are available (*conditioned simulation*);

- The **cdf** (histogram) of $Z_S(x)$ is equal to the imposed **cdf** $F(z)$. The $F(z)$ can be the experimental **cdf** or it can also be an arbitrary given **cdf** imposed by external information.

- One of the disadvantages of the simple simulation is that the variogram computed on the simulated value $Z_S(x)$ is not equal to the *a priori* variogram computed on the initial data.

- Moreover, the estimated indicators are smoothed because of the roughness criteria used in DSI.

To overcome these difficulties, especially the smoothness of the simulated values, a *sequential simulation* method was investigated below.

5.2 Sequential simulation

The *sequential simulation* consists in conditioning the simulated value by the original data and by the previous simulated values computed at each step. The algorithm is built as follows:

- define a random path to visit each grid point at which the simulation is performed;
- for a given grid point x_i, interpolate the indicator **cdf** $I(x_i;z) = F(x;z)$ using the original or previously simulated data points located in the neighbourhood;
- simulate a value $z(x_i,)$ using the **cdf** $F(x;z)$ by a simple Monte-Carlo method and save it in the simulated data set;
- transform the estimated **cdf** $I(x_i;z)$ into a step function computed on $Z(x_i)$, and put it in the initial data set;
- repeat the last three procedures for all grid points.

The sequential simulation provides less smooth simulation results, but it implies a reprocessing for each simulation. The computing time is then longer than in the simple simulation case. However, in the *sequential simulation*, the indicator vector can be coded in binary format (one bit per threshold), implying less memory. The sequential simulation does not take into account the variogram constraint. However, using a post-processing technique at the DSI iteration level, it is possible to constrain the simulation by the local variogram.

5.3 Accounting for the variogram

In order to take account of the variogram, we suggest applying the Monte Carlo simulation technique to a restricted interval of values in such a way that the local variogram is satisfied. More precisely, let α be the location at which the simulation is performed and $\Lambda(\alpha)$ the neighbourhood of location α. Two cases can be found:

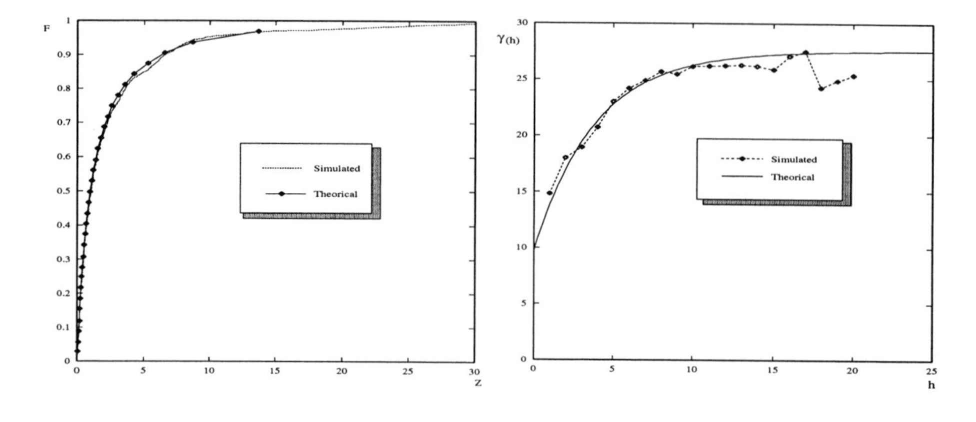

Figure 5. Comparison of the theoretical experimental cumulative distribution function of the data..

Figure 6. Comparison of the theoretical and experimental variogram of the simulated field.

- no point was simulated or observed in the neighbourhood $\Lambda(\alpha)$. In this case, the simulation is performed using $z(\alpha) = F^{-1}(\alpha; u)$ where $u \in [0;1]$ is a uniformly distributed random number;
- the neighbourhood $\Lambda(\alpha)$ is not empty. Firstly, the interval [0;1] is divided into a set of regularly spaced grid intervals: $V = \{v_1, v_2, \dots, v_i, \dots, v_k\}$; $v_i \in [0;1]$ for which the local variogram $\gamma^*(h)$ computed using the simulated values $\{Z(\alpha)\} = F^{-1}[V]$ in the neighbourhood $\Lambda(\alpha)$ is close to the *a priori* variogram $\gamma(h)$ for instance: $|\gamma^*(h) - \gamma(h)| < \delta$ where δ a tolerance factor (For a detailed study see Shtuka, 1994)

The above technique can be applied to the *simple* or *sequential simulation.* The algorithm can then be summarized as follows:

- define a random grid path;
- visit sequentially each grid point α
- **if** $\Lambda(\alpha)$ is empty simulate a new Z using the interval [0;1]; **else**
  - generate a list of values $Z_V(\alpha) = F^{-1}(\alpha; V)$ where V is a regular grid defined in [0;1];
  - compute the local variogram $\gamma^*(h)$ for each $Z_V(\alpha)$ using the neighbourhood $\Lambda(\alpha)$;
  - extract a sublist of values Z_W $w \in V$ for which $|\gamma^*(h) - \gamma(h)| < \delta$;
  - simulate a new Z_V by choosing at random in sublist Z_W;
- go to the next point in the random grid path;

5.4 Discussion

The accounting for the variogram constraint using this algorithm is relatively fast in terms of computing time compared to annealing methods, allowing interactive simulation as large as 2500 points (less than 10 seconds on a HP9000/710, including visualization).

The variogram computed on $Z_S(x)$ is quite the same as the imposed variogram (see Fig. 6). Because the variogram constraint restricts the interval on which the Monte Carlo method is performed, the final **cdf** computed on Z_S may differ slightly from the imposed **cdf** F(z).

Practically, the order of magnitude of the differences $|F(z) - F^*(z)|$ is small compared to the experimental errors (see Fig. 5). However, the above algorithm is sensitive to the extrema *zmin* and *zmax* used to build the indicator functions.

6 CASE STUDY

6.1 Data Set

The DSI simulation was tested on a data set from ESA published in the *"GSLIB: Geostatistical Software Library and Users' Guide"* by Deutsch and Journel (1992) using the files *true.dat* and *cluster.dat*.

The data are intended to represent a hazardous waste site. They were created using simulated annealing on a 50 x 50 regular grid. The variogram is isotropic and comprises two nested structures with an exponential and nugget components.

A set of 140 data values was selected at random as sample point values (Fig. 1). The *a priori* **cdf**, experimental variogram was then computed on the sample point values.

6.2 Simulating the Original Regular Grid

The simulation is performed on the 2500 nodes of the grid using the 140 initial sample point values as follows:

- compute the experimental variogram using the 140 samples;
- compute the experimental **cdf**;
- code the **cdf** by indicators using 30 thresholds;
- apply DSI with constraints to simulate the unknown values on the 50 x 50 grid;

6.3 Results

The results are reported in Figures 4 to 6. For comparison, the variogram and the **cdf** were computed using the simulated values, then compared to the true **cdf** and variogram (see Fig.5 and 7). The simulated values are reported in Figure 4. Figure 3 shows an interpolation of 50 x 50 maps using a simple DSI estimator given the sample values as control points. This example is interesting because it clearly shows the need for additional constraints. Indeed, one can notice that:

- the result obtained by DSI is a smoother field compared to the true image;
- some negative estimated values are produced by the DSI. This situation is common for estimators such as kriging, splines or least squares where the sample values present such a high spatial variability;

6.4 Comments

Regardless of the small differences observed between the true and the simulated **cdf** or variograms, those computed by sequential DSI simulation are quite similar in nature to the true ones, and the results fall within the precision of the measurements. The resulting simulation maps, which are obviously not exactly similar to the original, are quite acceptable. The challenge is in fact to satisfy contradictory constraints between the global **cdf** and the local variogram constraints.

7 DISCUSSION

The sequential DSI simulation seems to be a simple fast and flexible technique compared to the classical ones. The DSI method respects the **cdf** constraints (inequality imposed on the estimated indicator functions) which is not possible with kriging methods. In particular, Stochastic Simulation using indicators provides several advantages among which are:

- respect of the a priori distribution function and the spatial variability;
- possibility of producing probability maps of the estimated values which can be used in risk analysis protocols;
- possibility of using non-parametric estimators (mode, median, ..) for estimating the unknown parameters;
- possible extension of the Stochastic Simulation to multivariate problems (several parameters observed at the same location);
- possibility of including constraints or diffusion equations within the estimator;

8. CONCLUSION

The last point, together with the possibility of implementing additional constraints such as cross-correlation functions, would make the sequential DSI simulation a promising and interesting technique for future development, especially in environmental applications where the Stochastic Simulation by Indicator can be helpful for interpreting ecodata particularly in the following areas:

- risk analysis of natural hazards
- water resource monitoring
- overflow forecasting
- pollution monitoring
- waste disposal
- soil contamination problems

KEYWORDS

Risk assessment, stochastic modeling, mathematical model, geostatistic, simulation GOCAD, pollution, environment.

REFERENCES

[1] Deutsch C.V. and Journel A.G., GSLIB: Geostatistical Software Library and User's Guide. Oxford University Press, New York, 340p. (1992).

[2] Dowd P.A., A review of recent developments in Geostatistics. Computer and Geosciences, Vol. 17.10, 1481-1500 (1992).

[3] Isaaks E.H. and Srivastava R.M., *An Introduction to Applied Geostatistics.* Oxford University Press, 561p. (1989).

[4] Journel A.G., *Geostatistics: Models and tools for the earth sciences.* Mathematical Geology, 18(1):119-140 (1986).

[5] Journel A.G., *Fundamentals of Geostatistics in Five Lessons.* American Geophysical Union, Washington, D.C. (ISBN 0-87590-708-3) (1989).

[6] Journel A.G. and Huijbregts CH. J., *Mining Geostatistics.* Academic Press, New York (1978).

[7] Journel A.G., The indicator approach to estimation of spatial distributions. 793-806, AIME, 17th APCOM Symposium (1982).

[8] Journel A.G., Non-parametric estimation of spatial distributions. Math. Geol., 15:445-468 (1983).

[9] Journel A.G., *mAD and conditional quantile estimates.* Geostat-Tahoe, Reidel Publishing Company, Dordrecht, vol. 1, 261-270 (1984).

[10] Journel A.G., *The place of non-parametric geostatistics.*, Geostat-Tahoe, Reidel Publishing Company, Dordrecht, vol. 1, p. 307-335 (1984).

[11] Journel A.G., *Constrained interpolation and qualitative information - the soft kriging approach.*. Math. Geol., 18:269-286 (1986).

[12] Matheron G., *Les Variables Regionalisées et leur estimation.* Masson, Paris (1968).

[13] Mallet J.L., Discrete Smooth Interpolation in geometric modeling. Computer Aided Design Journal, Vol. 24, No. 4, 178-191 (1992).

[14] Royer J.J. and Shtuka A., Indicator based simulation using DSI. GOCAD Meeting, June, 1994, Nancy, ENSG, 16p. (1994).

[15] Royer J.J., Gérard B., Le Carlier de Veslud C. and Shtuka A., "3D modeling of complex natural obbjects" *In "Modeling Complex Data for Creative Information"* Ed. J.E. Dubois et N. Gershon, Springer Verlag, Paris, Vol. 2, 12 p. (1995).

[16] Shtuka A., Simulation, traitement et visualisation des images numériques: apport du codage par indicatrice en géostatistique. Thèse I.N.P.L., Nancy, 73-101 (1994).

Chapter 2

Global Atmospheric Evolution Impact of Anthropogenic Activities

Modeling Paleoclimatic Changes Using a Feedback Energetic System. Prediction of Mean Ice Global Variations.

Paul **ALLÉ**

C.R.P.G - C.N.R.S, 15 rue Notre Dame des Pauvres, BP 20, 54501 Vandoeuvre-lès-Nancy Cedex, France.

Abstract

It has been noted that several distinct ice ages existed during the past million years. Among other things, the marine sediment $\delta^{18}O$ record, interpreted as the variation through time of the ice mass, indirectly confirms the Milankovitch hypothesis that orbital variation governs climatic changes, and leads to a great number of sophisticated models. A simplified feedback model for the global variations in time of ice volume, mean surface temperature and atmospheric carbon dioxide concentration is constructed. The transfer function of the system came from our differential equation system solved by using Laplace and Z-transform techniques. The system must conserve energy, the input being the incident solar energy and the output the energy radiated by the earth. The atmosphere and surface of the earth are assumed to be a fast responding climatic system, while the ocean and ice shell have a slower response. The difference in the time constants of the main reservoirs explains the phase shift between the input and the output of the system and its transient behaviour. As a hypothesis the glaciers have a non linear response. This simplified approach makes possible *global energy transfer studies of the earth simultaneously including parameters such as the ice shell, the deep oceans, the atmosphere and surface, and the greenhouse effect. The ice mass over the last 800 K years computed by this approach, is in good agreement with the calculation using marine sediment* $\delta^{18}O$. *We suggest that the* CO_2 *variation over this period of time could result from feedback effect (governed by orbital variations), rather than from a direct geological effect. The same approach could be used to investigate the variations of the mean surface temperature, the mean ocean temperature and* CO_2 *atmospheric concentration.*

Modélisation des Changements Paléo-climatique à l'Aide d'un Système Énergétique à Contre-Réaction: Prévision des Variations au Cours du Temps du Volume Moyenne de la Couverture Glaciaire

RÉSUMÉ

Plusieurs cycles glaciaires se sont succédés durant le dernier million d'années. La théorie de Milankovitch, suivant laquelle, les variations des paramètres orbitaux terrestres provoquent ces changements climatiques a été confirmée par la mesure du $\delta^{18}O$ *dans les sédiments marins, interprétés comme traduisant l'enregistrement de la variation au cours du temps du volume des glaces terrestres. Ces mesures ont permis l'élaboration de nombreux modèles sophistiqués.*

Un modèle simple à contre réaction calculant les variations temporelles du volume total des glaces, la température moyenne de surface et la concentration atmosphérique en CO_2 *est présenté. L'algorithme de calcul est obtenu en traitant le système d'équations différentielles, décrivant le système, par les techniques de Transformées de Laplace, de transformée en Z et d'inversion matricielle. Le système doit conserver l'énergie solaire incidente, et la sortie d'énergie rayonnée par la terre. L'atmosphère et la surface terrestre forment la composante à réponse rapide du système, tandis que l'océan et les glaciers ont des constantes de temps plus importantes. Ces différentes constantes de temps expliquent les déphasages entre la sortie et l'entrée du système et son comportement transitoire. Par hypothèse les glaciers ont un comportement non linéaire. Cette approche simplifiée rend possible l'étude énergétique globale de la machine thermique terrestre incluant simultanément des paramètres comme les couvertures glaciaires, les océans profonds, le système atmosphérique et de surface, et l'effet de serre. La masse totale de glace calculée pour les derniers 800 000 ans par ce modèle corrèle bien avec l'enregistrement du* $\delta^{18}O$ *des sédiments marins. Nous suggérons que les variations de* CO_2 *durant cette période puissent être dues à la contre réaction du système thermique initiée par les variations orbitales, plutôt qu'un effet géologique direct. La même approche pourrait être utilisée pour étudier les variations de la température moyenne de surface, de la température océanique moyenne et du* CO_2 *atmosphérique.*

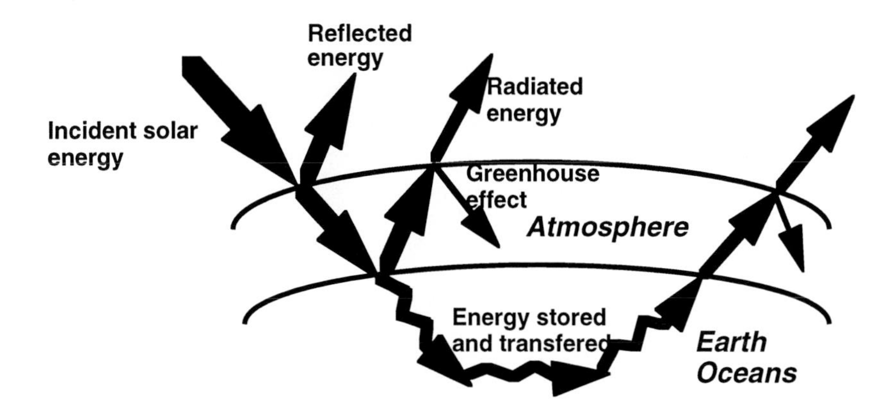

Figure 1. Energy received and radiated by the Earth. Row thickness are proportional to the account of energy respectively received and emitted.

1 THE EARTH'S CLIMATIC MACHINE

The earth receives solar energy and radiates it back into space (Fig. 1). This energy is radiated according to the black body law. Because of the temperature difference between the sun and the earth, this energy is in the visible light domain for the sun, and in the infrared domain for the earth. As some gases such as H_2O and CO_2 present in the atmosphere absorb preferentially infrared radiation, the earth's surface temperature is about 33°C higher than it would be without them. That is what is called the greenhouse effect. The full system is globally energy balanced, but there is spatial and temporal imbalance.

The excess energy received in the inter-tropical zone is conveyed to the polar zones by atmospheric and oceanic transport. The kinematics of these fluids is driven by the convection processes and by the Coriolis force. The input energy of the climatic system depends on the orbital parameters of the earth, and these have secular variations. Since some of the sub-system, such as oceans or ice sheets, have their own time response, the response of the full system is not in phase with the input. This means that energy can be stocked for centuries in these reservoirs and be exchanged between them.

2 PALEO-CLIMATOLOGIC MODELING

Two kinds of modeling have generally been done. Those called G.C.M. (General Circulation Models) calculate in terms of latitude, longitude and altitude the energy exchanges with respect to oceanic and atmospheric circulation. These very sophisticated models are used to calculate an instantaneous state of the climatic situation and cannot allow, because of the time required for calculation, the introduction of new parameters and the calculation of the long-term evolution of the climate (Broccoli and Marrabe, 1987). The most recent versions of such models have been run over a period covering a glacial cycle (Gallée et Al., 1991). The second type of model calculates the evolution of the climate over a period of about one million years, but without taking into account the general circulation. Such models use the Milankovitch theory: it is the insolation variations, driven by secular variations in the earth's orbital parameters (the astronomical signal), that cause the climatic changes.

This theory was confirmed in the seventies by the correlation between the spectrum of the marine sediment $\delta^{18}O$ record and that of the astronomical signal (Hays et Al., 1976). The marine sediment $\delta^{18}O$ record represents the oxygen isotopic variations of oceanic water. Knowing that in the evaporation process the heavy oxygen-18 isotope remains preferentially in the liquid phase, we can conclude that the formation of big ice sheets from precipitation during glaciation periods leaves in ^{18}O an enrichment in the oceanic water isotopic composition. That is why the marine sediment $\delta^{18}O$ record is generally considered as a first approximation to represent the global ice volume variation record.

Since at approximately the same time, the earth's orbital parameters were precisely recalculated (Berger, 1976; Berger, 1978), models could then be developed because of the availability of an input astronomical signal and an output sedimentary signal. The principal classes of such models are: (i)those based on rock and ice rheology and elasticity (Pollard, 1983), (ii) those based on the local energy balance (Hyde et Al., 1990), (iii) those coupling both approaches (Ghil and Le Trent, 1981) and those of a more mathematical nature (Saltzman and Sutera, 1984).

3 PROPOSED MODELING

The model presented here is a very simple one. It deals only with mean values and their evolution as a function of time. This simplification allows the introduction of new parameters in the model, such as CO_2 feedback and deep ocean time constant, and the use of a closed loop version of the model.

Doing a preliminary energy balance, one can see that the energy stored, for each square meter of the planet, between a glacial and an interglacial period, is around 30 Giga joules for the ice crystallization latent heat, and 10 Giga joules per degree Celsius in the oceans. These values have to be compared with the variations of incident energy caused by secular variations of earth orbit that are of the order of 10 Giga joules for 1000 years. This energy balance shows the importance of such considerations, the excess energy received during a few thousand years being necessary to prepare the climatic system to be rocked from the glacial to the interglacial state, and it will be the basis of the model presented in the next section.

The incident energy Ei is absorbed by the earth and the energy Er is emitted according to the black body law. The difference between these two energies ε is used by the climatic system. This energy ε modifies the surface temperature (atmosphere, superficial layers of oceans and continents) and is modeled by an integrator of capacity C_3. This energy can then be stored in the two main energetic reservoirs that are the ice sheets and the oceans (C_1 and C_2).

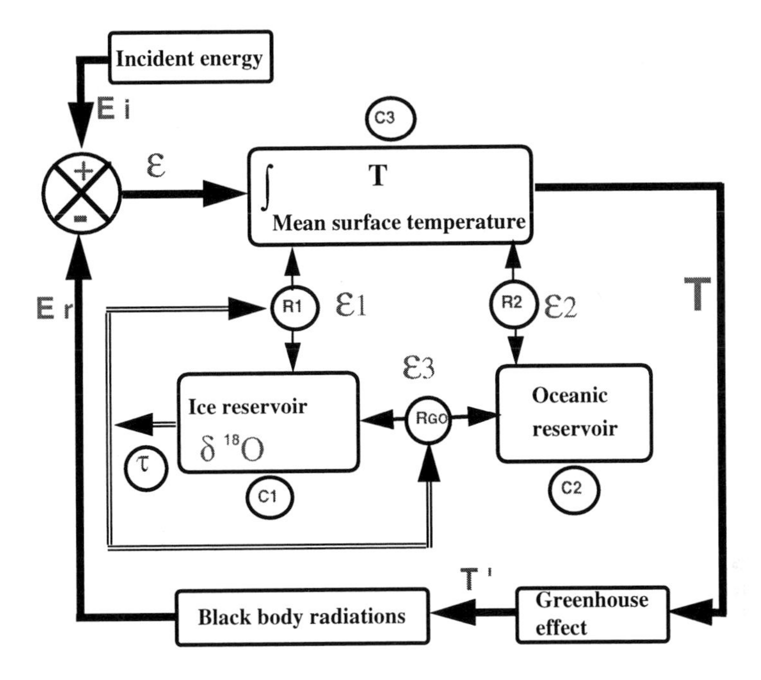

Figure 2. Closed-loop mean energy balance model (see text)

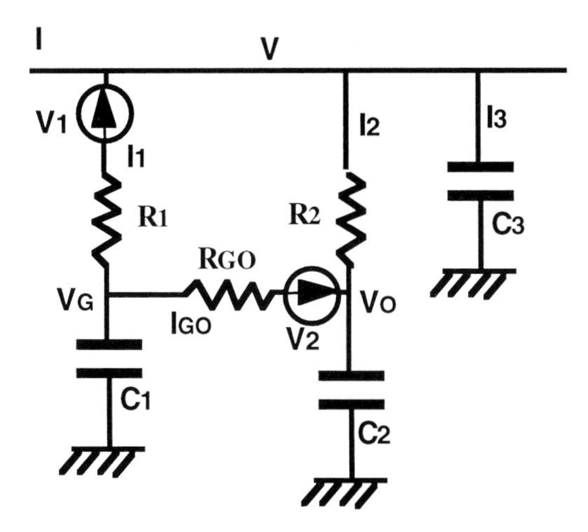

Figure 3. Equivalent electrical circuit of the main part of the closed loop energy balance model

These reservoirs can be filled more or less easily from the atmosphere, which is modeled by the resistance R_1 and R_2, and can directly exchange energy between them, resistance R_{GO}. For the generation of the 100 000 year dominant cycle, we use the main results of the rheology and elasticity based model, commuting R_1 and R_{GO} values to $R_{1'}$ and $R_{GO'}$ with a phase delay τ after the last glacial maximum. The system is made closed-loop by the emission of an energy Er, depending on the temperature T', according to the black body law. The temperature T' is shifted from the surface temperature T by a value itself a linear function of T (modeling of the greenhouse effect with the concentration of CO_2 and H_2O considered as positive feedback).

4 MATHEMATICAL EXPRESSION OF THE PHYSICAL MODEL

Such a physical model can be represented by a system of differential equations. The main part of this system is equivalent to the one describing the following electrical circuit (Fig. 3) :

Where current I represents energy ε, voltage V represents temperature T. Voltage sources V1 and V2 represent the temperature shift existing between mean surface temperature, mean oceanic temperature and polar local temperature that drive energetic exchanges with the ice sheets.

The equations describing the electrical circuit are:

$$I=I_1+I_2+I_3 \qquad I_3 = C_3 \frac{\partial V}{\partial t}$$

$$I_1\text{-}I_{GO} = C_1 \frac{\partial V_G}{\partial t}$$

$$I_2\text{-}I_{GO} = C_2 \frac{\partial V_O}{\partial t} \qquad I_1 = \frac{(V\text{-}V_1)\text{-}V_G}{R_1} \qquad I_2 = \frac{V\text{-}V_0}{R_2}$$

$$I_{GO} = \frac{V_G\text{-}(V_0\text{-}V_2)}{R_{GO}} \qquad Q_G = C_1 V_G = \int (I_1\text{-}I_{GO})dt$$

This gives us, after returning to the physical model, replacing I by ε, V by T, adding two equations for closing the loop, and calculating the Laplace transform :

$$\varepsilon = \varepsilon_1+\varepsilon_2+\varepsilon_3 \qquad \varepsilon_3 = C_3 PT \qquad \varepsilon_1\text{-}\varepsilon_{GO} = C_1 PT_G \qquad \varepsilon_2\text{-}\varepsilon_{GO} = C_2 PT_0$$

$$\varepsilon_1 = \frac{(T\text{-}T_1)\text{-}T_G}{R_1} \qquad \varepsilon_2 = \frac{T\text{-}T_0}{R_2} \qquad \varepsilon_{GO} = \frac{T_G\text{-}(T_0\text{-}T_2)}{R_{GO}}$$

$$Q_G = C_1 T_G = \frac{\varepsilon_1\text{-}\varepsilon_{GO}}{P}$$

$$\varepsilon = E_i\text{-}\delta T'^4 \qquad T' = T(1\text{-}\alpha)\text{-}\beta$$

The last equation represents the greenhouse effect feedback. The black body temperature T' superior to the surface temperature by an amount tuned with α and β. The previous one describes the closure of the system ($E_r = \delta T'^4$).

This system could then be written as follows, and be partially solved by matrix techniques.

$$\varepsilon=\varepsilon_1+\varepsilon_2+\varepsilon_3 \qquad T=\frac{\varepsilon_3}{C_3 P}$$

$$\varepsilon=E-\delta T'^4 \qquad T'=T(1-\alpha)-\beta$$

$$Q_G=C_1 T_G$$

$$\begin{bmatrix} 1 & 0 & 0 & 0 & \frac{1}{R_1} \\ 0 & 1 & 0 & \frac{1}{R_2} & 0 \\ 0 & 0 & 1 & \frac{1}{R_{GO}} & -\frac{1}{R_{GO}} \\ 1 & 0 & -1 & 0 & -C_1 P \\ 0 & 1 & 1 & -C_2 P & 0 \end{bmatrix} * \begin{bmatrix} \varepsilon_1 \\ \varepsilon_2 \\ \varepsilon_{GO} \\ T_O \\ T_G \end{bmatrix} = \begin{bmatrix} \frac{T-T_1}{R_1} \\ \frac{T}{R_2} \\ \frac{T_2}{R_{GO}} \\ 0 \\ 0 \end{bmatrix}$$

After solving the system by matrix inversion and finding $\varepsilon 1$, $\varepsilon 2$ and T_G as functions of T, we can find, using the other equations, T and Q_G as a function of T and E_r. Then we can calculate the Z transform using the bilinear approximation $P = \frac{1-Z^{-1}}{R_e}$ and deduce an algorithmic expression that allows us to calculate T and Q_G as a function of time. We can simplify the expressions by expressing the RC products in units of T_e. It gives the following equations :

$$T(R_1R_2R_{GO}C_1C_2C_3 + R_1R_2C_2C_3 + R_1R_2C_1C_3 + R_{GO}R_1C_1C_3 + R_{GO}R_2C_2C_3 + R_{GO}R_2C_1C_2 + R_1R_{GO}C_1C_2 + (R_{GO} + R_1 + R_2)(C_1 + C_2 + C_3))$$

$$= TZ^{-1} (3R_1R_2R_{GO}C_1C_2C_3 + 2(R_1R_2C_2C_3 + R_1R_2C_1C_3 + R_{GO}R_1C_1C_3 + R_{GO}R_2C_2C_3 + R_{GO}R_2C_1C_2 + R_1R_{GO}C_1C_2) + (R_{GO} + R_1 + R_2)(C_1 + C_2 + C_3))$$

$$- TZ^{-2} [3R_1R_2R_{GO}C_1C_2C_3 + (R_1R_2C_2C_3 + R_1R_2C_1C_3 + R_{GO}R_1C_1C_3 + R_{GO}R_2C_2C_3 + R_{GO}R_2C_1C_2 + R_1R_{GO}C_1C_2)]$$

$$+ TZ^{-3} R_1R_2R_{GO}C_1C_2C_3$$

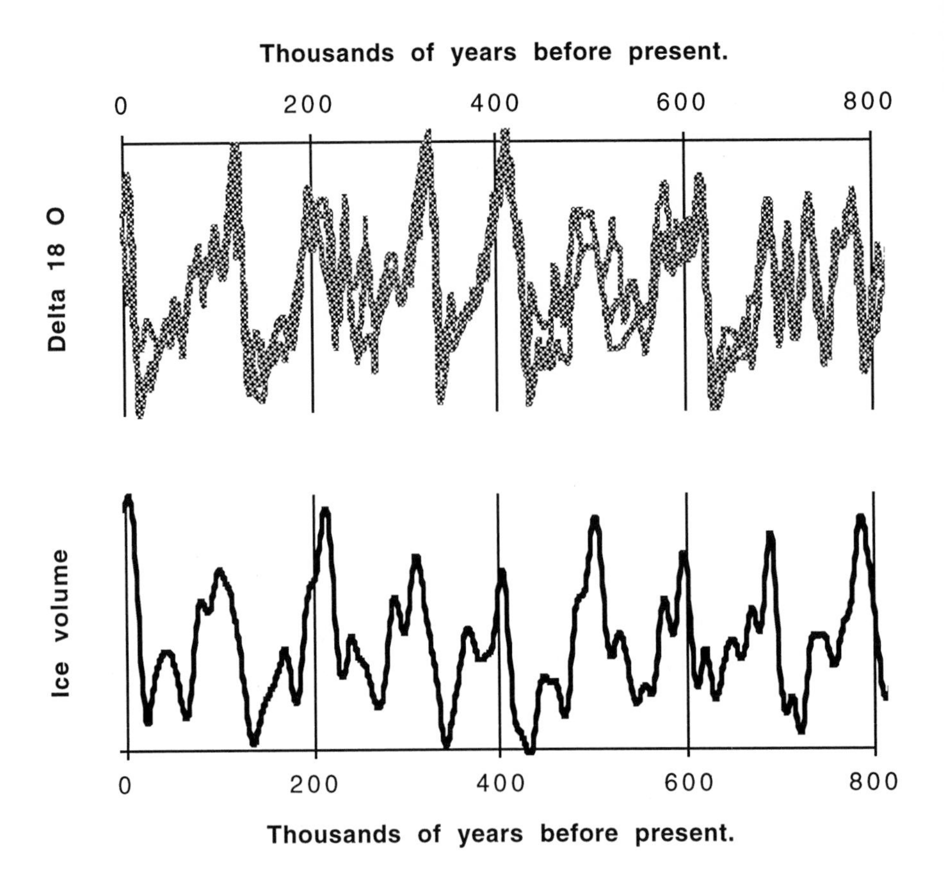

Figure 4. The model estimates time ice volume variations which are in good agreement with those recorded by $\delta^{18}O$ measured in sediments.

$$+ [E - \sigma((1 - \alpha)TZ^{-1} - \beta)^4] * [R_1R_2R_{GO}C_1C_2 + (R_1 + R_2)(C_1 + C_2) + R_{GO}(R_1C_1 + R_2C_2) + R_1 + R_2 + R_{GO}]$$

$$+ [EZ^{-1} - \sigma((1-\alpha)TZ^{-2} - \beta)^4] * [2R_1R_2R_{GO}C_1C_2 + (R_1 + R_2)(C_1 + C_2) + R_{GO}(R_1C_1 + R_2C_2)]$$

$$+ [EZ^{-2} - \sigma((1-\alpha)TZ^{-3} - \beta)^4] * [2R_1R_2R_{GO}C_1C_2]$$

$$Q_G \left(C_1C_2 + \frac{C_2 + C_1}{R_{GO}} + \frac{C_1}{R_2} + \frac{C_2}{R_1} + \frac{1}{R_1R_2} + \frac{1}{R_1R_{GO}} + \frac{1}{R_2R_{GO}}\right)$$

$$= C_1T_GZ^{-1} \left(2C_1C_2 + \frac{C_2 + C_1}{R_{GO}} + \frac{C_1}{R_2} + \frac{C_2}{R_1}\right) + C_1T_GZ^{-2} + C_1T\left(\frac{R_1 + R_2 + R_{GO}}{R_1R_2R_{GO}} + \frac{C_2}{R_1}\right)$$

$$+ C_1TZ^{-1}\left(\frac{C_2}{R_1}\right) - T_1 \left(\frac{R_2 + R_{GO}}{R_1R_2R_{GO}}\right) - T_2\left(\frac{1}{R_2R_{GO}}\right)$$

5 DISCUSSION

From this expression of the Z transform of the solution, we can write the software that calculates the ice volume from the astronomic signal and adjusts the model's parameters by successive trials. The input of the model is the variations of insolation (Berger, 1978), and the output (ice volume) is compared to the $\delta^{18}O$ sedimentary record (Raymo, et A.L. 1990). Figure 4 represents the $\delta^{18}O$ record and the ice volume calculated by the model.

Comparison between these two figures shows that we can get reasonable results with such a model, taking into account global mean energy balance, deep ocean time constant, and the greenhouse effect.

6 CONCLUSION

The originality of such a model is the simple and schematic vision that allows the introduction of parameters usually neglected in complex models, and the closed-loop approach. With further development, this approach could be used to bring a new point of view to more sophisticated models. We could test for example, the probable mean global influence of parameters such as deep ocean and greenhouse feedback effects, which could explain some differences between modeling and observation (Berger et Al., 1990). This approach could help resist the temptation to use the CO_2 palaeo record as an input signal (Genthon et Al., 1987; Pisias and Shackleton, 1984), thus modeling in an open-loop way a system known to be closed-loop.

KEYWORDS

Global change, climatic change, mathematical model, paleoclimatic evolution, CO_2, $\delta^{18}O$, green house effect.

REFERENCES

[1] Berger A., Gallée H., Fichefet T., Marsiat I. and Tricot C., "Testing the astronomical theory with a coupled climate ice sheet model". *Palaeogeography Palaeoclimatology Palaeoecology*. **89**:125-141 (1990).

[2] Berger A.L., "Obliquity and precession for the last 5000000 years". *Astron and Astrophys* **51**:127-135 (1976).

[3] Berger A.L., "Long-term variations of caloric insolation resulting from the earth's orbital elements". *Quaternary Research*, **9**:139-167 (1978).

[4] Broccoli A.J. and Manabe S., "The influence of continental ice, atmospheric CO_2, and land albedo on the climate of the last glacial maximum". *Climate Dynamics*, **1**:87-99 (1987).

[5] Gallée H., Van Ypersele J.P., Fichefet T.H., Tricot C.H. and Berger A., "Simulation of the last glacial cycle by a coupled sectorially averaged climate - ice sheet model 1. The climate model". *Journal of Geophysical Research*, **96**:13139-13161 (1991).

[6] Genthon C., Barnola J.M., Raynaud D., Lorius C., Jouzel J., Barkov N.I., Korotkevich V.S. and Kotlyakov U.M., "Vostok ice core: climatic response to CO_2 and orbital forcing changes over the last climatic cycle". *Nature*, **329**:414-418 (1987).
[7] Ghil M. and Le Treut, "A climate model with cryodynamics and Geodynamics". *Journal of Geophysical Research*, **86**:5262-5270 (1981).
[8] Hays J.D., Imbrie J. and Shackleton N.J., "Variations in the Earth's orbit: pacemaker of the ice ages". *Science*, **194**:1121-1132 (1976).
[9] Hyde W.T., Kim K.Y., Crowley T.J. and North G.R., "On the relation between polar continentality and climate: studies with a nonlinear seasonal energy balance model". *Journal of Geophysical Research*, **95**:18653-18668 (1990).
[10] Pisias N. G. and Shackleton N. J., "Modeling the global climate response to orbital forcing and atmospheric carbon dioxide changes". *Nature*, **310**:757-759 (1984).
[11] Pollard D., "A coupled climate-ice sheet model applied to the quaternary ice ages". *Journal of Geophysical Research*, **88**:7705-7718 (1983).
[12] Raymo M. E., Ruddiman W. F., Shackleton N. J. and Oppo D. W., "Evolution of Atlantic-Pacific $\delta^{13}C$ gradients over the last 2,5 m.y ". *Earth and Planetary Science Letters*, **97**:353-368 (1990).
[13] Saltzman B. and Sutera A., "A model of the internal feedback system involved in late quaternary climatic variations". *Journal of the Atmospheric Sciences*, **41**:N°5 (1984).

Gaseous Toxic Emission from Plastic Materials during their Thermal Decomposition

C. ARFI, C. ROTIVAL-LIBERT, E. RENACCO, J. PASTOR, A. M. PAULI and H. PORTUGAL

Laboratoire de Chimie Analytique, Faculté de Pharmacie, 27 Bd J. Moulin, 13385 Marseille, Cedex 5, France.

ABSTRACT

Because of their increasing use in modern technology, plastic materials undergo heating on various occasions. Thermal analysis (TG-DTA) provides knowledge of their thermal behaviour under precise conditions of atmosphere and heating. Emitted gases are collected during analysis, between limits of temperature corresponding to slope changes on the TGtrace. Then all the methods of analytical chemistry may be applied to qualitative and quantitative analysis of these gases.

Various plastic samples were studied: an acrylic varnish, a polypropylene film, a polyurethane glue, a polyvinylchloride bottle. Every sample showed a particular behaviour, justifying thereby such analyses.

Emission par les Plastiques de Toxiques Gazeux lors de leur Décomposition Thermique

RÉSUMÉ

Les matériaux plastiques subissent l'effet de la chaleur à diverses occasions en raison de leur utilisation accrue dans la technologie moderne. L'analyse thermique (TG-DTA) permet de connaître leur comportement thermique pour différentes conditions d'atmosphère et de chauffage. Les gaz diffusés sont captés pendant l'analyse pour une gamme de températures correspondant aux changements de pentes de la courbe TG. Les techniques de la chimie analytique peuvent alors être mises en oeuvre pour l'analyse quantitative et qualitative de ces gaz.

Plusieurs échantillons de plastiques ont été analysés: un vernis acrylique, un film en polypropylène, une colle polyuréthanne, une bouteille en polyvinychlore. Chaque échantillon montre un comportement spécifique qui justifie a posteriori ce type d'analyse.

1 INTRODUCTION

The substantial development of plastic materials in modern technology induced us to consider their behaviour in the many circumstances that make use of temperature raising:

- The production of manufactured objects necessarily uses glues, varnishes, paints which are polymers. Manufacturing processes use heating, therefore polymers are also heated (a varnish laid on metallic packaging is subjected to the temperature necessary to its formation).
- Conflagration is a risk that has to be known in order to be fought. Thus knowledge of the participation of plastic materials in the spread of fire and in the emission of gases and fumes is essential.
- Destruction has to be anticipated, because an important part of industrial, domestic, and hospital wastes is composed of plastic materials. Most often, these wastes are incinerated, but they can also be treated in order to constitute raw materials for other industrial activities.

Thus, before manufacture, incineration or recycling, the best heating conditions have to be determined, and the risks for the environment due to the release of gases formed have to be known.
Many authors were interested by this problem. We can mention some: Renman et al [1] determined isocyanate and aromatic amine emissions from thermally degraded polyurethanes. Chaigneau et al studied pyrolysis of polyurethanes [2] and polyvinyl chloride [3] between 500°C and 800°C and characterized the toxic gases emitted. Barabas et al [4] studied the volatile products of polypropylene thermal oxidation.
The aim of our study is the analysis of some toxic gases (carbon oxides, carbonyl compounds and other hydrocarbons, hydrogen cyanide, isocyanates, hydrogen chloride) emitted during the thermal analysis of various samples of plastics: an acrylic varnish, a polypropylene film, a polyurethane adhesive, a polyvinyl chloride bottle.

2 MATERIALS AND METHODS

2.1 Samples

Four samples were studied:

- the acrylic varnish was liquid and white. It was composed of:
  n Butyl Methacrylate (24%)
  Ethylcellulose (3%)
  Saccharose acetobutyrate (0,8%)
  Polyethylene (1%)
  Methyl Ethyl Ketone (71,2%)
  For the thermal analyses in the laboratory, it was necessary to spread it in a thin coat over a glass slab. After solvent evaporation in a drying oven at 40°C, the varnish film was removed. An approximately 30 mg sample was used for each test.

- the polypropylene film was meant to cover a metallic packaging. It was transparent.
- the polyurethane adhesive was meant to paste the polypropylene film on the packaging. It was delivered as a semi rigid block with small cavities.
- the polyvinyl chloride came from bottles of mineral water. It was rigid, transparent and lightly tinted.

2.2 Thermal analyses

The thermal behaviour of these samples was studied in a TG-DTA SETARAM apparatus, with a platinum crucible and a platinum rhodium thermocouple. It was possible to study simultaneously weight loss of the sample (TG) as a function of temperature and to observe exothermal and endothermal phenomena (DTA). (Fig. 1).
The experimental conditions were a heating rate of 5°C/mn to a maximum temperature of 650°C. An air flow drew the emitted gases towards impingers at an average rate of 10 ml a minute. We associated those techniques (TG-DTA) with those of analytical chemistry to analyze the gases emitted during the thermal decomposition within the temperatures ranges corresponding to different decomposition rates.

2.3 Gas sampling

Gases emitted inside the apparatus were led to an evacuation fitted with a three way valve. This made it possible to selectively feed the gas emission into one of the three impingers through a "Pulse pomp environment INC" pump.

The selection differed according to the nature of the collected gas.

- For acid gases (C02, HCN, HCl) the collection flasks contained a sodium hydroxide solution without carbonate in which was placed a glass bubbler [5].

- Carbonyl compounds were collected on silica cartridges (Sep-Pack C18) coated with 2-4 dinitrophenyl hydrazine in acetonitrile [6].

- Isocyanates were collected in impingers charged with 1-(2-methoxyphenyl) piperazine in acetonitrile [6].

- Other hydrocarbons were collected in flasks containing ethyl acetate.

- For carbon monoxide assays, gas was collected in a teflon collection bag equipped with a valve [6].

2.4 Gas determination methods

- Carbon dioxide fixed as sodium carbonate, as well as hydrogen chloride fixed as sodium chloride were assayed by a potentiometric method in the presence of excess sodium hydroxide.

- Hydrogen cyanide fixed as sodium cyanide was assayed by a potentiometric method with a silver nitrate solution in the presence of ammoniac.

- Carbonyl compounds fixed as hydrazones were separated and assayed by high performance liquid chromatography.

- Isocyanates converted to stable derivatives by 1-(2- methoxyphenyl) piperazine were also analyzed by high performance liquid chromatography.

- Hydrocarbons were separated by gas chromatography, and mass spectrometry allowed for qualitative and quantitative analyses.

- Carbon monoxide reacted on palladium and cacotheline; the formed derivative was determined by visible spectrophotometry.

3 RESULTS AND DISCUSSION

3.1 Acrylic varnish

DTA showed a decomposition endothermal peak at 392°C and another thermal peak at 485°C, corresponding to an exothermal phenomenon. TG showed a two-step decomposition. The first step was steep.

From 240°C to 425°C the weight loss was 82%, or a rate of weight loss of 2,23% min. The second step corresponded to the exothermal phenomenon: the weight loss was about 15% from 425° to 509°C, i.e. a rate of weight loss of 0,88 %/min.

Results of carbonyl compound assays (Table 1) showed the prevalence of acetaldehyde (69%), acrolein (13%) and formaldehyde (10%). They formed 92% of the total amount of aldehydes and ketones.

These results prevailed on the industrial users to choose a lower heating temperature for their thermal treatment. Therefore, thermal analyses had been realized up to 350°C, with a heating rate of 90°C/min (Table 2).

DTA showed only two endothermal phenomena at about 167°C and 340°C but no exothermal peak. TG presented only one stage of weight loss, beginning at 230°C, with a rate of 1,65%/min.

Analyses of aldehydes and ketones showed the predominance of acetaldehyde and formaldehyde which together represented 91,5% of all carbonyl compounds, but only a very low amount of acrolein was found (Table 2).

It was very important to avoid both an exothermal phenomenon and a highly irritant gas emission.

Table 1. Relative proportion of individual carbonyl compounds emitted during thermolysis up to 650°C of an acrylic varnish sample.

CARBONYL COMPOUND	RELATIVE AMOUNT %
Formaldehyde	10
Acetaldehyde	69
Acrolein	13
Acetone	0
Propanal	5
2- trans - Butenal	0.85
2- Butanone	0.75
3- Methyl butenal	0
Methyl propylketone	1.40
Pentanal	0

Table 2. Carbonyl compound emitted from 100 mg acrylic varnish sample during thermolysis up to 350°C.

CARBONYL COMPOUNDS	AMOUNT mg/100mg SAMPLE	RELATIVE AMOUNT %
Formaldehyde	0.186	14.4
Acetaldehyde	0.996	77
Acrolein	0.012	0.93
Acetone	0.035	2.7
Propanal	0.05	3.86
2 - trans - Butenal	0.0025	0.19
2 - Butanone	0.0081	0.62
3 - Methylbutenal	0	0
Methylpropylketone	0.0012	0.09
Pentanal	0.0035	0.27

Table 3. Assay results for Polypropylene film.

COLLECTION ZONE °C	WEIGHT LOSS %	DECOMPOSITION RATE % / mn	CO mg/100 sample	CO^2 mg/100 sample	CARBONYL mg/100 mg sample
200 - 500	98	1.88	8.82	39.2	0.105

Table 4. Aldehydes and ketones emitted during the thermal analysis of the polypropylene film sample.

CARBONYL COMPOUNDS	µg/100 mg PP
Formol	70
Acetaldehyde	10
Acrolein	7
Acetone	0
Propanal	13
2 - trans-butenal	5
Isovaleraldehyde	0.1
Pentanal	-
Pentanone	-
Total	105

3.2 Polypropylene film

The thermal analysis showed just a one-step decomposition between 200°C and 500°C. Carbonyl compounds and carbon oxides were collected during the thermal analysis between these two temperatures. Assay results are shown in Table 3 and 4.

The ratio of aldehydes and ketones was very low.

Carbon monoxide and carbon dioxide formed simultaneously. A relatively important quantity of carbon monoxide was emitted.

Table 5. Gas emission from the polyurethane adhesive sample.

COLLECTION ZONE °C	WEIGHT LOSS %	DECOMPOSITION RATE % / min	CO mg/100 mg of sample	CO2 mg/100 mg of sample	CARBONYL COMPOUNDS mg/100 mg of sample	HCN mg/100 mg of sample	ISOCYANATE mg/100 mg of sample
260 to 450	76	2.0	2.35	23.5		0.027	0.079
450 to 650	22	0.53	1.22	57.9		0.142	0.026
Total	98		3.57	81.4	0.059	0.169	0.105

3.3 Polyurethane adhesive

The thermal analysis showed a two-step decomposition. The first step was in a temperature zone included between 260°C and 450°C. The weight loss was 76%, therefore the rate of weight loss was 2%/ min. The second step, between 450°C and 650°C, was less steep since the rate of weight loss was 0,53% / min. Gas emission was collected in the two times corresponding to the two temperature zones. Carbon oxides, carbonyl compounds, hydrogen cyanide and isocyanates were assayed. Results are reported in Table 5. Carbonyl compound emission was very small and insignificant. The most important emission was carbon dioxide. Yet we have to consider carbon monoxide, hydrogen cyanide and phenylisocynate because of their toxicity. We can observe that carbon monoxide and phenylisocyanate were particularly emitted in the first collection zone, while carbon dioxide and hydrogen cyanide were particularly emitted in the second collection zone.

Other tests were realized with samples composed of the polypropylene film coated with the polyurethane adhesive. The experimental results obtained were higher than the total calculated from experimental results obtained from the original compounds.

3.4 Polyvinyl chloride bottle

The thermal analysis enabled us to delimit four collection zones. Hydrogen chloride was collected and assayed in these four zones. Results are reported in Table 6. We can observe that before 330°C more than half of the sample was decomposed and the hydrogen chloride emission was 92,7%. This high ratio and the dangerous feature of this gas have to be considered in case of incineration and conflagration.
Thus, the incineration rooms must resist chemical attacks and must be equipped with absorbent systems for stopping toxic emissions. In case of conflagration, we have to note that the most important part of hydrogen chloride is emitted while the fire is smouldering. That explains the legislative texts which control the use of polyvinyl chloride in the building trade.

Table 6. Hydrogen chloride emitted during the thermal analysis of the polyvinyl chloride bottle.

COLLECTION ZONE °C	WEIGHT LOSS %	DECOMPOSITION RATE % / min	HCl mg/100 mg of sample	Relative amount
240 to 330	55	3.05	45.8	92.7 %
330 to 420	5	0.28	1.99	4.0 %
420 to 490	22	1.57	0.45	0.9 %
490 to 590	18	0.90	1.19	2.4 %
total	100		49.4	100.0 %

4 CONCLUSION

These results show the value of determining gas emission during thermal analysis. One can thereby know the thermal behaviour of any plastic, under defined conditions, and the constitution of the gaseous emission during decomposition, within temperatures limits shown by the TG trace and corresponding to slope changes. It is then possible to modify the thermal processing conditions in order to avoid conflagration risks. Gas emission analyses have to be determined under identical conditions to assess the toxic risks and the consequences for the environment. Evacuation systems and recycling processes may then be adjusted to decrease these risks.

KEYWORDS

Air pollution, gaseous toxic emission, thermal decomposition, toxic risk, toxicity.

REFERENCES

[1] Renman L., Sangö C. , Skarping G., Am. Ind. Hyg. Assoc. J., 47, 621 (1986).
[2] Chaigneau M., Le Moan G., Ann. Pharm. Fr., 30, 402 - 414 (1972).
[3] Chaigneau M., Le Moan G., Ann. Pharm. Fr., 27, 97 - 101 (1969).
[4] Barabas K., Iring M., Lazlo-Hedvig S., Kelen T., Tudos F., Eur. Polym. J., 14, 405 - 407 (1978).
[5] Isvy-Arfi C., Thèse de Doctorat d'État en Pharmacie - Marseille (1977).
[6] Rotival-Libert C., Thèse de l'Université de Provence, Specialité Chimie de l'Environnement et Santé - Marseille (1994).

INVENTORY OF POLLUTANT EMISSIONS INTO ATMOSPHERE IN EUROPE - CORINAIR PROJECT

R. BOUSCAREN, J. P. FONTELLE and J. P. CHANG[1]

CITEPA, 10 rue du Fg Poissonnière, 75010 Paris, France

ABSTRACT

What is an air emission inventory? An air emission inventory may be considered formally as a data system composed of four dimensions: pollutants, sources of pollutant, spatial components (territorial units) and temporal components. Generally, the temporal component is reduced to a reference year, but there is some attempt to provide shorter resolutions. Air emission inventories, what for? The different possible users of such inventories are the following:

- *scientists for use in models ...*
- *economists for use in forecasting models ...*
- *politicians for responsibility evaluation, respect of international conventions, regulation orientation ...*
- *public for general information.*

Air emission inventory, how to proceed? The CORINAIR project tried to provide the most relevant way for achieving a consistent, transparent and harmonized inventory which may be a multi-purpose data bank.

Inventaire des Émissions Polluantes Rejetées dans l'Atmosphère en Europe - Le Projet CORINAIR.

RÉSUMÉ

Qu'est-ce qu'un inventaire des émissions de polluants atmosphériques? Un inventaire des émissions de polluants atmosphériques peut être conçu formellement comme un système de données comportant quatre dimensions: les polluants, les sources de pollution, les composantes spatiales (unités territoriales) et une composante temporelle. En général, la composante temporelle se réduit à une année de référence, mais parfois des résolutions plus fines sont nécessaires. A quoi servent les inventaires des émissions de polluants atmosphériques? Voici la liste d'utilisateurs potentiels:

---

[1] To whom correspondence may be addressed

- *les scientifiques pour les utiliser dans des modèles,*
- *les économistes pour leur utilisation dans des modèles de prévision,*
- *les politiciens pour évaluer les responsabilités, le respect des conventions internationales, et l'orientation des lois,*
- *le public pour l'information générale.*

Comment réaliser les inventaires des émissions de polluants atmosphériques? Le projet CORINAIR a tenté de fournir les moyens les plus appropriés afin d'obtenir des inventaires homogènes, transparents, harmonieux et qui, en outre, puissent être utilisés comme une banque de données à usages multiples.

1 BACKGROUND

The CORINAIR project is one component of the CORINE programme which followed the European Council Decision 85/338/EEC (27 June 1985) on a Commission work programme for gathering, coordinating and ensuring the consistency of information on the state of the environment and natural resources in the European community. The different projects within CORINE are biotopes, land cover, water resource projects and CORINAIR for the air emission project.

A first prototype CORINAIR inventory was achieved for the reference year 1985 and for the twelve EC Member states. Three pollutants were considered, SO_2, NOx and VOC (Volatile Organic Compound), see Annex 1.

The present CORINAIR 90 project for the reference year 1990, which was in its final step in September 1994, refers to more pollutants (height pollutants), and is extended to most European countries :

- the twelve EC Member States,
- five EFTA countries: Austria, Finland, Norway, Sweden and Switzerland,
- three Baltic States: Estonia, Latvia and Lithuania,
- Nine Central and Eastern European countries: Albania (*), Bulgaria, Croatia (*), Czech Republic, Hungary, Poland, Romania, Russia (*), Slovakia and Slovenia.

(*) Only potentially involved in the CORINAIR inventory.

Since the beginning, CITEPA was the main contractor for the CORINAIR project from the European Commission DG XI.

2 CORINAIR 90 SPECIFICATIONS

2.1 Objective

As presented in the introduction, the objective of the CORINAIR inventory is to be a multi-purpose inventory. Due to this objective, an important effort of collaboration and harmonization has been made with international organizations, such as UNECE for EMEP programme and OECD for IPCC programme.
As another example, at the European community level, the Large Combustion Plant Directive inventory may be provided within the CORINAIR system.

2.2 Required qualities

• Completeness: two aspects are covered. The geographic completeness relates to the availability of the CORINAIR system to almost all European countries including subnational divisions. The emission source completeness relates to the availability of an exhaustive list of relevant activities generating emissions for the considered height pollutants (SNAP 90 nomenclature).

• Consistency: this quality is achieved by application of the CORINAIR methodology and CORINAIR software.

• Transparency: this quality is reached by the provision (according to the CORINAIR methodology) of activity data and emission factors for area sources, and for point sources the provision of either activity and emission factor, or measured emissions and estimation methods.

2.3 Inventory specifications of CORINAIR 90

• The reference year: 1990

• The pollutants

Sulfur dioxides (SO_2) [SO_2 + SO_3 expressed as SO_2]
Oxides of Nitrogen (NOx) [NO + NO_2 expressed as NO_2]
Non-methane volatile organic compounds (NMVOC)
Methane (CH_4)
Carbon monoxide (CO)
Carbon dioxide (CO_2)
Nitrous oxide (N_2O)
Ammonia (NH_3)

• The generating sources of emissions

Relating to the eight considered pollutants, the relevant list of possible generating activities are given in the SNAP 90 nomenclatures: over 260 activities grouped in two hierarchical levels of sub-sectors and eleven main sectors which relate to the report level for UNECE inventory (see Annex 2).

• The spatial resolution

According to the idea of collecting data on the available format, the spatial resolution in CORINAIR is not a grid system but the administrative territorial unit system. The lowest level chosen is the NUTS level 3 or equivalent area (in France the NUTS level 3 relates to a Department).

• Specifications for Large Point Sources (LPS)

The large and localized pollutant emitters are inventoried individually in the CORINAIR system. Criteria for considering localized emitter as a point source in CORINAIR are given in Annex 3.

3 CORINAIR SYSTEM

3.1 CORINAIR "Tool Box"

The different components of the CORINAIR tool box for achieving a CORINAIR inventory are :

- nomenclatures and inventory specifications
- the handbook of default emission factors
- the CORINAIR software (and manual) for data input and emission calculation
- the COPERT software module (and manual) for detailed methodology on road traffic activities.

This "tool box" was distributed to all national CORINAIR experts, generally during training sessions.

3.2 CORINAIR organization

Under the supervision of DG XI EEA - Task Force, the different technical groups of the CORINAIR organization are structured as follows :

- the CORINAIR Technical Unit (less than ten persons including a representative of DG XI EEA - Task force) is in charge of methodology and general technical problems,

- the Project Leader team (CITEPA) is in charge of "Tool Box" development, assistance to national experts and checking/validation of national inventories,

- Working Groups covering specific sources/pollutant questions (stationary NOx, stationary VOC, mobile sources, natural VOC, ammonia) for improvement of knowledge in difficult topics,

- the CORINAIR Expert Group including all national experts and experts from international organizations interested in the CORINAIR programme.

4 CORINAIR METHODOLOGY

4.1 General principles

• Technology oriented inventory

The chosen type of inventory for CORINAIR is the technology oriented inventory (rather than economy or product oriented) because pollutant emissions are basically technology dependent.

• Mixture of Bottom-up and Top-down approach

When the information is known at the lowest territorial unit level, information at the upper level is then deduced (Bottom-up approach). But when information is known at a high level of the territorial unit and is unknown in some lower territorial units: CORINAIR methodology proposes a procedure to split values from top to bottom level according to well correlated data (surrogate data).

• Area sources and point sources

In CORINAIR methodology sources emitting pollutants are divided into two classes :

- area sources relate to all diffuse or mobile sources, and to minor localized sources.

- point sources relate to important localized sources (see criteria in "inventory specifications").

The area sources are treated collectively within territorial units, while point sources are treated individually with many details.

• Emission estimation principle

- For area sources, the emissions are estimated for one given activity and one given territorial unit with the simple equation :

  E = A x EF
  (E: emissions, A: activity value for the reference year, EF: emission factor).

- For point sources the emissions are either directly input when they are known from measurements, mass balance, ... or they are estimated by way of emission factors (E = A x EF).

4.2 Overview of CORINAIR process

To summarize, one can distinguish five main successive steps in the CORINAIR process :

1) Specifications/definition step: concerning common specifications and definitions, the CORINAIR technical unit is in charge of this initial part (pollutants, emitting sources, LPS specifications, ...). Concerning definitions specific to each country (e.g. territorial unit definition ...), each national expert is in charge as the first step of this work.

2) Data collection, treatment and emission calculation: each national expert is in charge of and responsible for this task for his country.

   During this step, continuous assistance is provided to national experts by the Project Leader team (CITEPA).

3) Checking/validation step: the Project Leader (CITEPA) is in charge of this task. This step does not start necessarily at the end of the data collection because of several possible feedbacks, between steps 2 and 3 (cf. annex 4).

4) Transfer of national databases into the single CORINE database: after checking/validation, all national databases (xBase) are transferred into the single CORINE database (ORACLE) in the European Environment Agency (EEA).

5) Diffusion of data from CORINE system: before a wide diffusion of CORINAIR data to different users, a confidentiality treatment extracts from the full CORINE database a free access database (without restricted national data). Then tables, maps and diskettes may be distributed.

5 State of CORINAIR 90 Project - Examples of Results for France

5.1 State of CORINAIR 90

As of September 1994, most EC Member States, EFTA countries and Central and Eastern European Countries provided final inventories. CORINAIR 90 inventories were to be definitively frozen by December 1994.

According to the steps previously defined, a provisional step 4 (transfer to CORINE) was run. Large diffusion of CORINAIR data was carried out in 1995.

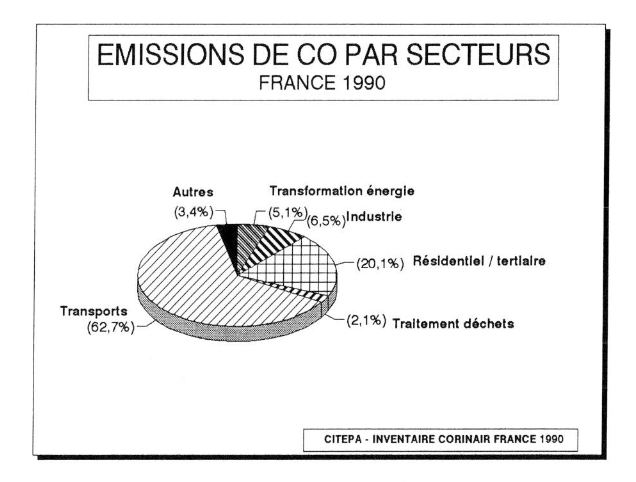

Figure 1. Emissions of CO by sectors (France 1990).

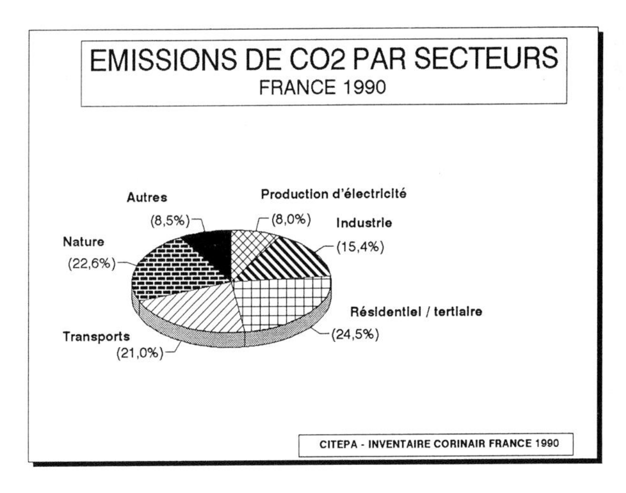

Figure 2. Emissions of CO_2 by sectors (France 1990).

5.2 Examples of French results

CITEPA was also in charge of the French CORINAIR 90 inventory with financial support from the French Ministry of the Environment. This inventory is now complete and a report is available.

Thereafter in Annex 5, some pages of CORINAIR 90- FRANCE report are given to illustrate results from CORINAIR inventory.

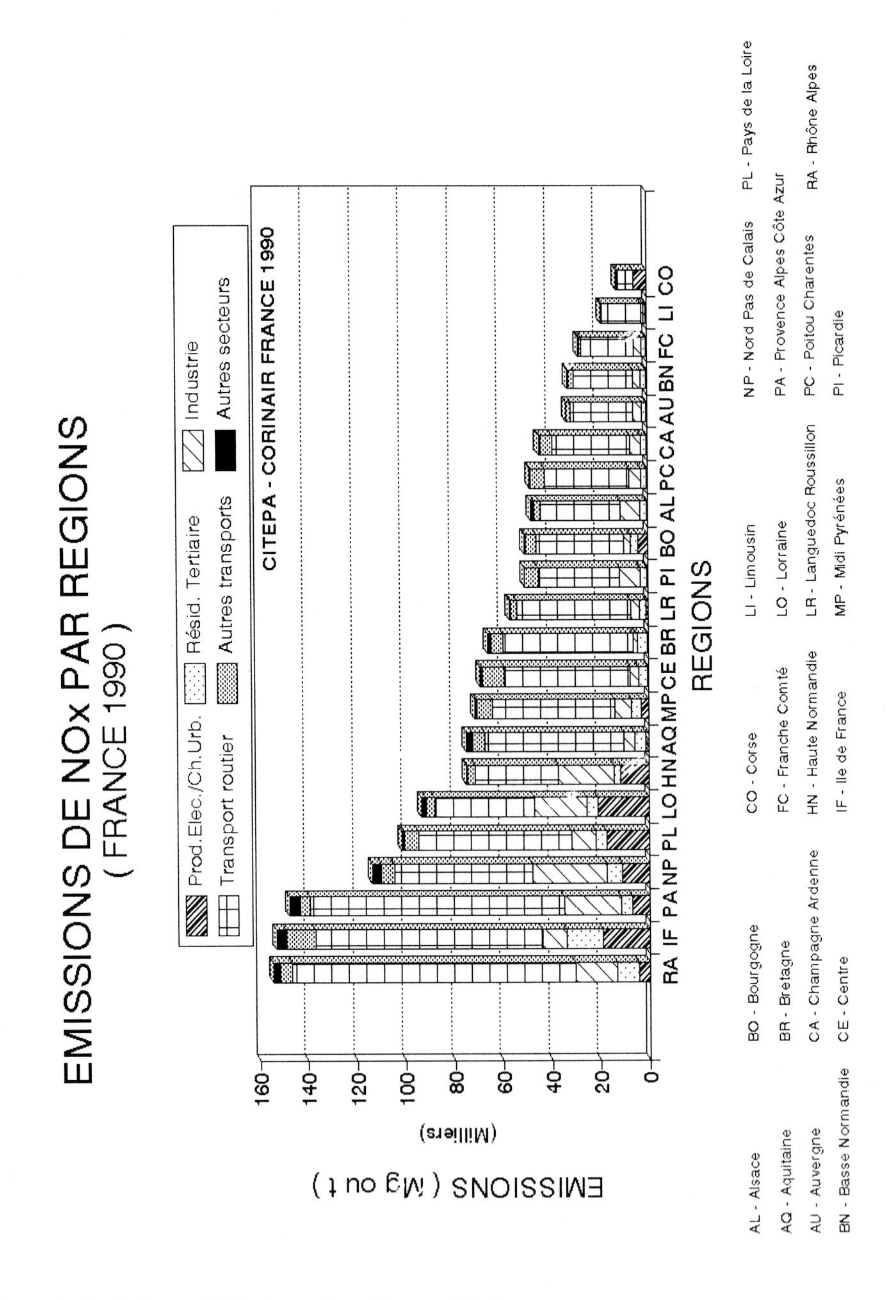

Figure 3. Emissions of NOx by regions (France 1990)

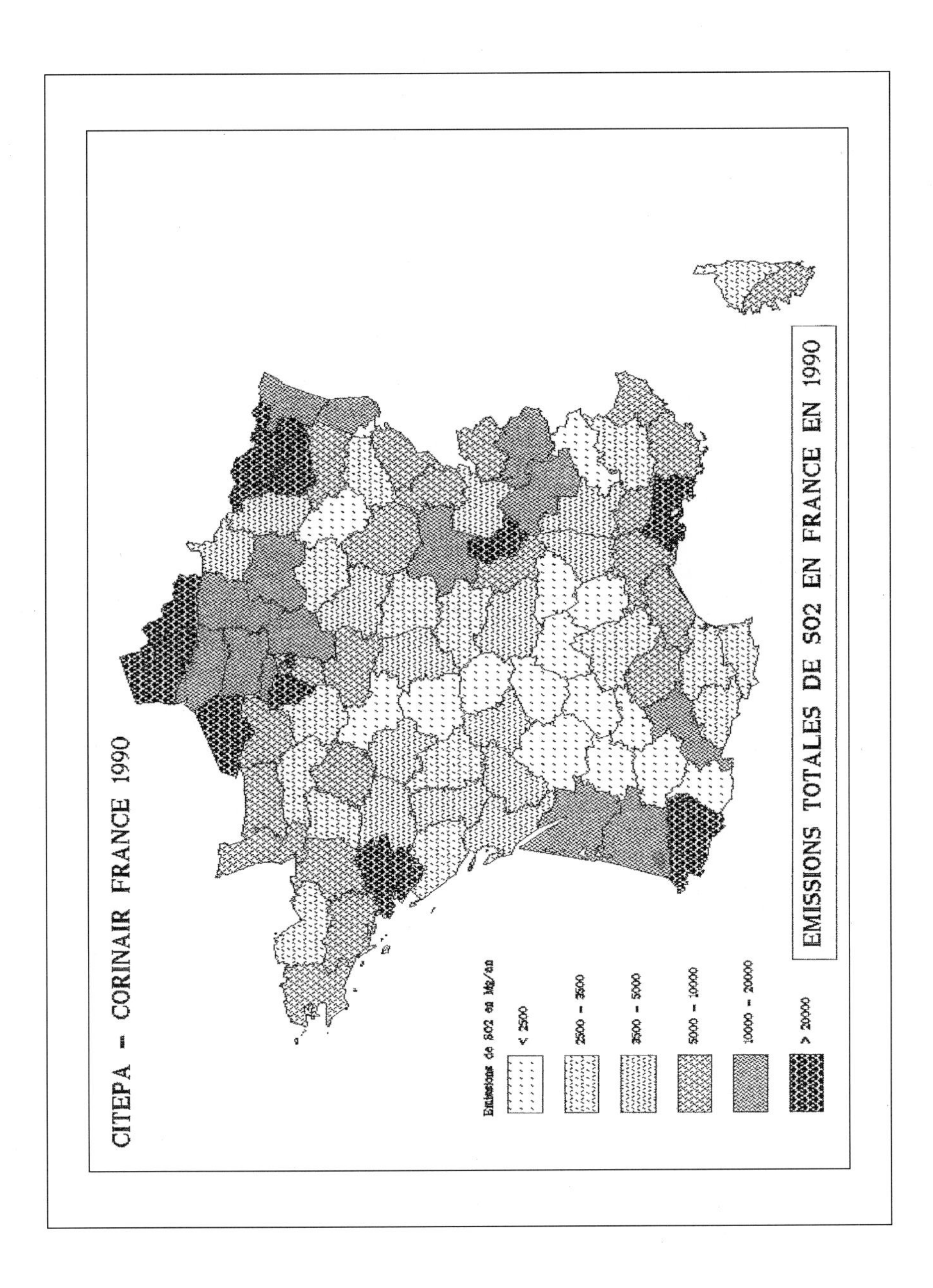

Figure 4. Total SO_2 emissions in France in 1990

KEYWORDS

Atmospheric pollution, European Community, CORINAIR, CO_2, CO, N_2O, NH_3, emissions.

REFERENCES

[1] *Official Journal of the European Communities* (OJ) n L 76, 6.7. 1985, Decision 85/338/EEC.
[2] Bouscaren R., Veldt, Zierock K.H. "CORINE Emission Inventory Project - Feasibility Study", (1986).
[3] Bouscaren R. et al. "European Inventory of Emissions of Pollutants into the Atmosphere" (CORINAIR project realized jointly by CEC and OECD). *Proceedings of the 8th World Clean Air Congresss*, The Hague, Vol. **5**, pp. 163-168 (1989).
[4] Bouscaren R., Veldt, and all CORINAIR working Group experts, "CORINAIR Inventory - Default emission factors handbook (Second edition)", January (1992).
[5] Zierock K.H. et al. "CORINAIR Working Group on Emission factors for calculating 1990 Emissions from road traffic", Vol. **1**, "Methodology and Emissions Factors", Vol.**2**, COPERT-users manual. ISBN 92-826-5770-1 (1990).
[6] Fontelle J.P. and Chang J.P. (CORINAIR 90 software) - Instructions for use - Version 5.13., (September 1992, July 1993).
[7] Fontelle J.P., Audoux N., Chang J.P. CORINAIR FRANCE 1990, "Estimation des émissions de polluants atmosphériques", Juin (1994).

ANNEX 1: CORINAIR 85 RESULTS

Note: These notes correspond to the final result of the CORINAIR 1985 inventory, some of them have been reviewed since 1985, especially for France.

CORINAIR-1985 Emissions of COV UNIT Mg/AN

	Combustion industry excepted	Combustion industry (sous chaudière)	Industrial processes (*)	Solvent evaporation	Road	Mines, décharges, épandage, distribution combust. gazeux	Nature and agriculture (**)	TOTAL
B	17782	2615	39312	82448	192528	70748	28069	433502
DK	14020	1635	5095	58468	96460	25008	7102	207788
D	100480	20869	120479	1119686	1196723	3069154	253773	5881164
GR	1751	541	10422	27925	114568	262550	195807	613564
E	35847	2226	42944	327055	489239	367707	876191	2141209
F	189105	6609	97090	437597	1188894	432388	423875	2775558
IRL	16621	568	1727	21182	24300	24374	20817	109589
I	46647	5331	83173	396554	989231	1203034	221037	2945007
L	487	81	2066	2636	8163	4321	2700	20454
NL	6513	10284	22551	163747	216573	0	13981	433649
P	637	658	20332	52149	53235	7106	65000	199117
UK	89704	56028	272841	668000	791296	1964497	79998	3922364
EUR 12	519594	107445	718032	3357447	5361210	7430887	2188350	19682965

CORONAIR-1985 Emissions of NOX Unit Mg/AN

	Combustion industry excepted	Combustion industry (sous chaudière)	Industrial processes (*)	Road	TOTAL
B	64504	37109	33096	182395	317104
DK	148642	12738	5379	102830	270589
D	833617	227322	169418	1484817	2715174
GR	147837	8429	31891	120050	308207
E	265667	39507	80527	453531	839232
F	257920	124162	127711	1095037	1604830
IRL	35744	8184	5718	35451	85097
I	441960	115765	152833	863366	1573924
L	1728	2589	9186	8633	22136
NL	134764	28258	37175	271036	471233
P	13968	10917	14492	56948	96325
UK	855083	299874	53430	916166	2124553
EUR 12	3201434	914854	721856	5590260	10428404

ANNEX 1: CORINAIR 85 RESULTS (continued)

Note: These notes correspond to the final result of the CORINAIR 1985 inventory, some of them have been reviewed since 1985, especially for France.

CORONAIR-1985 Emissions of SO2 Unit Mg/AN

	Combustion industry excepted	Combustion industry (sous chaudière)	Industrial processes (*)	Road	TOTAL
B	188576	98993	88962	15920	392451
DK	241387	61383	19652	11164	333586
D	1546609	416203	293640	59179	2315631
GR	373365	81139	45237	0	499741
E	1699180	263345	160139	67274	2189938
F	609694	444062	328511	99289	1481556
IRL	79065	55412	2289	4377	141143
I	1185891	549934	278082	75770	2089677
L	2817	4954	8309	489	16569
NL	71248	14521	102919	11070	199758
P	85945	68918	36348	6667	197878
UK	2948527	558466	216766	43480	3767239
EUR 12	9032304	2617330	1580854	394679	13625167

(*) including oil refinery combustion under generation
(**) CH_4 not included

ANNEX 2: EMEP/CORINAIR 90 SOURCE SECTOR SPLIT

SOURCE CATEGORY[a]	SO_x	NO_x	NMVOCs	CH_4	NH_3	CO
1. Public power, cogeneration and district heating plants.	XX	XX	X	X		X
2. Commercial, institutional and residential combustion plants.	X	X	X	X		XX
3. Industrial combustion plants and processes with combustion.	XX	XX	X	X		X
4. Non-combustion processes.	X	X	XX	X	X	(XX)
5. Extraction and distribution of fossil fuels.	X	X	X	XX		X
6. Solvent use.			XX			
7. Road transport.	X	XX	XX	X		XX
8. Other transport.	X	(XX)	X	X		X
9. Waste treatment and disposal.	X	X	X	XX	X	X
10. Agriculture.			X	XX	XX	
11. Nature.			(XX)	X		

[a] Relevant source categories are given by "X" and major source categories are shown by "XX". The parentheses indicate that the given source category may be a major one for some countries.

ANNEX 3 - CORINAIR 90 POINT SOURCE SPECIFICATIONS: DEFINITION OF LARGE POINT SOURCES (LPS)

Large Point Sources to be considered must be at least (whatever the stack height):

- **Power plants** with thermal capacity input $\geq$ 300 MW (items of SNAP 90: 1, 2 and 3.1) whatever the fuel

- All **oil refineries** (items: 3.1, 3.2.1, 4.1)

- All **units of production of sulphuric acid**, including installations for abating SO_2 emissions in non-ferrous metal or elsewhere (item 4.4.1)

- **All units of production of nitric acid** (item 4.4.2)

- **Integrated iron and steel plants** with a capacity of production higher than 3 millions tons per year. An integrated iron and steel plant generally includes: iron ore reception, coke production, sinter plant, blast furnace, steel conversion, rolling mill (grey iron foundry, lime plants, galvanizing, surface treatment are not included) (items 3.1, 3.2.2, 3.2.3, 3.3.1, 3.3.2, 4.2)

- **Paper pulp plant** with a production capacity higher than 100 000 t/y of pulp (9 % dryness condition, ready for use and shipping) (items 3.1, 3.3.21, 4.6.2, 4.6.3)

- **Vehicule painting units** with a production capacity higher than 100 000 veh./y (or equivalent). Some painting lines are presently equipped for using low-VOC emission paints; they must be taken into account (item 6.1.1)

- **Airports** (item 8.5) with LTO cycles > 100 000
  National and international air traffic above 3 000 feet is not covered by the present emission inventory. Only LTO cycle (landing-take off) is concerned.
  Airport activities should also include land operations (road traffic, boilers, etc...); nevertheless these emissions are generally considered much lower than LTO emissions.

- **LPS with large flows of pollutants** (whatever the item)

. for SO_2, NOx, VOC	plants emitting $\geq$ 1 000 Mg/y
. for CO_2	plants emitting $\geq$ 300 000 Mg CO_2/y
. for CH_4, N_2O, CO	no large emitter defined for these pollutants.

N.B.: Whatever the criteria of large emitters, for a defined LPS, all eight pollutants must be taken into account.

ANNEX 4 - CROSS COUNTRY EMISSION FACTOR CHECKING (CORINAIR 90)

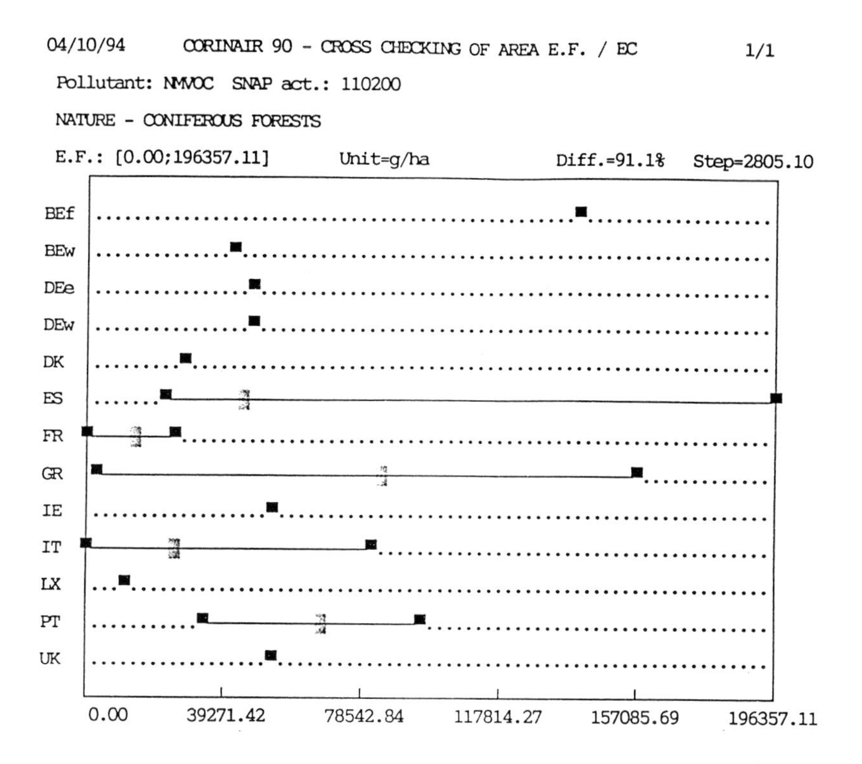

N_ISO	Or	EF mini	EF maxi	diff%	Emissions (Mg)	Act. (ha)	Average EF
BEf	T	140832.00	140832.00	0.0	8061	57240	140832.00
BEw		42400.00	42400.00	0.0	9905	233600	42400.00
DEe	T	47500.00	47500.00	0.0	106875	2250000	47500.00
DEw	T	47500.00	47500.00	0.0	243200	5120000	47500.00
DK		27569.00	27569.00	0.0	7802	283000	27569.00
ES		21586.73	196357.11	89.0	347659	7741999	44905.53
FR		0.00	24752.07	100.0	73552	4854500	15151.30
GR		2000.00	157000.00	98.7	81758	966393	84601.60
IE		52500.00	52500.00	0.0	16238	309295	52500.00
IT		0.00	82343.66	100.0	39147	1557518	25134.22
LX		12468.00	12468.00	0.0	395	31700	12468.00
PT		33699.80	94002.07	64.1	96663	1423685	67896.54
UK		52560.00	52560.00	0.0	77000	1465000	52560.00

N.B.: Or = 'T' means the presented EF are not the original ones, but the Transformed ones because reference units were not used.

ANNEX 5 - CORINAIR 90 FRANCE / SOME RESULTS

Emission versus emissions categories

These values can be revised later on according to the improvement of knowledge and of the methods of estimation. The reader is requested to test if more recent data are available.

CITEPA - CORINAIR FRANCE 1990 17/06/94 resfraf.wq2

EMISSION GROUPS	SO_2 10^3 M g	NO_x 10^3 M g	COVNM 10^3 M g	CH_4 10^3 M g	NH_3 10^3 M g	CO 10^3 M g	CO_2 10^3 M g	N_2O 10^3 M g
1. Production of electricity cogeneration and district heating plants.	343.7	105.9	1.3	1.1	AZ	21.0	40.0	1.1
2. Commercial, institutional and residential combustion plants. (biomass included)	116.2	88.6	214.2	149.9	AZ	1892.0	117.6	3.8
3. Industrial combustion plants.	514.1	164.3	7.3	6.6	AZ	598.1	86.7	2.1
4. Industrial processes, non-combustion processes.	110.9	30.8	100.1	5.9	16.6	668.3	6.8	93.6
5. Extraction and distribution of fossil fuels.	23.8	3.5	122.5	352.2	AZ	0.0	0.2	0.0
6. Solvent use.	0.0	0.0	635.6	0.0	0.0	0.0	0.0	0.0
7. Road transport.	145.3	1037.8	1169.8	22.3	0.8	6812.2	97.4	3.8
8. Other transport.	24.6	129.1	121.8	0.6	AZ	512.2	11.7	0.4
9. Waste treatment and disposal.	19.2	23.9	19.0	729.0	1.8	232.2	11.9	1.3
10. Agriculture.	0.0	0.0	10.5	2410.2	625.9	NE	NE	60.8
11. Nature.	2.5	5.5	461.8	191.1	AZ	193.9	108.8	55.7
TOTAL	1300.4	1589.4	2864.0	3868.8	645.1	10930.1	481.0	222.7

value = 0 signifies emissions < 0.05 10^3 — NO_x expressed in equivalent NO_2, CO_2 in CO_2 effective

AZ: approximately zero or negligible — NE: not estimated

Chapter 3

SPACE AND EARTH REMOTE SENSING GIS Application to Environmental Impact Studies

ENVIRONMENTAL IMPACT ANALYSIS USING REMOTELY SENSED DATA

Piero BOCCARDO

Dipartimento di Georisorse e Territorio Politecnico di Torino, C.so Duca degli Abruzzi, 24, 10129 Torino, Italy

ABSTRACT

This paper presents results from an activity undertaken at the Dipartimento di Georisorse e Territorio - Politecnico di Torino, to analyze remotely sensed data in order to generate thematic maps suitable for environmental impact studies. The aim is to identify optimal procedures and methodologies, from data acquisition, through to the final restitution of the product. Several types of data were processed and different algorithms were implemented in order to compare their performances. These methods were then used to classify digital images acquired at different times of the year and in different locations. Particular attention was paid to the elaboration of a new multi spectral digital image analysis capable of standardizing the procedures used for the different problems studied. This standard analysis includes: remotely sensed data acquisition, pre-elaboration of data, spectral signature extraction, definition of the training samples, fully automated and/or supervised classification and thematic map generation and editing. Three different case-studies are presented here: (i) an experimental test to obtain a Bormida valley land use map using Landsat 5 Thematic Mapper data (ii); the effects of multiple look direction radar and back scatter enhancement from flooded terrain using airborne SAR imagery; (iii) moisture content analysis using Landsat 5 Thematic Mapper data and generation of a water-table contour map.

Utilisation des Données de Télédétection pour les Études d'Impact sur l'Environnement

RÉSUMÉ

Ce travail présente les résultats d'une recherche entreprise au Politecnico de Turin dans le Département de Georisorse e Territorio, dont le but est la réalisation de cartes thématiques à partir de données de télédétection pour les études d'impact sur l'environnement. L'objectif est d'identifier des procédures et méthodologies optimales, à commencer par l'acquisition des données, jusqu'à la restitution finale du produit. Différentes catégories de données ont été traitées et divers algorithmes ont été testés afin de comparer leurs performances. Ces méthodes ont ensuite été utilisées pour segmenter des images numériques obtenues à différentes époques de l'année et situées dans des régions différentes. Une attention toute particulière a été consacrée sur la mise au point d'une méthode d'analyse pour les images digitalisées multispectrales, capable de standardiser les procédures employées en fonction des problèmes étudiés. Cette analyse des standards comprend: l'acquisition des données par télédétection, le pré-traitement des données, l'extraction des signatures spectrales, la définition des échantillons de référence, la classification automatisée et/ou supervisée, ainsi que l'élaboration et la publication de cartes thématiques. Nous présentons dans ce travail trois études de cas: (i) un test portant sur la réalisation pratique d'une carte d'occupation des sols de la vallée de Bormida, à partir des données Landsat 5 Thematic Mapper; (ii) l'utilisation des données radar à directions multiples et des données de réflexion pour l'étude des terrains inondables à partir des images SAR; (iii) l'analyse de l'état hydrique du sol à partir des données Landsat 5 Thematic Mapper et le dessin de cartes isobathes des eaux souterraines.

1 INTRODUCTION

In recent years, the large amount of data derived from multi spectral satellite images created the problem of how to use this data and, above all, which data should be taken into account in environmental analysis. The problem essentially is to extract the largest quantity of information from the analyzed data, possibly by avoiding redundancies and discrepancies.

Remotely sensed data fusion, or in general digital image data fusion, is a quite common problem, and essentially regards the integration of data acquired by the same sensor or by different sensors. The first case generally deals with multi spectral data integration, where the fusion occurs, by means of synthetic addition techniques of radiometric components, generating true or false color image composites. The second case deals with digital images acquired, either by different sensors mounted on board the same platform (for example, all the data obtained by Landsat's MSS and T.M. sensors), or using different images (such as imagery in the microwave spectral band, digital conversions of aerial photographs in the near infrared spectral band, etc.). The two different methods of integration, although extremely different, present several common characteristics that are analyzed in the following paragraphs.

2 DIGITAL IMAGE PROCESSING TECHNIQUE

2.1 Data acquisition

Data acquisition essentially depends on two factors: scale and spectral bands. Scale is intrinsically related to the acquisition system (sensor) geometrical resolution and allows one to obtain a quantitative measure of the minimum ground details that can be identified. IFOV (Instantaneous Field Of View) is not always that which can be detected on the ground; in fact, some features (such as highways, railroads, mirrors and corner reflectors) though smaller than the minimum sensor geometric resolution, register a consistent radiant flux on the sensor (a radiant flux that is considerably different from that of the surrounding area), leaving a "trace" of their presence on the digital image.

Nowadays, the geometric resolution of the satellite acquisition systems does not allow large scale applications, being suitable for medium scale surveys (1:25.000, 1:50.000). Several new sensors are due to be launched in the near future, thus improving geometric resolution to a few meters (bib Washington).

To overcome resolution shortcomings, one can often use aerial imagery, acquired both in a direct digital form (scanner mounted on board of airborne platform) and hardcopy scanned (aerial photograph).

Spectral resolution (the possibility of acquiring signals in different bands of the electromagnetic spectrum) is particularly important in the feature recognition process and in the classification of environmental phenomena. It is obvious that the greater the spectral resolution, the greater the possibility of extracting information from the data. This extraction is also related to the radiometric resolution concept, that is the capacity to (quantify) the radiant flux that comes from the ground in a certain number of radiometric classes, with the subsequent storage of these classes in digital data fraction (bits).

2.2 Data pre-processing

The data available must be previously corrected in order to eliminate two types of distortion: radiometric and geometric distortion.

Radiometric distortion refers to the scanners' calibration instability and to the interactions between the electromagnetic energy, the atmosphere and the ground.

Geometric distortions refer to both the relative position of the acquisition platform with respect to the atmosphere and the ground (terrestrial rotation and curvature effect, reflection and refraction phenomena due to the different atmospheric layer density and to the presence of suspended particles), and the sensor operating devices (rotating mirror, line or area CCD array).

It is absolutely necessary to perform an image georeference by determining the geometric position in image and ground coordinates of a sufficient number of ground control points; then, by using desirable geometric transformation algorithms, to rectify the digital data through repositioning and radiometric resampling.

2.3 Spectral signature extraction

If the remotely sensed data environmental analysis goal is to generate thematic maps, the main task is to extract different spectral feature signatures in order to classify the whole image.

A spectral signature is the radiant behavior (reflection or emission) of a certain feature, this behavior being detected exactly where the feature is located. The extraction has to be carried out by direct surveys, by means of spectrometers, or by choosing some training samples on the multi spectral images.

A spectral signature is the minimum knowledge of the studied area that one can obtain, and it is particularly important to understand how the analyzed phenomenon are distributed on the ground and to know the relation with the physical processes that generate them.

2.4 Definition of training samples

In order to compare the physical phenomena representation and the real evidence of the same phenomena, it is necessary to extract some training samples that are small portions of the studied area where a theoretical model can be built. Training samples have to be both statistically meaningful and statistically representative, because they are the only evidence one can distinguish.

2.5 Fully automated and supervised classification

Classification is the process that allows one to extract information from the raw data.

Classification algorithms are essentially based on a radiometric distribution statistical analysis of the electromagnetic signal in a whole data set (the digital image). Such an analysis is performed both by evaluating the number of radiometric classes in the digital image in order to select different data clusters (unsupervised classification), and by comparing the likelihood of the radiant flux that comes from the ground features collected from the sensor with the spectral signatures extracted from the training samples.

The two different algorithm usages depend on a simple assumption; if the degree of understanding of the studied area is low or non existent, it is necessary to perform an unsupervised classification in order to extract the information (the meaning of the different classes).

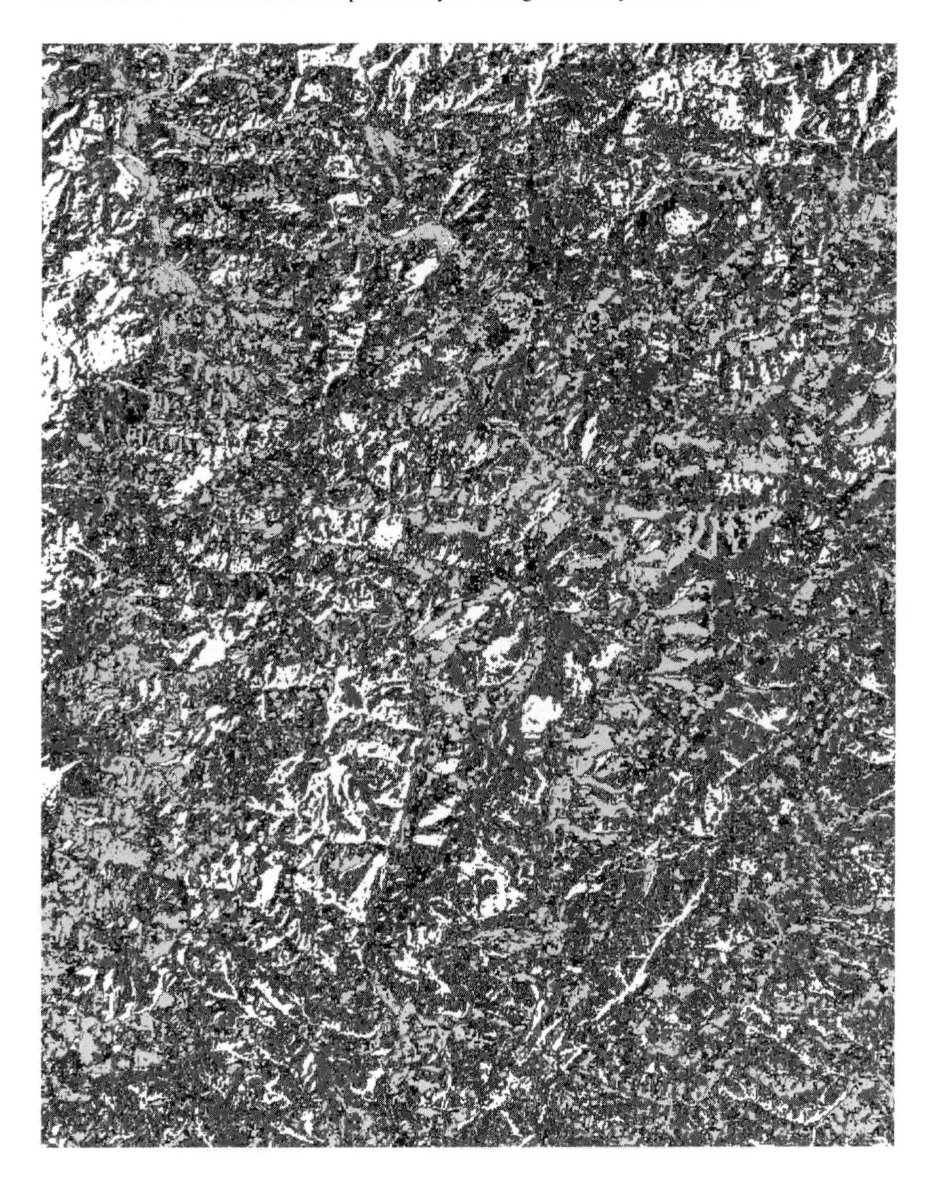

Figure 1. The classified studied area using the results obtained by different classification algorithms: Bormida Valley land use map using Landsat 5 data. (For color figure, see page 299)

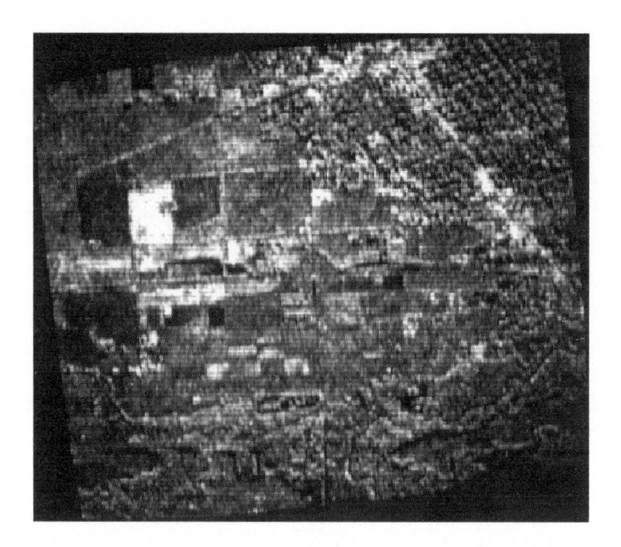

Figure 2. The Northern looking direction radar image, flooded rice fields

Figure 3. The classified infrared image of the ground moisture content: Bormida Valley

If the degree of understanding is sufficient or almost complete, it is useful to proceed with a supervised classification that can link the known information (the training samples and the derived spectral signatures) with the raw data stored in the digital image.

Sometimes the information can be extracted by using different procedures (such as radiometric processing) by means of low complex techniques and algorithms (contrast stretching, digital filtering, etc.).

The methodology presented here is a combination of the different classification algorithms (supervised and unsupervised) and of the simple radiometric processing.

2.6 Thematic map generation and editing

The last step of this digital image analysis methodology is the generation of thematic maps, that is the spatial representation of the analyzed phenomena. A thematic map is the classification result corrected from the interpretation errors, and plotted at the right scale.

In this map the information should not be redundant (otherwise it will be necessary to reprocess the data redefining the spectral signatures), and the selected classes (the organization of the information) should be consistent, both statistically (classes should represent the analyzed phenomena) and geometrically (spatial positioning should be precise).

3 Case Studies

3.1 An experimental test to obtain a Bormida valley land use map using Landsat 5 Thematic Mapper data

The first results of a land use study of a valley in the Piedmont region are shown in this experimental test, using remotely sensed data. In particular, attention was paid to the data collected by the American satellite Landsat 5 in seven different spectral bands of the multi spectral Thematic Mapper scanner (T.M.).

The digital classification was conducted using different classification algorithms (both supervised and unsupervised) and then compared to previous classification data, carried out by using the classical photo interpretation technique on aerophotogrammetric color pictures and field data.

Due to the particular conditions of the analyzed area (a hilly valley with frequent small area agricultural crops), different digital image processing procedures were used; the results obtained using these procedures were then merged in order to extract as much information as possible from the different band images.

3.2 The effects of multiple look direction radar and backscatter enhancement from flooded terrain using airborne SAR imagery

Synthetic Aperture Radar (SAR) has many advantages over conventional optical/mechanical remote sensing systems. Some of these advantages are well known i.e. nearly all types of weather, day or night operating capabilities with synoptic scale imagery at relatively high resolution. The understanding of SAR characteristics is important for optimizing its use. In this case the effects of multiple look direction and the enhancement of the radar backscattered signal, from flooded terrain, were of special interest.

False color composite radar images from three look directions were generated and evaluated, using X-band Intera multiple look direction radar data of South Louisiana, collected under the auspices of the U.S. Geological Radar Mapping Program.

The enhancement of a radar signal from a flooded surface masked by a vegetation canopy provides valuable information for both Earth and biological scientists. This signal enhancement was corroborated by field data for an X-band radar of flooded rice fields.

3.3 Moisture content analysis using Landsat 5 Thematic Mapper data and generation of a water-table contour map.

This study regards the extraction of moisture content data using conventional Landsat 5, Thematic Mapper images. First, a statistical data analysis was conducted with the aim of evaluating the data distribution in the seven different band images; secondly, a clustering procedure was developed in order to compare the retrieved classes with a moisture content ground survey. The spectral signatures extracted from the training samples, chosen according to their supposed humidity condition, were then processed band by band evaluating the information content for every image. All information redundancy was thus avoided, and an infrared optimum band was selected as being typical of the analyzed phenomena.

A water table contour map was generated from the radiometry of the pixel that makes up the image, and then verified by measuring the piezometric surface depth in different sample areas using ground surveys.

KEYWORDS

Environmental impact study, GIS, Thematic maps, remote sensing, risk analysis in flooded terrain, moisture content in soil, ground water.

REFERENCES

[1] Boccardo P., Lewis A. J., Hanson B. C., "Radar Back Scatter Enhancement Using Multi-Look Direction X-Band Synthetic Aperture Radar: A South Louisiana Example", 1993 ACSM/ASPRS American Annual Convention & Exposition, February 15-18, New Orleans, Louisiana (1993).

[2] Boccardo P. and Lewis A. J., "Understanding SAR Imagery: The Effects of Multiple Look Direction Radar and Back scatter Enhancement from Flooded Terrain, ISPRS Commision I Workshop on Digital Sensors and Systems", June 21-25, Trento, Italy (1993).

[3] Boccardo P. and Garnero G., "Prove sperimentali per la redazione di una carta dell'uso del suolo della Valle Bormida mediante dati telerilevati", *Seminario A.I.G.R. sul tema del recupero dell'edilizia rurale nel contesto territoriale*, 13-16 giugno, Sassari (1994).

[4] Boccardo P., "Marginalit rurale e valorizzazione del territorio", in Belforte S. (edited by), "Oltre l'ACNA: identità e risorse per la rinascita della Valle Bormida", Franco Angeli, Milano (1993).

[5] Campbell J. B., "Introduction to Remote Sensing", *The Guilford Press*, New York (1987).

[6] Lewis, A. J. , ed., "Geoscience Applications of Imaging Radar Systems", *Remote Sensing of the Electromagnetic Spectrum (RSEMS)*, Remote Sensing Committee of the Association of American Geographers, Vol. 3, No.3, 1-152 (1976).

[7] Lewis A. J. and Hanson B. C., "Radar Mapping Program of Louisiana", *Proceedings of the Louisiana Aerospace Forum*, University of New Orleans, New Orleans, LA (1992).

[8] Slater P. N., "Remote sensing: optics and optical systems", Addison-Wesley Publishing Company, Reading, MA (1980).

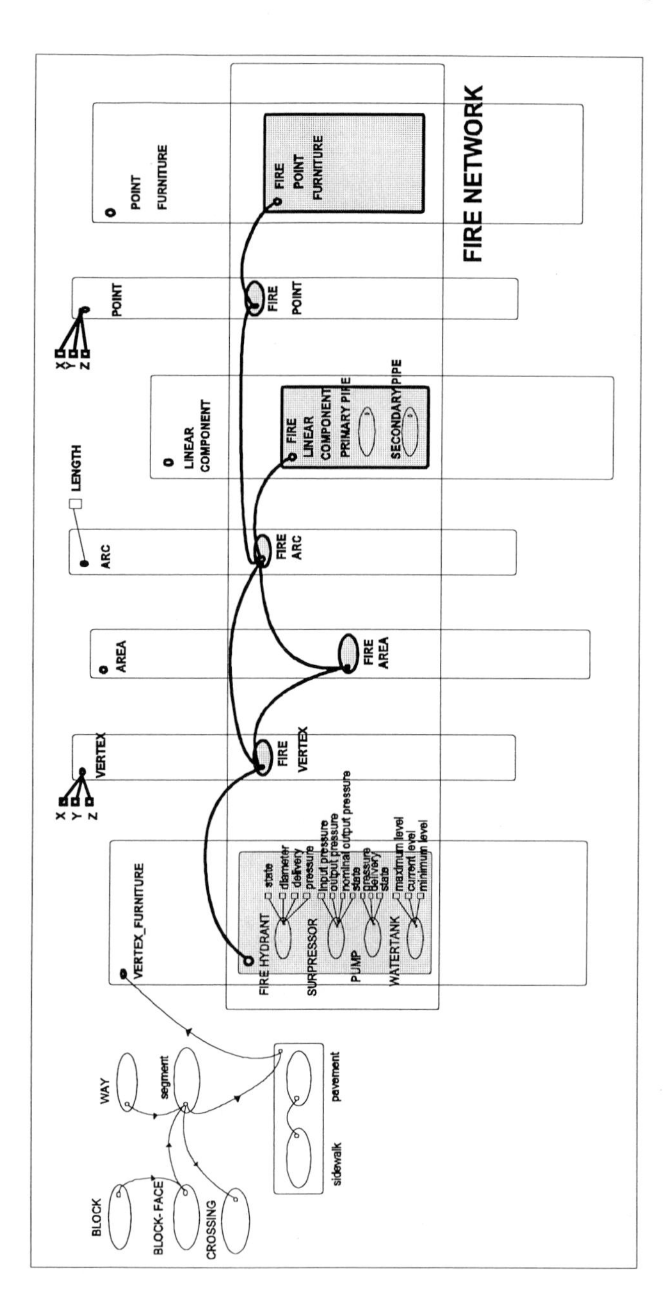

Use of Expert Systems and G.I.S. - This picture represent a part of the structure devoted to the "Fire Hydrant Network", connected to that of the Urban Texture, composed of blocks, ways, segments, etc. A few classes of this water network are shown with their specific attributes. (after Bouillé F., see page 245).

USE OF THE ROSE-DIAGRAM METHOD FOR VEGETATION FIRE PATTERN ANALYSIS ON A REGIONAL SCALE IN AFRICA

P. A. BRIVIO[a,1], J-M. GRÉGOIRE[b], B. KOFFI[b] and G. OBER[a]

[a] Remote Sensing Dept, IRRS - CNR, 56 Via Ampère, 20131 Milan, Italy
[b] MTV- Inst. Remote Sensing Applications, JRC-CEC, 21020 Ispra (Va), Italy

ABSTRACT

Biomass burning is a phenomenon of continental proportions, capable of causing large scale environmental changes. Satellite observations, in particular NOAA-AVHRR GAC data four kilometer ground resolution, are the only source of information currently useful to document burning patterns on a continental scale and over long periods of time.

In this paper we present the application of the rose-diagram method in the analysis of a set of NOAA-AVHRR data to characterize and quantitatively describe the evolution of spatial patterns of vegetation fire on regional and continental scales in Africa.

Utilisation de la Méthode du Diagrammes-Rose pour L'Analyse de la Distribution des Feux de Végétation à l'Échelle Régionale en Afrique.

RÉSUMÉ

Le phénomène des feux de biomasse qui atteint des proportions continentales, est capable d'entraîner des changements environnementaux à grande échelle. Les observations par satellite, en particulier les données de NOAA-AVHRR de quatre km de résolution au sol, représentent la seule source d'information actuellement exploitable à l'échelle continentale pour étudier la dynamique des feux de végétation sur de longues périodes de temps.

Nous présentons dans cet article une application de la méthode du diagramme-rose à l'analyse de données NOAA-AVHRR afin de caractériser et décrire d'une manière quantitative l'évolution spatiale des feux de végétation aux échelles régionales et continentales en Afrique.

---

[1] To whom correspondence may be addressed

1 INTRODUCTION

Global change is more than climate change or global warming; it also concerns changes in land use and land use practices over large areas. Vegetation fires are a key element of land use practices in the tropics. Much of the world's tropical Savannahs are periodically burned: in Africa alone an area of more than 300 million hectares, 75% of the Savannah, burns annually. Episodic fires in the tropical rain forests are not uncommon and are thought to be on the increase. The ecological, environmental and economic effects of biomass burning occur on all scales, from local to global, and systematic documentation and history of the phenomenon are not yet available.

While the subject of fire in the tropical environment has been widely studied, there is now a need to assess the characteristics of fire on unexplored scales. Biomass burning is a phenomenon of continental proportions, capable of causing large scale environmental changes (Malingreau and Tucker, 1988). This calls for a better understanding of burning on regional and continental scales, a better appreciation of historical trends and, if possible, improved capabilities to chart future trends and possible impacts.

Earth observations from space by remote sensing satellites provide systematic and consistent measurements of a series of parameters related to vegetation fires and their environmental impacts. Fire detection is normally done on the basis of the visible wavelength and surface brightness temperatures derived from Landsat, SPOT and NOAA-AVHRR sensors (Langaas, 1992; Grégoire, 1993).

The aim of this work is to explore spatial characteristics of vegetation fire on regional and continental scales of representation and to quantitatively describe the evolution of spatial patterns of vegetation fire on regional and continental scales in Africa.

2 BIOMASS BURNING DOCUMENTATION USING AVHRR-GAC TIME SERIES

Sub-continental patterns of fire distribution, both in time and space, may vary from year to year. These variations may be due to climate, interannual variations in rainfall distribution, or to anthropic factors, such as movements of populations and land use changes. Data from NOAA-AVHRR allow regular observation of biomass burning at regional and continental scales.

While the sampling procedure used to produce the AVHRR Global Area Coverage (GAC) data, 4 kilometers ground resolution, makes it difficult to collect quantitative information related to fire (Justice *et al*, 1989), GAC archives are very useful for getting information on the main regions of fire occurrence and its spatio-temporal variability, as an index of fire activity (Belward *et al*, 1993). Moreover, the availability for the research community of large volumes of these images, constituting a historical archive covering the last decade from July 1981, both of 4 kilometers (GAC data) and 1 kilometer (LAC data: Large American Coverage) ground resolution, can give a temporal perspective to the analysis of burning patterns (Fig. 1).

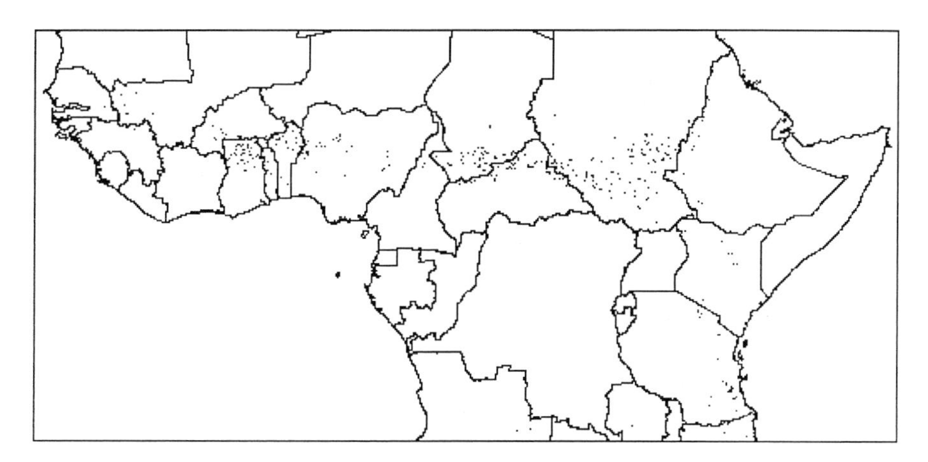

Figure 1. Fire distribution in Central Africa from AVHRR-GAC data of November 1987.

The FIRE project (Fire In global Resource and Environmental monitoring) of the Monitoring of Tropical Vegetation unit (MTV-JRC, CEC) is currently involved in the analysis of ten years of GAC archives to characterize the extent, status and variability of vegetation fire patterns at regional and global scales in order to understand the dynamic evolution of the terrestrial biosphere (Grégoire *et al*, 1994).

3 Vegetation Fire Pattern Analysis

3.1 Rose-diagram approach

A numeric approach, commonly used by geologists in the lineament analysis field where the main parameters are rose-diagrams of frequency and length, has been applied to this problem.

For each image, a *reference point* is assumed as origin of polar plane and each fire pixel is connected to the origin generating a series of linear elements. The fire pixel position then becomes a function of polar coordinates (r,q), where r is the distance of a generic fire from the reference point and q is the angle.

The spatial domain of the fires is subdivided into angular sectors 15° wide, and linear elements are examined in terms of length and angle, East being 0° and North 90°. Quantities considered are frequency and distance of the fires per angular sector (Fig. 2). Reference points can be selected to be representative of some particular feature of National Parks, Natural Reserves, equatorial rain forest.

The rose-diagram of *fire frequency* gives an immediate impression of the biomass burning phenomenon on a regional scale. Radius indicates the number of fire events and direction indicates the angular sector, 15° wide, where they occur.

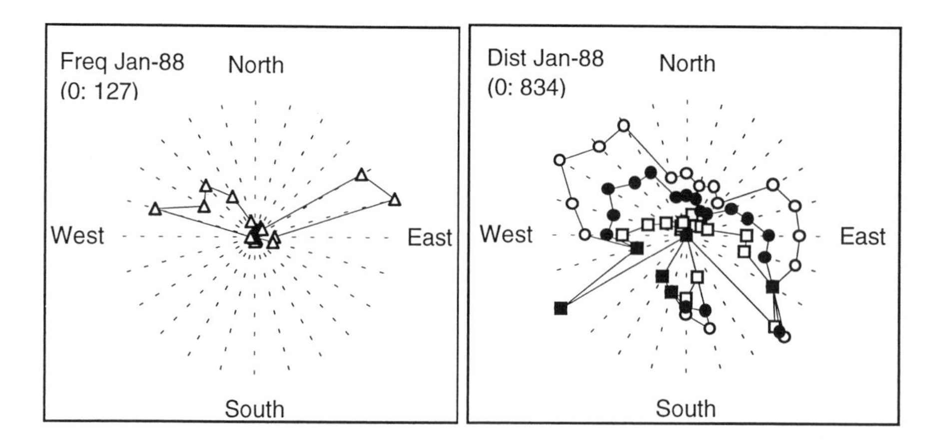

Figure 2. Rose-diagrams of fire frequency and distance for AVHRR-GAC monthly image of January 1988. Reference point: Enyele town (2°49'N, 18°06'E) in Congo.

Rose-diagrams of *fire distance* present in the same graph minimum, mean and maximum distances and show characteristic shapes very different from previous ones. Interpretation of this kind of graph is not straightforward: interesting elements are lower values highlighting angular sectors where fire events are nearer to the reference point, e.g. limit between Savannah and forest. Moreover if lower values coincide with sectors with high frequency, their importance becomes more evident, and they might coincide with some critical area or direction of the fire activity.

3.2 AVHRR-GAC data for 1988-89 burning season

This study concerns the African continent and data considered at GAC (4 km x 4 km pixel resolution) cover the period from September 1988 to August 1989. The data-set consists of monthly images, created by accumulating day-by-day pixels recognized as fire according to a methodology developed by the MTV unit (Kennedy *et al*, 1994, Koffi *et al.*, 1996) which improved thresholding criteria proposed by Kaufman *et al* (1990).

Analysis of these twelve monthly images by the rose-diagram technique cannot only give synthetic information about the spatial distribution of fire activity on a regional and continental scale in Africa, but can also enhance the temporal evolution of the fire displacements during the defined period and allow exploration of some characteristics of fire seasonality in the central part of the African continent (Fig. 1).
Results of *frequency* analysis are presented in Figure 3. One can observe how the highest frequences gradually move from Southern sectors to Northern ones, month by month from September 1988 to August 1989: the temporal path starts in the fourth quadrant (South-East), goes to first quadrant, where the data from December 1988 to May 1989, are clearly located to the third quadrant for July and August 1989.

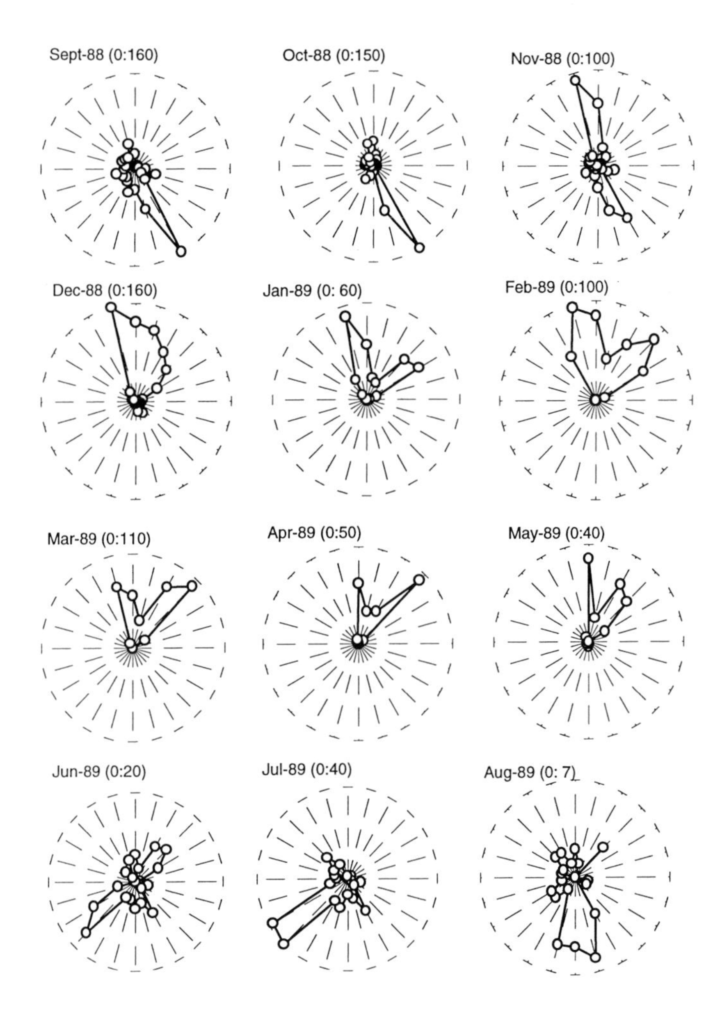

Figure 3. Spatio-temporal distribution of fire activity derived from AVHRR-GAC monthly images for the period September 1988 - August 1989 in the central part of the African continent. Reference point is located on the Equator.

Moreover some diagrams show a clear maximum, while others have two preferred sectors. Particular care must be taken in reading data with no prevailing direction. Analysis of *mean distance*, being obtained as an average value for linear elements per angular sector, is strongly influenced by the number of objects: a single fire occurring far away can have the same or higher impact than many fires occurring in the same direction. Nevertheless it is possible to distinguish several directions completely free from the presence of burning phenomena (first four months of the year, for example) and to distinguish isolated branches (January and July data).

the year, for example) and to distinguish isolated branches (January and July data).

4 CONCLUSION

The rose-diagram technique shows very promising results, particularly when the *reference point* is selected on the basis of meaningful consideration of the final users (Natural Preservation Area, rain forest, etc.). The results of this study are candidates for use in the EXPRESSO (EXPeriment for Regional Source and Sinks of Oxidants) experiment and are to be inserted in the VFIS-Vegetation Fire Information System of MTV. Definition of similar quantities, considering a *reference line* (Savanna-Forest transition line, etc.) requires further studies.

KEYWORDS

Vegetation fire, fire dynamics, Africa, AVHRR satellite imagery, remote sensing, risk analysis, rose-diagram method, spatio-temporal analysis.

REFERENCES

[1] Belward A.S., Kennedy P.J. and Grégoire J.M., "The limitations and potential of AVHRR GAC data for continental scale fire studies", *Int. J. Remote Sensing* , 15(11), pp. 2215-2234, (1994).

[2] Grégoire J. M., "Description quantitative des régimes de feu en zone soudanienne d'Afrique de l'Ouest", *Secheresse*, **4**(1), pp. 37-45 (1993).

[3] Grégoire J.M., Belward A.S. and Malingreau J.P., "The FIRE project: Fire In Global Resource and Environmental Monitoring", *1st TREES Conference*, Belgirate (Italy), 20-21 October (1993).

[4] Justice C.O., Markham B.L., Townshend J.R.G. and Kennard R.L., "Spatial degradation of satellite data", *Int. J. Remote Sensing*, **10**, pp. 1539-1561, (1989).

[5] Kaufman Y.J., Tucker C.J. and Fung I., "Remote sensing of biomass burning in the tropics", *J. Geophysical Research*, **95** pp. 9927-39, (1990).

[6] Kennedy P.J., Belward A.S. and Grégoire J.M., "An improved approach to fire monitoring in West Africa using AVHRR data", *Int. J. Remote Sensing* , 15 (11) pp. 2235-2255, (1994).

[7] Koffi B., Grégoire J-M. and Eva H.D., "Satellite monitoring of vegetation fires on a multi-annual basis at continental scale in Africa", in Biomass Burning and Global Change (Levine J.S., ed.) MIT Press, Vol. **I**, chapter 21 (1996 in press).

[8] Langaas S., "Temporal and spatial distribution of savannah fires in Senegal and Gambia, West Africa, 1989-90, derived from multi-temporal AVHRR night images", *Int. J. Wildland Fire*, **2**(1) pp. 21-36, (1992).

[9] Malingreau J.P. and Tucker C.J., "Large-scale deforestation in the southeastern Amazon Basin of Brazil", *AMBIO*, **17**(1) pp. 49, (1988).

INTEGRATION OF LANDSAT TM DATA AND SPOT DIGITAL ELEVATION MODEL APPLIED TO GEOSCIENTIFIC AND GEOTECHNICAL RESEARCH IN THE AREA NORTH OF KATHMANDU, NEPAL

G. GABERT[a,1]***, K. M. AMATYA***[b]***, C. BARDINET***[c,1]***, E. BOURNAY***[d]***, P. HOPPE***[e] ***and P. PRADHAN***[d,1]

[a]*CODATA-Germany, D-30916 Isernhagen, Germany*

[b]*Core Consultancy Ltd., P.O. Box 720, Kathmandu, Népal*

[c]*E.N.S., 45 rue d'Ulm, F-75230 Paris Cedex 05, France*

[d]*ICIMOD, P.O. Box 3226, Kathmandu, Népal*

[e]*BGR, Stilleweg 2, D-30655 Hannover, Germany*

ABSTRACT

The paper presents an initial study within the scope of a CODATA Project on the availability, acquisition and analysis of geo-environmental data on a regional scale in the Himalayas of Nepal and Tibet. In this project, digital multisatellite data and multidisciplinary environmental information will be connected by means of geographical information systems (GIS), in order to study the interdependence of problems like deforestation and soil erosion; lithology, morphology and landslides; snow-melting, glacial lake outbursts and flooding; socio-economic climatic changes and regional climate variations. For the case studies of the project, which will be carried out in interdisciplinary and international cooperation, two well defined areas in Central Nepal and in Southern Tibet have been selected. The project will advance environmental geo-indicators, facilitate risk assessment and contribute to cause-effect analysis of geo-hazards and management policy options of hazard prevention. In this paper, a digital 3-dimensional elevation model (DEM) derived from SPOT stereo-images is digitally connected with LANDSAT TM color composites and geological map compilations of the Melamchi area north of Kathmandu. The resulting 3-dimensional thematic and geological maps will facilitate the interpretation of complicated geological problems, for example at the northern margin of the Kathmandu Valley, where gneisses of the High Himalaya are in contact with metamorphic rocks of the Kathmandu Complex. The geoscientific interpretation will also contribute to the solution of problems in engineering-geological and geotechnical projects in this area.

---

[1] To whom correspondence may be addressed

Croisement des Données Landsat TM et des Modèles Numériques de Terrains Obtenus par SPOT pour des Recherches en Géosciences et en Géotechnique dans le Nord de Katmandou, Népal

RÉSUMÉ

Ce travail présente les résultats d'une première étude réalisée dans le cadre d'un Projet CODATA dont le but est d'évaluer la faisabilité, l'acquisition et l'analyse de données géo-environnementales acquises à une échelle régionale dans la région de l'Himalaya, du Népal et du Tibet. Il a pour objectifs de croiser les données numériques multisatellitaires avec les informations environnementales multi-disciplinaires à l'aide de systèmes d'information géographiques (SIG), afin d'étudier l'interdépendance des problèmes comme la déforestation et l'érosion des sols, la lithologie, la morphologie et les glissements de terrain, la fonte des glaces, les débordements des lacs d'origine glaciaire et les inondations, les changements climatiques socio-économiques et les variations climatiques régionales. En ce qui concerne les études de cas de ce projet qui sera conduit dans le cadre d'une coopération interdisciplinaire et internationale, deux zones bien définies ont été sélectionnées, l'un au centre du Népal, l'autre, dans le sud du Tibet. Ce projet a pour but de mettre au point des indicateurs géographiques environnementaux, de faciliter l'évaluation des risques, de contribuer à une analyse "de cause à effet" des risques naturels et de proposer une méthode de gestion pour la mise en place d'une politique de prévention des risques naturels. Dans ce travail, un modèle numérique de terrain tridimensionnel (MNT) construit à partir des images stéréo de SPOT a été superposé aux images en fausses couleurs dérivées de LANDSAT TM et aux cartes géographiques de la zone de Melamchi, située au nord de Katmandou. Les cartes thématiques tridimensionnelles et les cartes géographiques ainsi obtenues vont faciliter l'interprétation des structures géologiques complexes, comme par exemple, la zone se trouvant à la lisière nord de la vallée de Katmandou où le gneiss du Haut Himalaya est en contact avec les roches métamorphiques du complexe de Katmandou. Cette démarche géoscientifique peut également contribuer à trouver des solutions aux problèmes posés lors de travaux public ou de projets géotechniques dans la région.

1 INTRODUCTION

Geological research, mapping and exploration programs carried out in Nepal since 1950, resulted in vast accumulations of geoscience data stored in databases in analogue and digital form. These data cover scattered areas, as large parts of the country are devoid of infrastructure and remote areas are almost inaccessible. In the Nepal Himalayas, morphological conditions are extreme with regard to slope gradients and altitudes; they render field work most strenuous and time-consuming. Geological features are often difficult to decipher due to intense meta-morphism and tectonism; neotectonic faults, unfavourable lithology and strong erosion pose great problems to geological engineering and geotechnical safety.

Figure 1. SPOT - Stereo digital terrain model: Kathmandu valley and Shivapuri area.

In view of these facts, the application of remote sensing in general, and the integration of multisatellite data and geo-environmental data in particular, are of increasing importance for geological and thematic mapping at different scales for various purposes.

In 1992, the CODATA Commission on Data for Global Change conceived a research project in two critical zones in the Nepal and Tibet Himalayas to study the interdependence of natural conditions and geo-hazards by multisatellite mapping and integration of remotely sensed and geo-environmental data.

The present paper describes a case study within the scope of the CODATA project; it was commissioned by the department of Mines and Geology, HMG Nepal, and carried out in cooperation with Nepali, French and German scientists as an example of the application of remote sensing to the planning and management of the environment, natural resources and physical infrastructure in a high mountain area in Nepal.

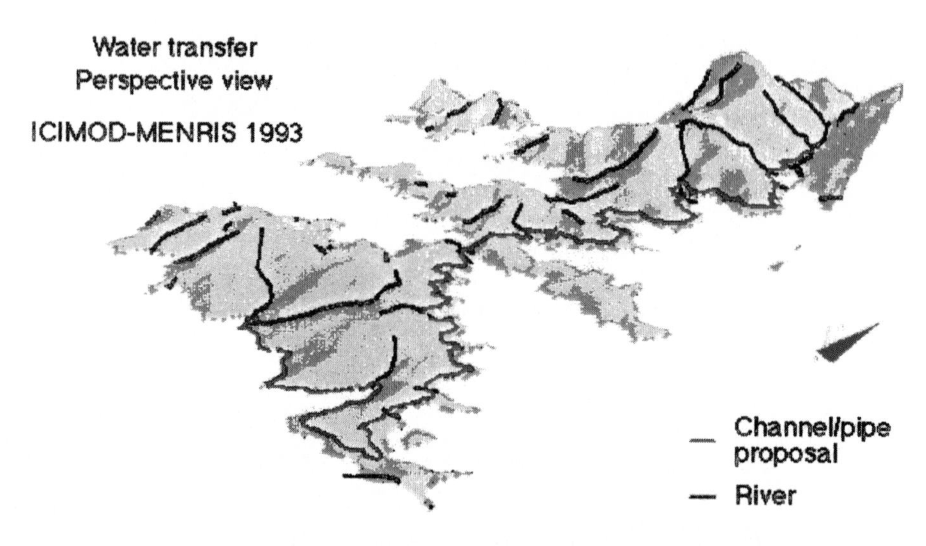

Figure. 2. Melamchi-Sundarijal Water Transfer: Spot Relief View Digital Elevation Model above 1600 m altitude overlaid by the Channel/Pipe Proposal. Sun elevation 35° and azimuth 100°, Observator elevation 20° and azimuth 150° (after ICIMOD). (For color figure, see page 303)

The study concerns the solution of specific geological and geotechnical problems, in particular the planned transfer of surface water from the glacier-fed Melamchi River in the High Himalaya to the Kathmandu Valley. As the study area is prone to geo-hazards, the geo-environmental conditions and constraints had to be analysed, and by means of stereo-relief construction, data integration and network analysis, an optimal ground route for the water transfer had to be determined.

2 CASE STUDY

According to a master plan by UNDP and WHO, the water transfer scheme will consist of an intake from the Melamchi River, a 27 km long tunnel, storage reservoirs and a treatment plant to provide drinking and irrigation water for the Kathmandu Valley. Supported by geological field and ground truth work, the search for an optimal ground route for the planned water transfer was based, first of all, on the evaluation and integration of existing and new databases; they comprised:

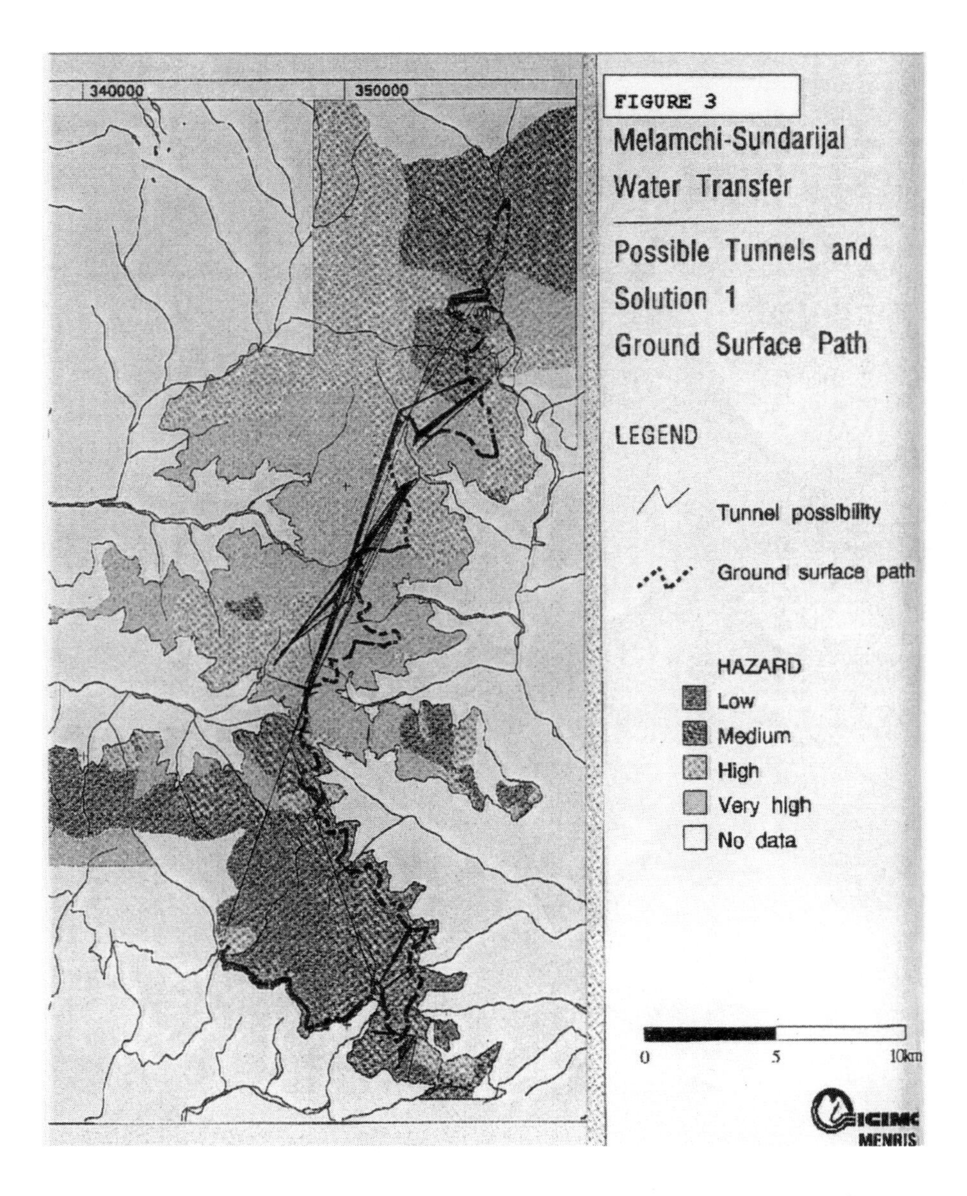

Figure 3. Melamchi-Sundarijal Water Transfer. Possible Tunnels and Solution 1: Ground Surface Path. (For color figure, see page 300)

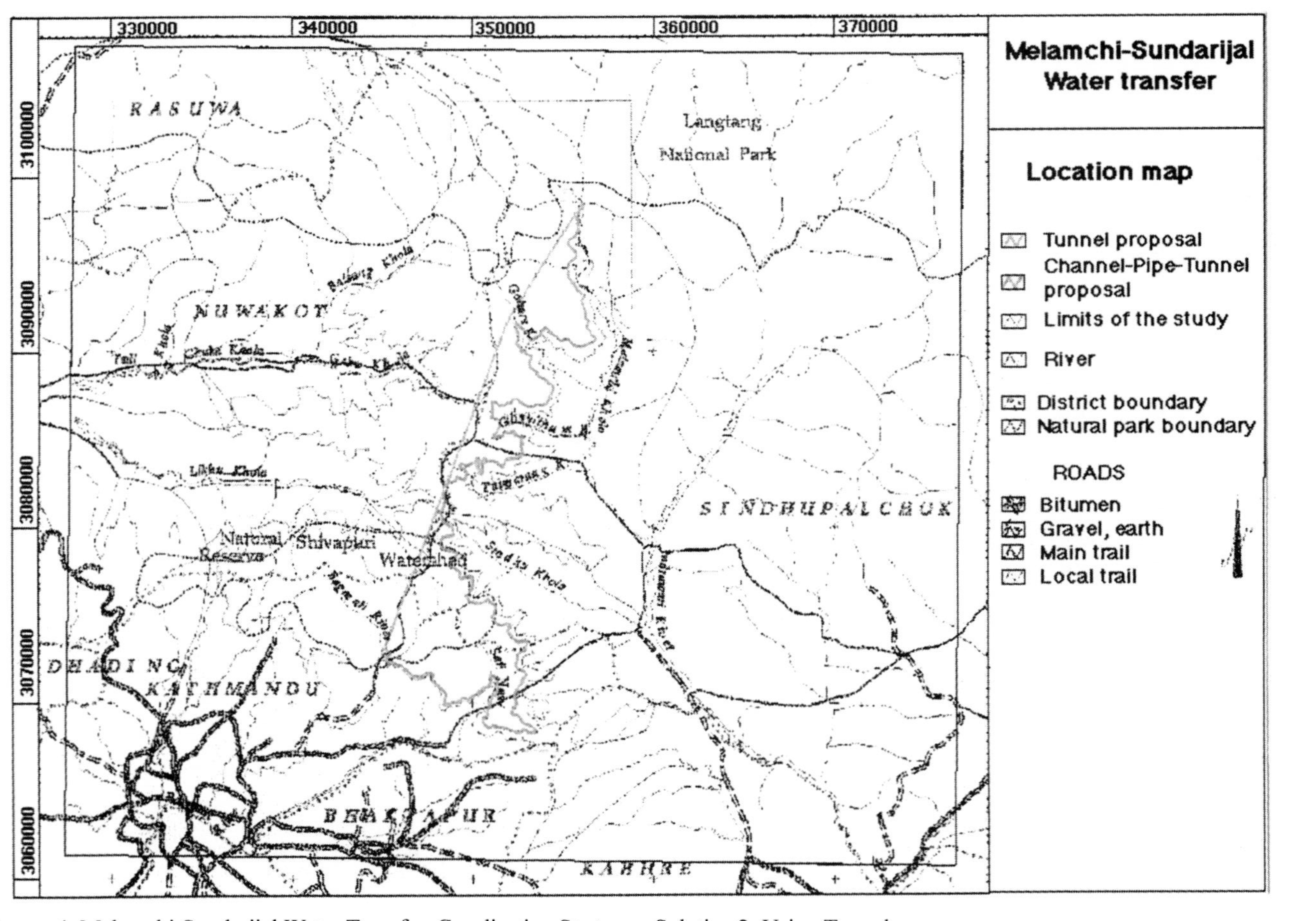

Figure 4. Melamchi-Sundarijal Water Transfer. Canalization Strategy - Solution 2: Using Tunnel (after ICIMOD - MENRIS). (For color figure, see page 301)

- geo-databases: topographical, hydrological, geological, tectonic and geo-hazard data
- environmental databases: land-use, population density, land accessibility, road and trail data
- remote sensing databases: LANDSAT TM, SPOT panchromatic and stereo.

2.1 Application of Geoscience Data

The main data sources were ICIMOD (International Centre for Integrated Mountain Development), DMG (Department of Mines and Geology, HMG Nepal), E.N.S. (Ecole Normale Supérieure, France), ISTAR (Imagerie Stéréo Appliquée au Relief, France), BGR (Federal Institute for Geosciences and Natural Resources, Germany). Those map data which still existed in analogue form, were digitized and included in the MENRIS (Mountain Environment Resources Information System) database of ICIMOD in Kathmandu.

In the laboratories of ISTAR and G-Sat/ENS in France, a high-quality Digital Elevation Model (DEM) of the study area was constructed from two SPOT stereo-images (Fig. 1); the earth surface was sampled every 20m, and the elevation value of each sample has an error interval of 10m. The DEM was then digitally connected with LANDSAT TM colour composites (TM scene 141-041, B=HV1, G=HV4, R=HV7) and with the digitized geological data of the map compilations of the study area. The resulting 3-dimensional geological and geomorphologic maps facilitated the interpretation of geological problems.

The geological interpretation had to take into account that the geology and tectonics of the study area were partly unknown and that the area is situated in a critical geological zone: it comprises partly the northern flank of the tectonic synclinorium of the Kathmandu Complex, and partly the southern Central Crystalline of the High Himalaya which is the root zone of the Kathmandu Complex. As revealed by LANDSAT TM and SPOT image evaluation, supported by DEM interpretation and ground truth, the basal gneisses of the Kathmandu Complex (Bimphedi Group) and the Sheopuri and Gosaikund gneisses of the Central Crystalline merge here; a major tectonic thrust fault, the Main Central Thrust (MCT), apparently does not separate these two geological units. This geoscientific result is also of importance for the geotechnical problems regarding the planned water transfer.

2.2 Application of SPOT and LANDSAT / TM Data to civil engineering

To tackle the geotechnical problems, first of all a slope map was derived from the DEM. Mainly in high altitudes, above 2500m, slopes exceeding 40 degrees were detected; in the major part of the study area, the slopes average 30 degrees. At this stage, geo-hazard data were added to the slope data like landslides and fault zones, in order to establish a sound basis for the assessment of the risk potential of the study area. The area belongs to the southern slope of Langtang Himal which forms a part of the High Himalaya range at the border of Nepal and Tibet. The landform is highly irregular with altitudes between 1500 and 5000m, comprising narrow ridges and steeply incised, V-shaped valleys.

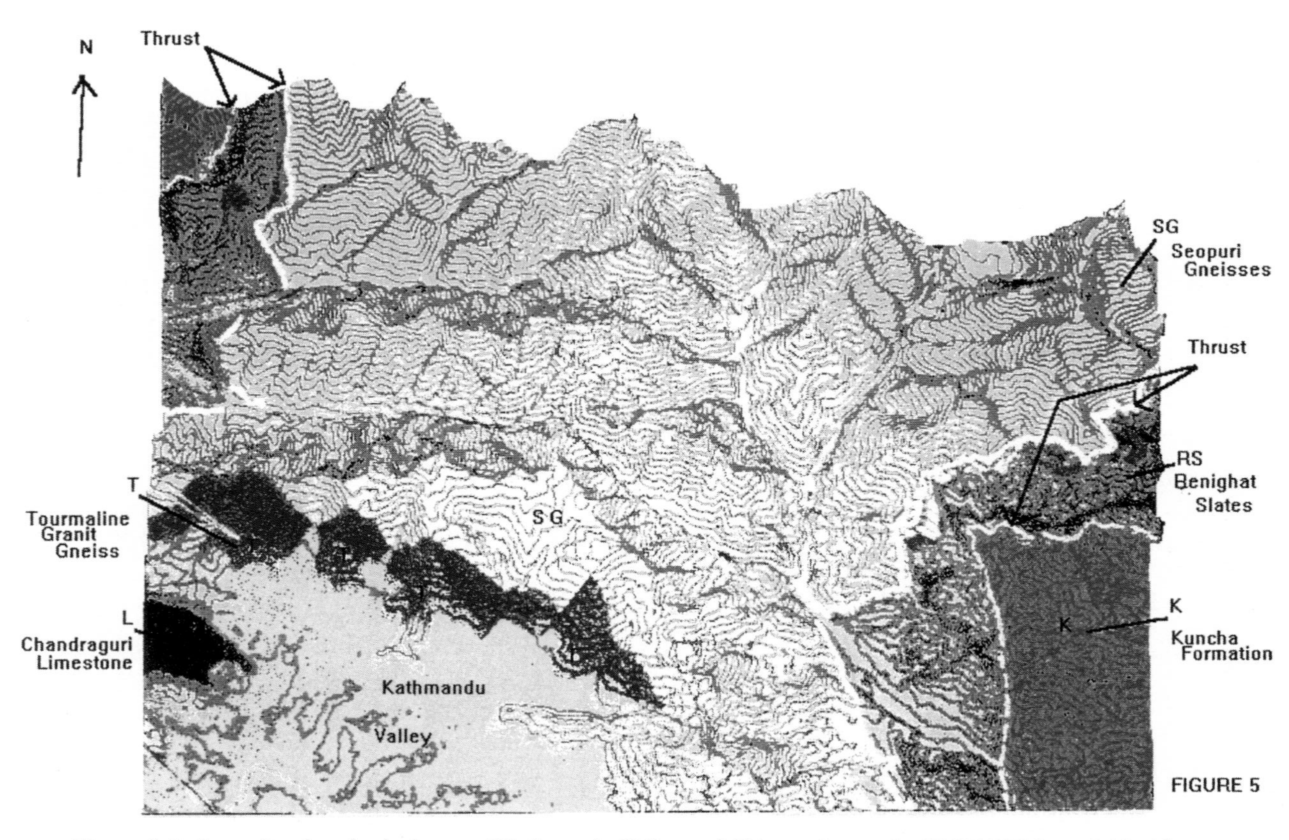

Figure 5. 3-dimensional geological map of Kathmandu Valley and Shivapuri area: the SPOT-DEM overlaid with the geological data. (see also page 302)

The ridges and valleys have steep slopes of 30 to 50 degrees and are intensively dissected by a drainage pattern of gullies; the ridges have a thin residual soil cover, while their flanks are made up of colluvium, debris and talus fans. The extensive erosion on the ridge slopes indicates their surface instability which is proven by frequent old, new and reactivated landslides of varying dimensions; they, in turn, have affected the overburden and the bedrock. Half of the study area shows a high risk potential and comprises, in addition, protected forest areas.

It was thus evident that from the geotechnical point of view, a geo-hazard map had to be established. In a first step, this was done by the interpretation of LANDSAT TM false colour composites using an ARC/INFO workstation, and by combining the digitized results with data from a digitized landslide map of the DMG, based on airphoto interpretation and field mapping. In a second step, tectonic data were added which were derived from lineament interpretation of the same LANDSAT TM false colour composites, on colour prints at 1:125 000 scale. Two main lineament systems were identified striking NNE - SSW and N - S. They represent a neotectonic fault pattern which dissects the entire project area, and to which some earthquake risk potential is connected.

2.3 Hazard mapping for water supply project

All the hazard data are displayed in a geo-hazard map from which "hazard values" or hazard parameters were derived. To this end, the following factors had to be considered: presence of landslides and erosion gullies, steepness of slopes, status of vegetation cover, nature of land-use, geological and hydrogeological conditions. For the solution of the geotechnical problems, regarding the determination of an optimal route for the planned water transfer (Fig. 2), adequate criteria had to be taken into account concerning the:

- natural conditions (extreme morphology, difficult accessibility)
- engineering conditions (geo-hazards, availability of building material)
- environmental conditions (protected areas, different forms of land-use, population density).

Using GIS overlay techniques, specific parameters for hazards, elevation, landslides and accessibility were introduced in order to optimize the route for the planned water transfer. To this end, a path-finding algorithm was used with the following constraints:

- gravity-driven water flow at moderate slope conditions
- minimum of natural obstacles
- minimum of the overall costs.

3 Conclusion

By means of network analysis and a path-finding program of ARC/INFO, an optimal route for a channel, with a few short tunnels in places, could thus be determined, and the water intake site could be moved from a high-hazard to a medium-hazard area (Fig. 3); potential tunnel routes were identified for follow-up ground checks. By minimizing the crossing of protected areas, forests and arable land, the environmental impact of the planned channel/tunnel combination will be greatly reduced (Fig. 4).

The original cost estimate by UNDP/WHO based solely on a tunnel construction for the water transfer could thus well be reduced by one third, provided a channel, pipeline and a few short tunnels will be used along the optimal route detected, local labour employed and local building material used on-site.

The research work for this case study utilized several GIS tools of ARC/INFO V6.1 on workstation IBM RISC 6000 Model 530; for image processing ERDAS 7.4 was used on IBM/RS 6000, and for geographic analysis, combination, visualization and map production ARC/INFO V6.1. The image integration functions of ARC/INFO were utilized to integrate the results of image processing modules in ERDAS 7.4 on the workstation; the combination of these two softwares provided a powerful tool.
The raster-processing GRID of ARC/INFO was used to display and merge large raster datasets, and to perform complex spatial analysis on raster data.

KEYWORDS

GIS, deforestation, soil erosion, hazard prevention, remote sensing data, ecosystems, geohazard, geotechnics, Himalayas, Nepal

REFERENCES

[1] Adikhary, T.P., Bannert, D. and Gabert, G., Compilation of Geoscience Data for the Display of Nepal Himalayan Geology. - First South Asia Geol. Congr., Islamabad 1992; Abstr. Vol. (1992).

[2] Amatya, K.M., Bournay, E. and Pradhan, P., Water transfer from Melamchi Khola to Kathmandu Valley - Research of a ground surface path in a hazard area using network analysis methods, SPOT DEM and multisatellite data. - 5. UN/CDG/ESA/ICIMOD Regional Training Course, Kathmandu 1993; Unpubl. Rep. (1993).

[3] Bardinet, C. and Gabert, G., Multisatellite Mapping and Stereo-Relief Construction for the International Geosphere-Biosphere Program in the Nepal and Tibet Himalayas: a CODATA Project. - Japan Soc. Geoinformatics (Special Issue, Ed.: N. Nishiwaki), Vol. 4, N° 3, p. 301, Kyoto (1993).

[4] Bardinet, C., Bournay, E., Shrestha, S. and Wang, C.-Y., Accessibility of Data for Global Change on Critical Zones in Nepal and Tibet: the Case of ICIMOD. -PRC. II. High Mount. Remote Sens. Cartogr. Sympos., Lhasa 1992; Eds.: Liu, J.-Y. and Buchroithner, M.; p. 1 - 19; Beijing (1993).

[5] Bardinet, C., Gabert, C., Monget, J.M. and Yu, Z., Application of Multisatellite Data to Thematic Mapping. - Geol. Jb. 367, p. 3 - 74, Hannover (1988).

[6] Bournay, E. and Pradhan, P., A Comparative Assessment of Water Transfer Modes in Hilly Terrain (Bagmati Zone).- In: Application of GIS to Rural Development Planning in Nepal; MENRIS Case Study Series N° 2 (ICIMOD), p. 47 - 69, Kathmandu (1994).

[7] Lavreau, J. (Ed.), Le programme GARS de l'Unesco/IUGS en Afrique; Vol. II: Spectrométrie à haute résolution in situ et analyse d'image LANDSAT et SPOT au Burundi. - Ann. Série Sci. Géol., Vol. 98, p. 1 - 155, Mus. Roy. Afr. Centr., Tervuren (1990).

Geographic Information System for Environmental Assessment of the Long-range Pollution Effects in Remote Areas (Himalayas, Nepal)

Gianni TARTARI[a,1], Massimo ANTONINETTI[b], Giancarlo BORTOLAMI[c], Pietro A. BRIVIO[b], Claudia DE VITO[c], Giorgio IABICHINO[d], Monica PEPE[b], Gabriele A. TARTARI[e] and Sara VALSECCHi[a].

[a]*CNR Istituto di Ricerca Sulle Acque. 20047 Brugherio, Italy*

[b]*CNR Istituto di Ricerca sul Rischio Sismico, Via Ampère 56. 20131 Milano, Italy*

[c]*Dipartimento di Scienze della Terra, Università di Torino. Via Accademia delle Scienze 5. 10123 Torino, Italy*

[d]*CNR Centro di Studio per i Problemi Minerari, C/o Politecnico di Torino, Corso Duca degli Abruzzi 24. 10129 Torino, Italy*

[e]*CNR Istituto Italiano di Idrobiologia, Largo Tonolli 50/52. 28048 Verbania-Pallanza, Italy*

ABSTRACT

In the framework of the Italian Ev-K^2-CNR Project (High Mountain Scientific and Technological Research) a Geographic Information System (G.I.S.) has been planned to model the hydrochemical characteristics of lacustrine environment due to interactions between meteoric waters and soils. The study area is located in the Khumbu valley, at the foot of Mount Everest in the Himalayan Nepal, where, at an altitude of 5,050 m a.s.l., a "Pyramid" of glass and aluminium has been installed to host the scientific laboratories and researchers. The G.I.S. will be organized with different layers containing geographical, geological, geochemical and hydrological information of the area, collected from available cartography, enhanced remotely sensed imagery and in situ field studies. A Kumbu-Himal lake cadastre, including hydrochemical and geo-lithological characteristics, the assessment of water quality and the inventory of phyto and zooplankton populations of monitored lakes, will be included in the G.I.S. to obtain a better definition of the principal components determining the basic chemical composition of the lake environment. The comparison and interpretation of the different data can be useful to identify the presence and the consequent influence of long-range pollution in this area, in an attempt to understand the processes by which many chemicals are transported, and the long-term effects on the lacustrine environment.

---

[1] To whom correspondence may be addressed

Système d'Information pour la Gestion Environnementale des Pollutions à Long Terme des Régions Éloignées (Himalaya, Népal)

RÉSUMÉ

Dans le cadre du Projet Italien Ev-K^2 - CNR (Recherches Scientifiques et Technologiques en Haute Montagne) un système d'Information Géographique (GIS) a été conçu pour modéliser les caractéristiques hydrochimiques d'un environnement lacustre dues aux interactions entre les eaux météoriques et les sols. La région où se situe l'étude est la Vallée Khumbu, au pied du Mont Everest dans le Népal himalayen où une "Pyramide" de glace et d'aluminium a été érigée à 5050 d'altitude, pour abriter les laboratoires scientifiques et pour accueillir les chercheurs. Le GIS est structuré en différentes couches, comprenant des informations sur la géographie, la géologie, la géochimie et l'hydrologie de la région étudiée. Ces informations ont été recueillies à partir des cartes disponibles, des photos satellites et des études in situ. Le cadastre du lac Khumbu, ses caractéristiques hydrochimiques et lithologiques, ainsi que les études de qualité de l'eau et l'inventaire des populations phyto et zooplanctoniques des lacs étudiés seront inclus dans le GIS afin de disposer d'une meilleure définition des principaux composants qui déterminent la composition chimique de base d'un environnement lacustre. La comparaison, ainsi que l'interprétation des différentes données peuvent s'avérer utiles pour identifier la présence et l'influence d'une pollution à long terme dans cette région. Ces études tenteront de mettre en évidence les processus de transport et les effets à long terme sur un environnement lacustre de plusieurs produits chimiques.

1 INTRODUCTION

Elevated remote areas, as well as polar and middle-oceanic areas, are very useful in defining global change effects (Birks et al., 1992). Worldwide, there are at least five to seven high altitude zones (Alaska, Andes, Alps, Kilimanjiaro, Himalayas, Karakorum, Ural, ...) that are useful in anthropogenic-pollution monitoring, when pollution is related to atmospheric long-term diffusion. The Himalayan zone has the peculiarity of being situated between India and China - the most populated nations on Earth - developing rapidly, and far away from most industrialized Western zones. So this area is helpful for observing developing pollution diffusion phenomena.

The Ev-K^2-CNR project is an interdisciplinary high altitude research program (Tartari et al., 1991), which is carried out in a 5050 m a.s.l. laboratory (Pyramid Laboratory) located in Khumbu Valley, Sagarmatha National Park, Nepal, at the foot of Mount Everest. Since 1989 environmental and geological research has been carried out near the laboratory in order to assess information, also using remote sensing techniques that are integrated by a Geographic Information System (G.I.S). This G.I.S. models the basic chemical composition in lacustrine environments, which is useful to identify the external polluting-element.

The initial aims of limnological research in high altitude lacustrine environments in the Khumbu Valley - as resumed in Gosso et al., 1993 - were an evaluation of the long-range pollution effect of rising anthropic influence, due to touring and trekking, as achieved from trophic alterations in aquatic environments.

The possibility to join multidisciplinary research results, added other aims: evaluation of the possibilities and potentialities of this tool in environmental modeling.

The aims of this paper are:

- a presentation of the lake cadastre in the study area and of hydrochemical frames in lacustrine environments, used for quality-characterization modeling;

- a presentation of the basic decisional stages in the creation of a large diffusion-allowing G.I.S., with low costs and small software/hardware tool dimensions.

2 GEOGRAPHICAL AND GEOLOGICAL DATA BASE

The study area (Fig.1) is located in the Khumbu-Himal region, between 27°50' and 28°01' latitude North and 86°45' and 86°59' longitude East and it is included in the High Imja Khola basin. This area, about 440 km^2 wide, has been arbitrarily chosen on the basis of the most recent available topographic map ("Mount Everest" map, 1:50,000, National Geographic Society, Washington D.C., 1991). It considers two large valleys formed by the Khumbu Glacier, starting from Everest (8846 m), Lhotse (8501 m) and Pumori (7145 m) Mounts, and by the Nupse Glacier, starting from homonymous Mount Nupse (7896 m). The study area is about 35% covered by glaciers, and it lies between the thermic zero summer limit (at about 5600 m) and an altitude of about 4100 m.

Localization, altitude, catchment limits, catchment area and lake area have been defined for each lake - presented on topographic maps - except for glacier tongue lakes. Three main morphological features, altitude distributions, lake and catchment areas, are summarized in Figure 2. Most of the 64 environments studied have an altitude ranging from 5200 to 5600 m a.s.l. Usually lake areas are quite small, at least 5 x 10^3 m^2, although their watershed areas are comparatively larger, 100-2500 x 10^3 m^2.

From a climatic point of view the Khumbu Valley is characterized by a monsoon seasonal type, less rainy than in other Himalayan regions. Indeed measurements taken in the Periche station, located in the central part of the lake cadastre zone (Fig. 1), indicate a mean annual precipitation of about 450 mm/y; moreover 75% of that value is related to the summer monsoon period only (Grabs and Pokhrel, 1993).

From a geo-lithological point of view there are three main formations involved in assessed lake drainage basins. Two of them are early Paleozoic-Precambrian in age - the Island Peak Complex and the Black Gneiss Formation - the other is characterized by Moraines formed in Quaternary times (Bortolami et al.,1976).

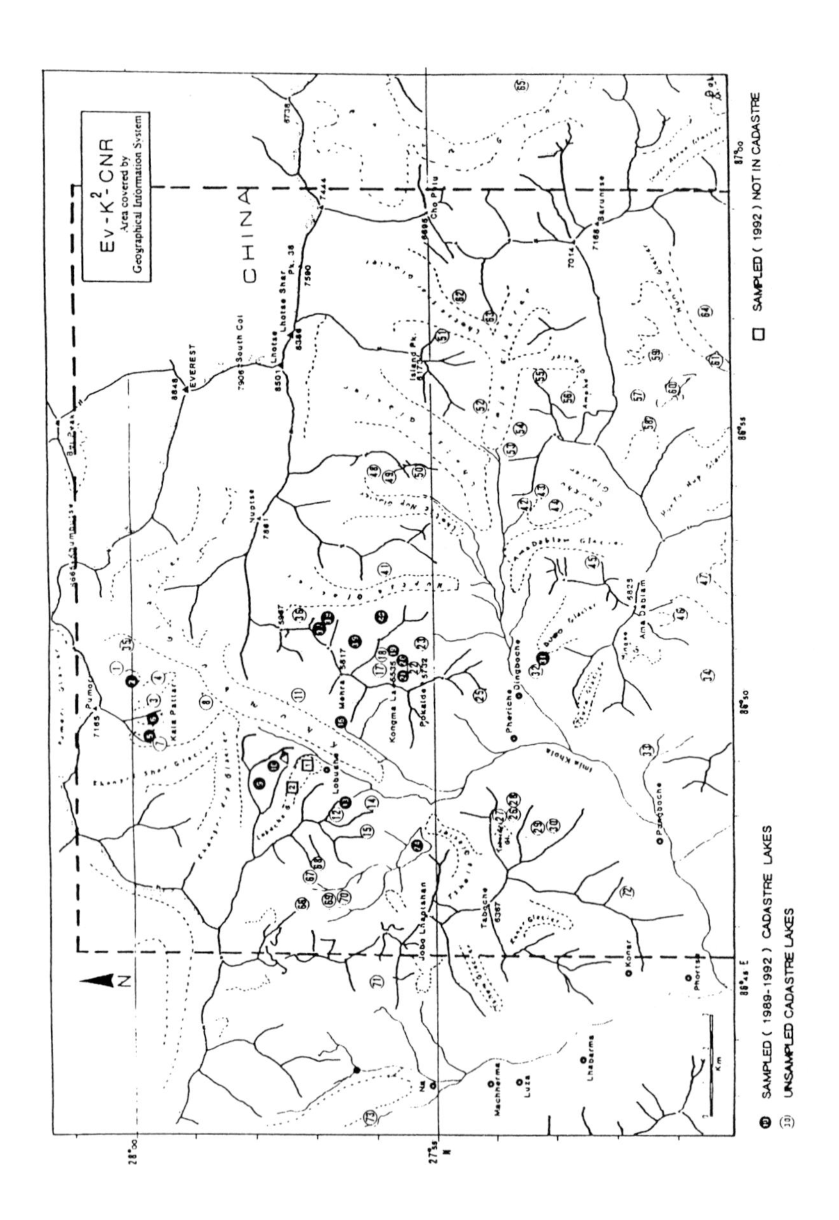

Figure 1. Ev-K^2-CNR lake cadastre. Dotted line bounds area considered in the G.I.S.

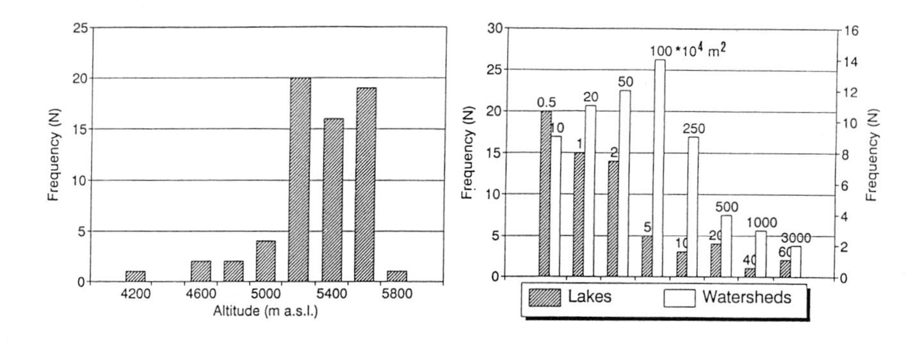

Figure 2. Lake altimetric distribution and lake and watershed surface frequency in the Khumbu-Himal area.

The Island Peak Complex contains augen gneiss, biotite gneisses and micaschists. The overlying Black Gneiss Formation contains biotite-sillimanite gneisses, biotite-muscovite gneisses, biotite-muscovite-staurolite-garnet micaschists and scarce metaconglomerates. This formation locally contains granitic dyke swarms. Moraines and moraine ridges, overlying the two previous formations, are composed of elements of both Island Peak and Black Gneiss lithologies.

3 HYDROCHEMICAL DATA BASE

Hydrochemical feature studies on lacustrine environments started in 1989, but only data collected since 1990 are efficient according to analytical quality controls.

Since 1990 sixteen lacustrine environments were sampled. As shown in Table 1 during these years sampling frequencies have changed according to the aims in limnological researches: a study on seasonal and long-term variations in lower and higher Pyramid lakes, and a definition of spatial hydrochemical variability, which allowed the creation of the data base used in the G.I.S. That table shows the mean concentrations of main dissolved species: hydrogen carbonate, ammonium, nitrate, sulphates, chlorides, calcium, magnesium, sodium and potassium.

Hydrochemical macroconstituents appear to be mainly related to atmosphere weathering catchment rocks as the direct precipitation from atmosphere on the lake surface is negligible.

In fact, mean concentrations in about 40 wet deposition events analyzed during the monsoon period from 1990 to 1992 (July-October) reported in Table 1, are two orders of magnitude lower than lake ones.

Table 1. Lake water and atmospheric wet deposition chemical features.

Table 1. Lake water and atmospheric wet deposition chemical features.

Lake cadastre numbers	Altitude m a.s.l.	Sampling period mm,yy	Sample N	pH 20°C	Conductivity 20°C µS/cm	HCO_3 µeq/l	SO_4 µeq/l	Cl µeq/l	NO_3 µeq/l	NH_4 µeq/l	Ca µeq/l	Mg µeq/l	Na µeq/l	K µeq/l	TP µgP/l	TIN µgN/l	TN µgN/l	Si mgSi/l
2	5300	IX;91	1	7.7	11	66	22	3	<DL	0	101	6	7	7	3	3	200	0.1
5	5460	IX;92	1	7.7	28	144	84	3	2	<DL	213	25	9	12	5	32	267	0.1
6	5460	IX;92	1	8.2	21	125	82	2	2	<DL	202	15	12	9	5	30	240	0.3
9	5213	VII-X; 90-93	5	7.7	23	110	82	4	8	<DL	169	14	15	12	2	112	175	0.1
10	5067	VII-X; 90-93	8	7.6	22	108	65	3	7	2	153	14	18	12	2	130	188	0.1
13	5420	IX;91	1	7.4	11	72	23	3	<DL	0	99	6	6	7	6	1	221	0.1
16	4939	X;91	1	7.8	34	134	148	2	8	<DL	246	36	24	13	6	105	346	1.1
19	5580	X;91-93	2	7.7	30	140	121	3	<DL	0	226	27	26	20	6	2	363	0.3
20	5580	X;91-93	2	7.7	27	158	67	3	3	0	197	29	19	23	2	40	167	0.5
21	5546	X;91-93	2	7.7	20	131	48	3	0	0	164	15	11	16	11	7	272	0.2
24	4532	X;91	1	7.4	28	101	138	5	3	0	212	31	26	16	10	51	638	0.2
31	4688	IX;92	1	7.4	15	82	47	2	5	<DL	112	7	17	4	3	64	118	0.1
37	5436	IX;92	1	8.0	51	422	29	2	5	<DL	450	16	13	14	9	67	751	0.1
38	5460	IX;92	1	8.1	49	382	17	2	5	<DL	406	10	7	8	2	68	160	0.1
39	5366	IX;92	1	7.7	21	145	27	1	3	<DL	159	18	12	12	4	40	249	0.1
40	5220	IX-X; 92-93	2	7.8	12	88	7	3	6	1	85	9	16	6	3	94	129	0.1
Wet atmospheric depositions		VII-IX; 90-92	39	6.4	2.4		1.0	2.5	1.3	1.1	2.4	0.6	2.0	1.0				

DL: detection limit.

The chemical composition of lake water is mainly characterized by calcium, among cations, and by hydrogen carbonate and sulphate, among anions. On the basis of these hydrochemical observations, the probable watershed lithology would be dominated by limestone and gypsum. This conclusion does not include complexity in hydrochemical phenomena nor other factors (morphometry, vegetation cover, etc.) responsible for solution processes inside catchments.

In fact, the chemical analysis discrepancies between different lake waters are probably due to different rock presences inside the catchment, to outcrop, plunge and steep conditions and to dealing between on and under-surface waters. The general occurrence of calcium values, higher in lake waters than in wet depositions, as well as of hydrogen carbonate, cannot be directly related to lithologies crossed by drainage basins.

This anomaly needs further geological field observations and analysis. The magnesium and potassium concentrations, higher in most of the sampled lakes than in wet deposition ones, are caused by the dominant presence of mica in the gneisses and micaschists of the basal complex and in some moraine elements.

The high content of sulphate ions in lacustrine waters is probably related to the presence of hydrothermal stage minerals, connected to the presence of granitic dykes in the Black Gneiss Formation.

Finally, dissolved silicate, total phosphorous and total nitrogen concentrations are also represented in Table 1. Collected values of the latter two parameters may indicate that the majority of lakes are oligotrophic.

4 GEOGRAPHIC INFORMATION SYSTEM SETTLEMENT

To create a broad distributable G.I.S. we used the Auto Cad (Auto Desk) software program, PC version, to digitize spatial data, and Access (Microsoft) software program, PC version, to create a data-base collecting other data.

To realize the G.I.S. we integrated information obtained from three maps: "Khumbu Himal", scale 1:50,000 and the extracted "Mahalangur Himal/Chomolonga Mount Everest", scale 1:25,000, edited by Schneider, Wien, (1957); "Mount Everest", scale 1:50,000, by National Geographic, Washington D.C. (1988). As a reference to digitized map we used the 1:25,000 "Mahalangur Himal" map integrated with a photostatic enlarged 1:25,000 "Mount Everest" map. We digitized all 100 m contour lines, and all 20 m contour lines near the mountain tops, creating a layer. Both U.T.M. and Geographic coordinates were introduced in it to allow an easier comparison between different maps. The available geological map, "Geological map of the Upper Imjia Khola", scale 1:25,000, C.A.I./C.N.R.(1981), was digitized as a separate layer.

Other layers were created to represent rivers, lakes, watersheds, elevations of lakes and mountains, toponyms.

A data base was created to collect all information concerning each lake (cartographic sketch, watershed, geographical framing, geological setting, chemical and hydrobiological characteristics) and presented it in the form of tables.

The use of the in-progress G.I.S. is necessary to handle a great amount of multi-source data and to allow the updating of information with field data that will be collected in future sampling periods.

5 DISCUSSION

The realization of a G.I.S. in the study area could lead to the evaluation of G.I.S. possibilities in environmental phenomena representation and could be helpful for interdisciplinary integrated modeling of information collected by different research groups. Moreover sampled lacustrine environments could be a reference for future studies concerning environmental modifications due to atmospheric long-range diffusion of polluting species. Pre-existing limnological information collected in 1964 (Loffler, 1969) concerning 18 lakes, 9 of them sampled during this project, could be a useful reference to evaluate anthropogenic alterations, by means of lacustrine biocoenosis analysis (Manca et al., 1994) and biomarker analysis in lacustrine sediments (Guilizzoni and Lami, in press).

Data collection carried out from 1990 to 1993 is still inadequate in relation to project aims because the sampled lakes are representative only of a partial area of our lake cadastre, although the G.I.S. software realization is partly complete. Consequently in 1994 further analyses are planned on unsampled lakes, associated with geological surveying in unsurveyed areas, to define all factors governing the basic hydrochemical composition of lacustrine waters. These analyses, integrated with remote sensing data - i.e. on vegetation cover- will enable development of mathematic modeling, which is the last step of the project.

6 CONCLUSION

The final form of the G.I.S. created in this Himalayan zone could also be applied, with a few modifications, to Italian Alpine environments. This application could redefine diffusion factors of acidifying elements, whose present definition is inconsistent with pollution consequences in high altitude environments.

ACKNOWLEDGEMENTS

This research activity was carried out within the framework of the Ev-K^2-CNR Project, promoted and sponsored by the Italian National Research Council (CNR), and partially supported by the EEC Project CI*-CT 90-0855 (DTEE).
The authors express their grateful thanks to Dr Simona Ramponi for support in planning the lake cadastre, and to Mr Luciano Previtali for kind attention to graphic production.

KEYWORDS

GIS, Himalayas, soil erosion, water quality, lake environment, long range hydrochemistry pollution, lacustrine ecosystem.

REFERENCES

[1] Antoninetti, M., De Vito C., Iabichino G., Tartari G., Valsecchi S. and Bortolami G., "Geographic Information System of the Khumbu valley (Himalayas, Nepal): integration of remote sensing data with the data collected during the Ev-K2-CNR Project expeditions", *Proc. 3rd Int. Symposium on High-Mountain Remote Sensing Cartography.* Mendoza, Argentina, (In press) (7-12 November 1994).

[2] Bortolami, G., Lombardo B. and Polino R., "The Higher Himalaya and the Tibetan series in the Lothse area (Eastern Nepal)", *Boll. Soc. Geol. It.,* **95**: 489-499 (1976).

[3] Birks, J. W., Calvert J. G. and Sievers R. E. (Ed.s), *The Chemistry of the Atmosphere: Its Impact on Global Change. CHEMRAWN VII. Perspectives and Recommendations.* International Union of Pure and Applied Chemistry. Agency for International Development Publ.. Washington, D.C. 163 pp. (1992).

[4] Gosso, E., Tartari G., Valsecchi S., Ramponi S. and Baudo R., "Hydrochemistry of remote high altitude lakes in the Himalayan Region", *Verh. Internat. Verein. Limnol.*, **25**: 800-803 (1993).

[5] Grabs, W. E. and Pokhrel A. P., "Establishment of a measuring service for snow and glacier hydology in Nepal - Conceptual and operational aspects", In: G. J. Young. *Snow and glacier hydrology.* IAHS Publ., no **218**: 3-16 (1993).

[6] Guilizzoni, P. and Lami A. (In press), Chemical and biomarker analysis of sediments from one Himalayan lake (Khumbu Valley, Nepal), *Ev-K^2-CNR Environmental Workshop. "Results of the 1989-1992 research at Pyramid Laboratory. What kind of future?".* Milano, March 16th.

[7] Lofler, H. "High altitude lakes in the Mt. Everest region", *Verh. Internat. Verein. Limnol.,* **17**: 373-385 (1969).

[8] Manca, M., Cammarano P. and Spagnuolo T., "Notes on *Cladocera* and *Copepoda* from high altitude lakes in the Mount Everest Region (Nepal)", *Hydrobiologia,* **287**: 225-231 (1994).

[9] Tartari, G., Gosso E., Valsecchi S. and Nino P., "Atmospheric deposition and lake chemistry program in the Himalayan region", *Documenta Ist. ital. Idrobiol.*, **32**: 121-126 (1991).

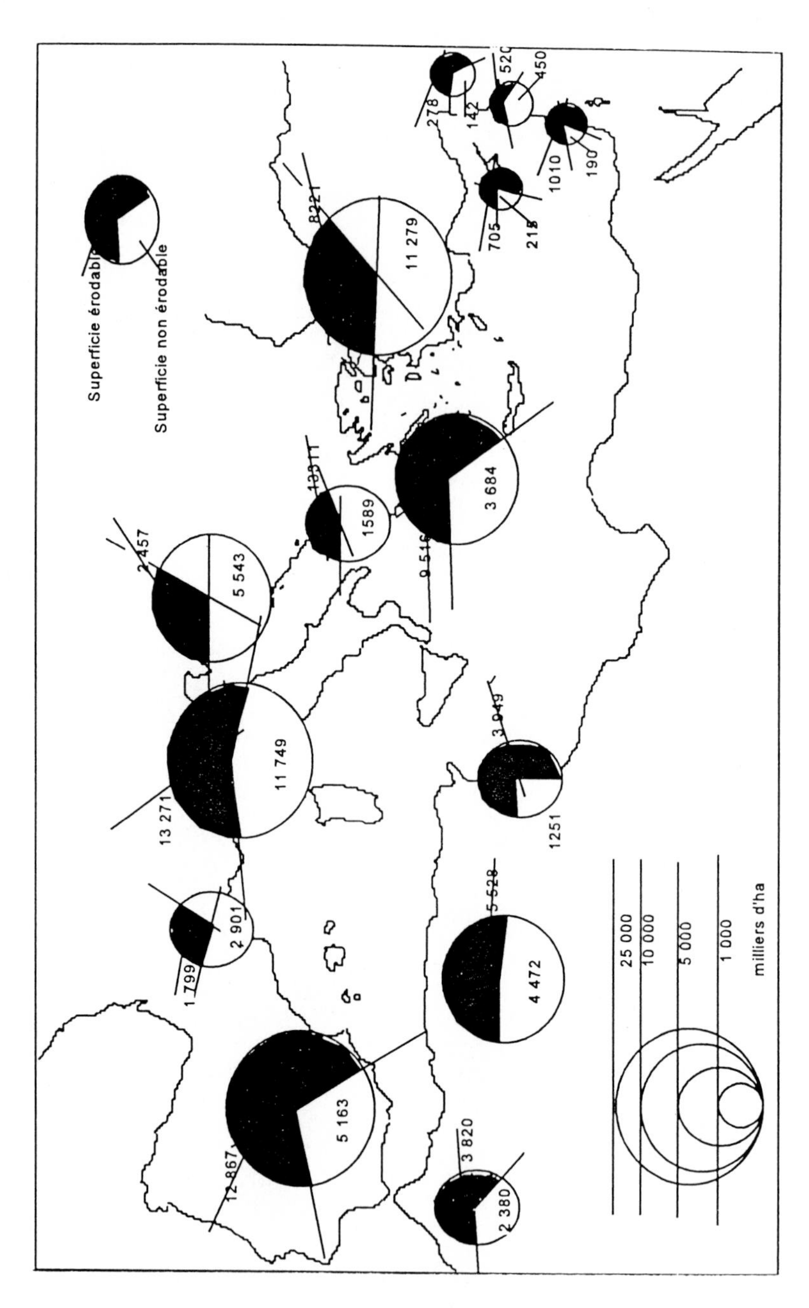

Soil Erosion risk in the Mediterranean (Blue Plan, 1989) (After Dubost, see page 188).

Chapter 4

REGIONAL ENVIRONMENTAL CHANGES
Application to Soil Erosion Modeling

DESERTIFICATION OF MEDITERRANEAN AND MOUNTAINOUS REGIONS

Michel DUBOST

ICALPE, Casa pastureccia, Riventosa, 20250 Corte (Corsica) - France

ABSTRACT

Mountain areas are characterized by a delicate balance between ecosystem productivity and human use of natural resources. This is particularly true in many mountain regions of the developing countries like the Himalayas, the Andes or the African ranges, where the growing demographic pressure initiates vicious circles of degradation. Serious problems also concern mountain areas in developed countries.

Deforestation, forest fires, erosion, agricultural decline, air pollution, massive tourism, traffic, etc. are but a few examples which critically affect soil, vegetation, water and other vital resources of mountainous areas. Good descriptions and case studies of resource degradation are available from the Mediterranean, but an overall understanding is currently missing, as is a theoretical framework to help analyze the multiple-sided issue of desertification. Some research projects offer an opportunity to compare several study areas with different methods throughout the Mediterranean region. The study of desertification phenomena requires combining various sources of information at different time and space levels.

A major objective of such studies is the identification of potential indices of desertification in diversified historical and natural contexts and study sites Remote sensing techniques offer a series of promising achievements in this direction. The analysis of desertification in the context of regional environmental change requires various study-scales in time and space, from the local to the regional level, and local case studies must be completed with regional synthesis. Reconstructing past vegetation dynamics is a good example of such an approach, and methods derived from historical botany, like pollen analysis, dendro-chronology or pedo-anthracology are particularly relevant for discussing the wide range of methods and techniques that must be combined in any integrated research effort to elucidate desertification processes in the Mediterranean mountains and regions.

Bilan de l'État de Désertification des Régions Méditerranéennes et Montagneuses.

RÉSUMÉ

Les régions montagneuses se caractérisent par un équilibre fragile entre le renouvellement des écosystèmes et l'exploitation par l'homme des ressources naturelles. Cette remarque s'applique tout particulièrement aux régions montagneuses des pays en développement, comme l'Himalaya, les Andes ou les chaînes africaines, où la pression démographique croissante provoque des cercles vicieux de dégradation. De sérieux problèmes se posent aussi dans les régions montagneuses des pays développés.

La déforestation, les feux de forêts, l'érosion, le déclin agricole, la pollution de l'air, le tourisme de masse, le trafic etc..... voici quelques exemples qui affectent d'une manière critique le sol, la végétation, l'eau et d'autres ressources vitales des zones montagneuses. Il existe de bonnes descriptions et des études de cas de la dégradation des ressources dans les pays méditerranéens. Toutefois, la compréhension générale du processus de désertification, ainsi qu'un cadre théorique permettant son analyse par une approche pluridisciplinaire, restent à faire. Quelques projets de recherche offrent l'opportunité de comparer plusieurs régions d'étude des pays méditerranéens au moyen de méthodes différentes. L'étude du phénomène de désertification exige l'association de plusieurs sources d'informations à des moments différents dans le temps et dans l'espace, à des niveaux d'échelle différents.

L'objectif principal de ces études est d'identifier des indices potentiels de désertification dans des contextes historiques et naturels divers. Cependant l'analyse de la désertification dans le contexte de modifications environnementales régionales demande des études à échelles variées, dans le temps et dans l'espace, du niveau local au niveau régional. La reconstruction de la dynamique de la végétation d'autrefois est un bon exemple d'une telle approche et les méthodes dérivées de la botanique historique, comme l'analyse des pollens, la dendro-chronologie ou la pédo-anthracologie sont particulièrement intéressantes pour modéliser et contrôler les processus de désertification dans les montagnes et les pays méditerranéens.

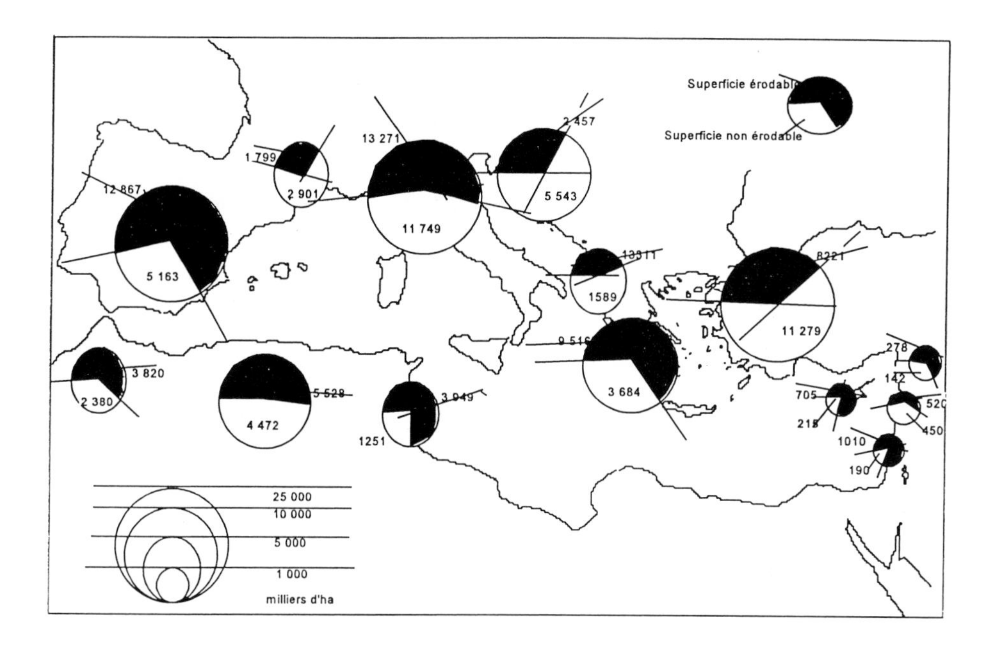

Figure 1. Soil Erosion risk in the Mediterranean (Blue Plan, 1989).

1 Threats to the Mediterranean and Evolution of Mountainous Areas

Among the most critical issues for the future of the Mediterranean region are the protection of coastal areas, the prevention of forest fires, soil conservation, and demographic growth.

All Mediterranean countries are already affected by an extension of severe erosion risks. In 1980, a loss of 5 to 50 tons/ha/yr was estimated for 35% of the cultivated lands, and not only in the Southern rim, but also in European countries like Greece, Italy or Spain (see Fig. 1, Blue Plan, 1989).

1.1 Water and demographic growth

Water availability is another major problem for the future, even without a climate change.

Mountains are the wellsprings of any continent. But this is particularly critical in a region where water scarcity is widespread and population is more and more concentrated along the coasts. Since water availability is a major issue for its future, considering the hydrological consequences of ecological change in the mountains is essential for the whole Mediterranean region.

Table 1. Water resources and use in the Mediterranean (Margat, 1992).

Countries (Mediterranean part)	Renewable water resource	index of exploitation or consumption		Regular water resource	index of exploitation or consumption	
		iex (%)	ic (%)		iexr (%)	icr (%)
Spain	31,1	64,3	39,2	7,5	266,6	162,6
France	74	23,2	2,7	35,2	48,8	5,68
Italy	187	24,8	8	30,5	152	48,7
Malta	≈ 0,07	49	42,3	≈ 0,03	113,3	100
Yugoslavia	77.5	1.9	0.4	11.5	3.04	2.43
Albany	50	5.9	2.2	6.5	45.7	16.92
Greece	58.65	11.9	6.5	7.7	90.9	48.7
Turkey	≈ 67	10	5	15.6	42.9	20.5
Cyprus	0.9	42	28	0.27	140.7	92.6
Syria	4	50	25	2.3	86.95	43.49
Lebanon	≈ 4	20	8	≈ 2.8	28.57	11.4
Israel	≈ 1	150	92	≈ 1	150	92
Egypt	57.3	97.5	663	55.8	100.17	68.1
Libya	≈ 0.7	157	120	0.2	800	425
Tunisia	3.1	64.5	42	≈ 1.5	133.3	86.66
Algeria	10.9	15.6	8	2.5	68	34
Morocco	3.8	29	15	0.9	122.22	64.44

At the regional level, in terms of water resources and exploitation or consumption rates, apart from two very extreme and opposite situations (Libya, Israel, Tunisia, vs. Albania, ex-Yugoslavia), there is a rather clear distinction between the North and the South and East of the Mediterranean basin, with more favourable situations in France, Italy, Greece, Turkey, than in Algeria, Egypt, Cyprus, Morocco, Spain, Syria and Lebanon. In all these countries water requirements reach a maximum when resources are at the minimum level, during the dry summer season, which causes many quantitative and qualitative problems. Therefore, the ratios indicated in Table 1 are based not only on the total or potential natural resources, but also on the minimum available or regular resources (see Table 1, Margat J., 1992).

The above classification of Mediterranean countries with regard to water availabilities and needs should not significantly change up to the years 2000-2025. But discrepancies must increase between the North and the South and East with more or less stabilized situations in the North and grave deterioration in the South and East.

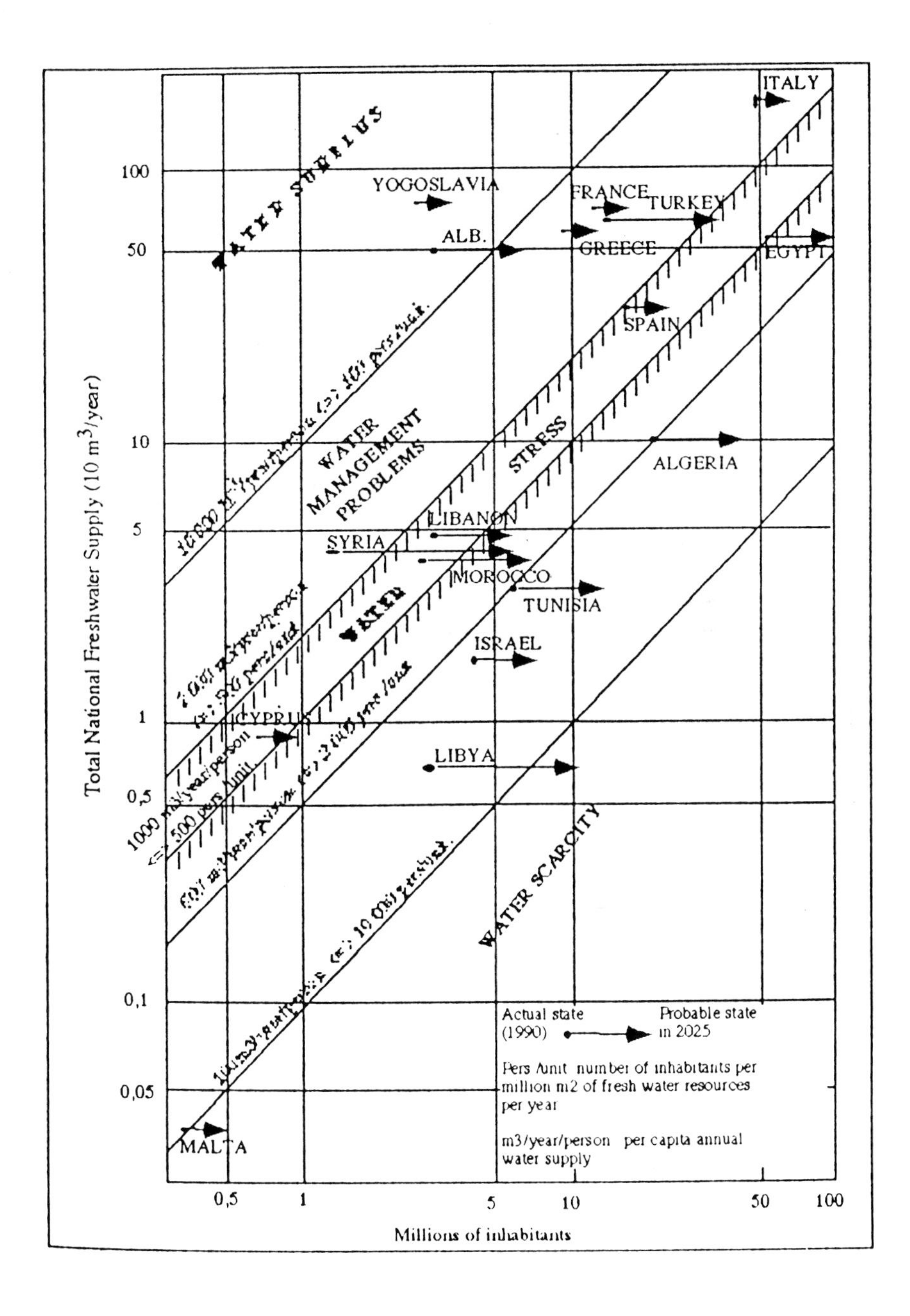

Figure 2. Water resources and population in the Mediterranean Beyond 2000 (Margat, 1992).

For most of the countries in the North quantitative pressures would remain similar to present ones with some possibilities for adapting to extreme events and for preserving water quality and the environment.

But in the countries of the South and East quantitative pressures and water shortage risks would increase. Demands should be adapted to decreasing availability and natural environments would be reduced dramatically (see Fig. 2. Margat 1992).

1.2 Probable impacts of a climate change

A climate change will have a major impact on the availability of water, in terms of feeding the terrestrial ecosystems as well as the fresh water systems. It is largely agreed that a probable climate change in the Mediterranean basin and particularly the Western part would result in an increase of temperature and decrease of precipitation (Hulme H. 1989).

This will aggravate the water shortage during the sensitive summer season, reduce the total water availability for vegetation and crop growth as well as for human population, and certainly affect the capacity of the vegetation cover to protect the soil from erosion.

Thus, with regard to the protective role of mountains for water and soil resources, the possible impact of a climate change on mountain ecosystems is a critical issue for all the Mediterranean regions, with many social, economic and environmental implications.

Some available data illustrate the dimension of problems, like the indicative hydrological shifts due to a changing climate for the European countries based on the UKMO scenarios (see Fig. 3, Brouwer, Falkenmark, 1989). Countries like Spain, Italy or Greece would experience a large-scale and permanent water stress problem.

The ability of mountain ecosystems to protect their vegetation cover, soil and water resources will be crucial, to an extent that is probably unexpected today. This is why understanding their past and present evolution is so important in order for us to be able to adapt integrated strategies for the future.

1.3 Recent changes in the Mediterranean mountains

The recent historical evolution has generated two very distinct situations around the Mediterranean basin.

In the North, the non management of abandoned rural landscapes is responsible for severe environmental threats, such as primarily forest fires, but also erosion, depletion of water resources, loss of biodiversity and landscapes.

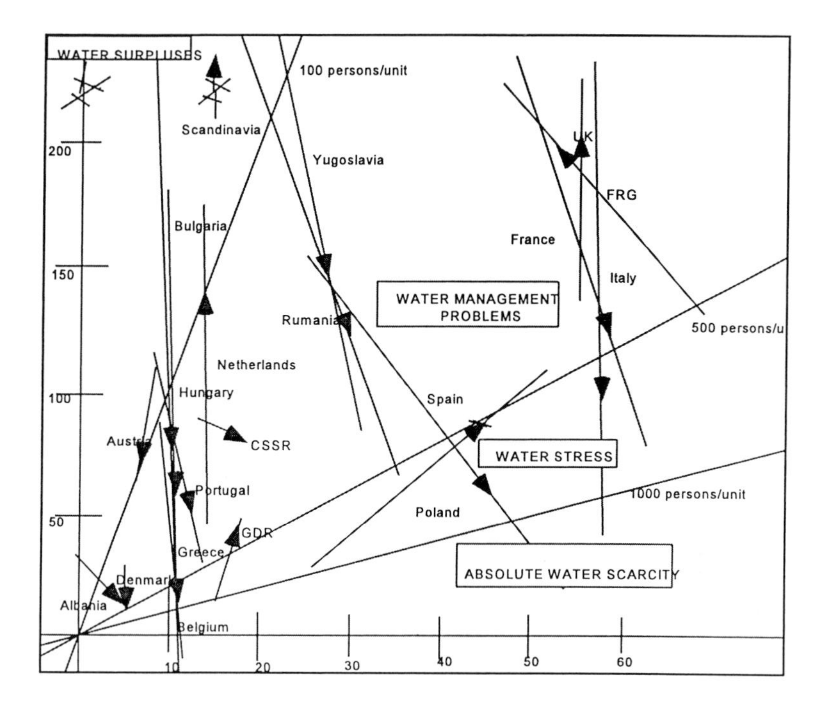

Figure 3. Hydrological shifts after a climate change for the European countries (Brouwer and Falkenmark, 1989).

In some sense, the abandonment of ancient agricultural or pastoral lands is a recovering process, recreating forest or semi-natural landscape, where natural fires could contribute, to some extent, to the regeneration of ecosystems and maintenance of bio-diversity at the landscape level.

However, the high-frequency of forest fires in many Mediterranean regions, by pure negligence or actual voluntary actions, produces wide-spread and serious degradations in the environment.

In the South, for opposite reasons, mainly relating to demographic pressure, with a parallel and dramatic increase of stocking densities and consumption of wood for domestic purposes, the environment is also undergoing severe degradation processes. This is jeopardizing the future of many mountain communities, particularly in the mountain areas of the North-African countries.

Statistical data available to compare the evolution of the forests and arable lands among countries of the Mediterranean basin provide good evidence of these two very opposite trends in the North and the South (see Table 2, Quezel, 1956).

Table 2. Evolution of forests and arable lands in the Mediterranean, from 1965 to 1976 (in Km^2) (Quezel, 1956).

	FORESTS			ARABLE LANDS		
	1965	**1976**	%	**1965**	**1976**	%
Portugal	3 1650	36 410	+ 15	43 320	36 000	- 16.9
Spain	131 600	153 330	+ 16.5	207 090	206 590	- 0.2
France	119 050	145 760	22.4	210 670	187 300	- 11.1
Italy	59 840	63 130	+ 5.5	154 540	123 480	- 19.1
Greece	24 790	26 180	+ 5.6	38 000	38 850	+ 3.0
Turkey	201 700	201 000	0	257 750	276 990	+ 7.5
Syria	4 460	4 570	+ 2.7	65 230	56 720	- 13.1
Lebanon	920	780	- 15.2	2 760	3 480	+ 26.1
Tunisia	6 740	5 300	- 18.8	44 060	44 100	+ 0.1
Algeria	25 490	24 240	- 4.9	62 610	71 100	+ 13.6
Morocco	53 020	51 640	- 2.1	70 660	78 300	+ 10.8

2 THE APPROACH TO DESERTIFICATION ISSUES IN THE MEDITERRANEAN MOUNTAINS

Addressing the role of mountainous areas in the desertification process around the Mediterranean basin is a key issue for this region.
As evidenced by the presentation of major threats to the Mediterranean future and the present and active degradation processes in the mountainous areas, maintaining the capacity of mountains as primary water sources is highly critical.

2.1 The concept of desertification

The concept of desertification is a matter of controversy. A broad definition would include the concept of non-reversibility or non-recovering capacity in the degradation processes that affect primary resources vital to human life, such as vegetation, soil and water.

This conception implies the interaction between human and natural processes, and a wide range of desertification types could be described according to the factors involved in human societies and natural environments.

Good descriptions and case studies of resource degradation are available from the Mediterranean region, but an overall understanding is actually missing, as is a theoretical framework to help analyze the multiple-sided issue of desertification. Some research projects offer an opportunity to compare several study areas by different methods throughout the Mediterranean region.

2.2 The example of the MEDIMONT research project

MEDIMONT is an example of such a project. It is a multi-national and multi-disciplinary research program on the role and place of the mountains in the desertification process of the Mediterranean regions, selected for support by the European Community Directorate General for Research, within the frame of its Environment program. This program includes a specific chapter on desertification.

The main objectives of MEDIMONT are to understand the desertification process in the Mediterranean mountains, under various natural and human conditions, and to produce guidelines for a rational management of such areas, including prevention, monitoring and appropriate policies for sustainable development. The project is a combination of local and regional-scale investigations.

At the local scale a series of multi-disciplinary case studies is carried out in pilot zones, to study the phenomenon, identify criteria, develop guide lines, but also give evidence of the diversity, the similarities and dissimilarities among the study areas.

At the regional scale, the work is completed by some general approaches to include the local case studies in a consistent overview, extrapolate findings, possible applications and recommendations from the local to the regional-scale level.

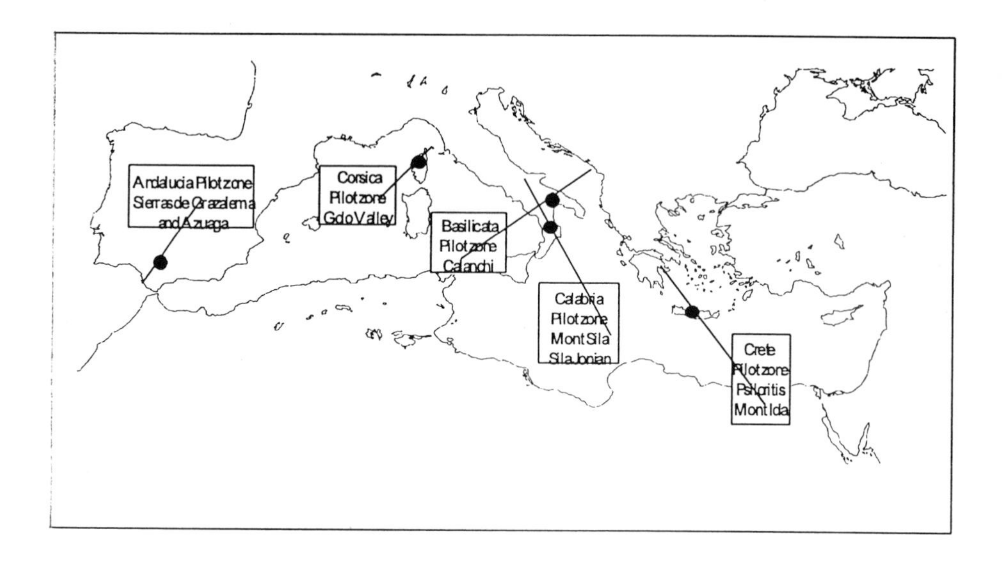

Figure 4. The five selected regions and pilot zones of Medimont.

Pilot zones were selected in five Mediterranean regions undergoing a severe desertification process: Andalusia, Corsica, Basilicate, Calabria and Crete (see Fig. 4).

Regional-scale approaches consist in a series of parallel investigations ranging from sensitivity studies to climate change, through the use of remote sensing techniques and macro-economic scenarios, to historical botanic reviews.

3 THE DATA ISSUE: MAJOR OBJECTIVES AND METHODS FOR RESEARCH ON DESERTIFICATION IN THE MEDITERRANEAN CONTEXT

As in the MEDIMONT project, the study of desertification phenomena requires the combination of various sources of information at different time and space scale-levels.

A major objective of such studies is the identification of potential indices of desertification in diversified historical and natural contexts and study sites.
Remote sensing techniques offer a series of promising achievements in this direction.

However, remotely sensed data must be analyzed in relation to other data derived from field measurements, statistical records and historical archives.

The use of GIS is essential to overlap so much information on a geographical basis, and developments in this field are of major importance for such studies.

But analyzing desertification in the context of regional environmental change requires various study-scales in time and space.

Local case studies must be completed with regional syntheses, and there is also a need for both retro-and prospective analysis, including social and natural aspects.

For instance, prospective analyses based on demographic data compared to the availability of critical natural resources provide a basis for more detailed scenarios on the future of the region. Also there is a need to carry out sensitivity studies on the consequences of climate variability and change by comparing climatic scenarios to the possible response of key biotic and abiotic parameters.

A retrospective approach is also necessary to provide information on the state of the environment before and after some critical change in human settlement and history and to assess its capacity to adapt or recover after major disturbances.

Reconstructing past vegetation dynamics is a good example of such an approach, and methods derived from historical botany, like pollen analysis, dendrochronology or pedo-anthracology are particularly relevant.

4 Conclusion

A wide range of methods and techniques must be combined in any integrated research effort aiming at modeling and monitoring desertification processes in the Mediterranean mountains and regions.

Keywords

Ecosystem, mountains, Mediterranean regions, soil erosion risk, climate change, desertification, water resources, global change.

References

[1] Brouwer F., Falkenmark M., "Climate induced water availability changes in Europe. Environmental monitoring and assessment". Kluwer Academic Publisher, Printed in the Netherlands (1989).

[2] B.A.H.C., Climate-Hydrology-Ecosystems Interrelations in Mountainous Regions (CHESMO). An International Initiative for Integrative Research. IGBP - BAHC / UNEP Workshop, Report n°2, St Moritz, Switzerland, December 1993 (1994).

[3] Hulme H., Climate Scenarios. Models and Monitoring. Proceedings of the First European LICC Conference, Lunteren, the Netherlands (1989).

[4] Margat J., "L'eau dans le Bassin Méditerranéen - Situation et Prospective", in Les fascicules du Plan Bleu, Ed. Economica (1992).

[5] Plan Bleu, "Avenirs du Bassin Mediterranéen", Les fascicules du Plan Bleu, Ed. Economica (1989).

[6] Quezel P., "Contribution à l'étude des Forêts de Chênes à feuilles caduques d'Algérie", Mem. Soc. Hist. Nat., Afrique du Nord (1956).

LAND COVER CHANGE DETECTION IN A MEDITERRANEAN AREA (MARITIME SYRIA) FROM REMOTELY SENSED DATA

Jean-François KHREIM and Bernard LACAZE[1]

Centre d'Ecologie Fonctionnelle et Evolutive, CNRS-BP 5051, 34033 Montpellier Cedex, France

ABSTRACT

In order to detect land cover changes in maritime Syria over the past 16 years (1977-1993), a multi-temporal normalized difference vegetation index (NDVI) analysis computed from MSS-Landsat data was performed. A RGB-NDVI image composite technique was applied to visualize NDVI changes between three acquisition dates (March 1977, April 1986, March 1993). Qualification of change was obtained by a combination of this method with a simple classification comparison procedure. Our results are summarized in a synthetical map revealing a trend towards a global degradation of vegetation cover.

Détection d'un Changement du Couvert Végétal des Zones Méditerranéennes par Télédétection (Syrie Maritime)

RÉSUMÉ

A partir de trois scènes MSS-Landsat acquises entre 1977 et 1993, une analyse multidate de l'indice de végétation normalisé (NDVI) a été réalisée afin de caractériser le changement des grandes unités de l'occupation des sols dans la région de Syrie maritime. La visualisation en composition colorée rouge vert et bleu (RVB) combinée à une méthode simple de classification a permis de suivre l'évolution du NDVI et de quantifier l'importance du changement observé. Les résultats obtenus font l'objet d'une carte synthétique montrant une dégradation globale de la végétation dans la région considérée.

1 INTRODUCTION

Nowadays, everyone is aware that the future of the Earth is threatened by the degradation of ecosystems, the loss of biodiversity and the irrational use of natural resources. These processes are very pronounced in the Mediterranean Basin where fragile ecosystems result from the confluence of natural and anthropogenic factors, leading to erosion, desertification, deforestation, overgrazing, over-

---

[1] To whom correspondence may be addressed

population and urban industrial pollution. Such changes are coupled with a global warming due to increasing levels of carbon dioxide and other greenhouse gases. Remote sensing technology through a repetitive coverage of data is a good tracer for the evolution of ecosystems at regional and global scales. The Landsat MSS data acquired since 1972 represent an important basis of temporal, spectral and spatial information for the study of global change phenomena in terrestrial ecosystems. In this paper, which is part of a current research project, we illustrate the usefulness of multitemporal remotely sensed data in deciphering the land cover changes of the Mediterranean region. In this study, the normalized vegetation index (NDVI) is used as an indicator for land cover changes. The area under investigation is part of the Eastern Mediterranean Basin and concerns maritime Syria located North-West of this country and with a typical Mediterranean climate, characterized by dry summers and moist winters (Khreim and Lacaze, 1991). The studied zone precisely pertains to the humid to sub-humid bioclimatic stage according to the Emberger climatogram applied to Lebanon by Abi-Saleh (1978). It constitutes the more vegetative region in Syria (Quezel and Barbero, 1985).

2 Multitemporal NDVI Analyses

The techniques used in land cover change detection are based on the comparison of remotely sensed data and map data, or the comparison of remotely sensed data taken at two or more different times. Numerous methods are available, each one offering its own advantages and drawbacks: univariate image differencing, image regression, image rationing, vegetation index differencing, principal component analysis, post-classification comparison, direct multidate classification, change vector analyses... (Singh, 1989). Among these methods, we chose to consider the vegetation index for multitemporal comparison. Since vegetation is considered to be the functional equivalent of terrestrial ecosystems (Graetz, 1990), then vegetation is a prime indicator of the type of ecosystem present (i.e. grasslands, forests, deserts). The characterization of vegetation generally offers the observer a great amount of information about its geographical location, current ecological conditions and may even yield information on previous changes that have occurred in a given region. The vegetation index which is the ratio between the reflected radiation in near-infrared (NIR) and red (R) bands measured by the satellite sensor, is directly linked to the density and state of green vegetation (LAI: Leaf Area Index). The index used in this study is the normalized difference vegetation index: (NDVI) = NIR - R / NIR + R (Rouse et al. 1974) as it can be easily calculated and allows the relief effects of illumination to be reduced. Land cover changes over time may be distinguished through the analysis of multitemporal vegetation index imagery.

3 Data Processing

The data we used are 800x800 pixel extracts from three Multi Spectral Sensor (MSS) scenes of Landsat 1 and 2 satellites, acquired on March 21, 1977 (Landsat 2), April 5, 1986 and March 5, 1993. Produced by ESA (European Space Agency, Fucino, Italy), these scenes allow a 2200 km^2 coverage of the region. The absolute radiometric correction can be accomplished by converting the different grey levels to physical values of radiance and then to apparent reflectance at the sensor.

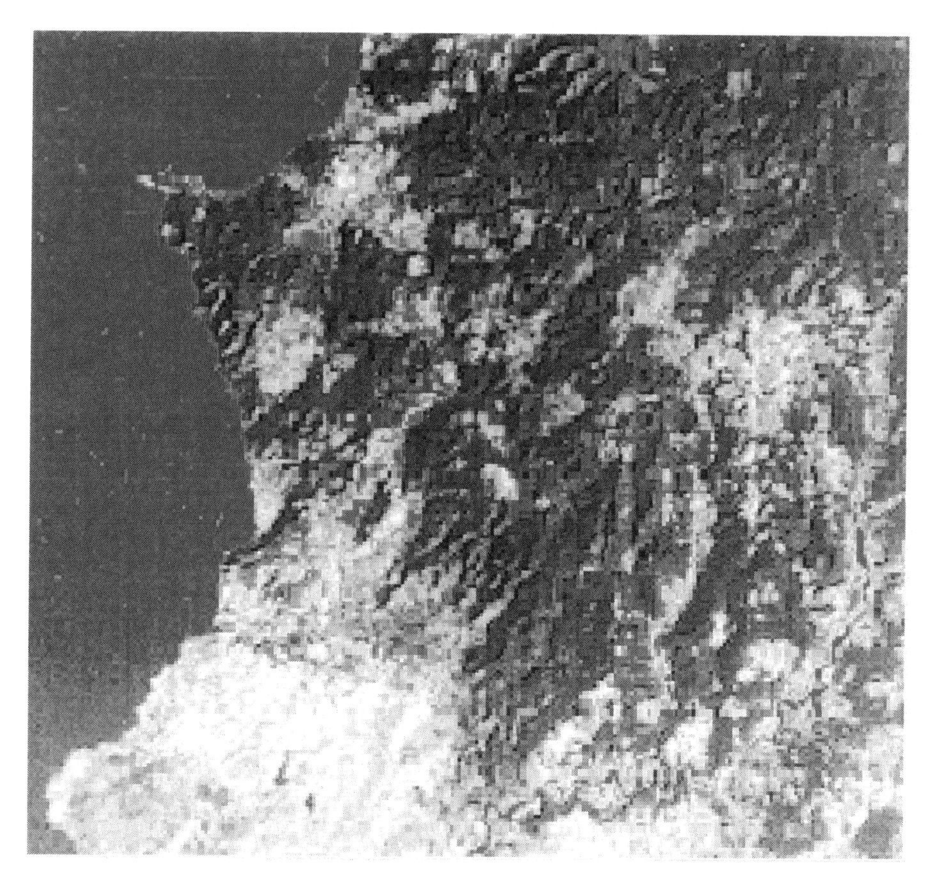

Fig. 1. Landsat MSS false color composite image (bands 7,5,4 in RGB) acquired on March 21, 1977. This scene represents the northern part of maritime Syria. (see also page 303)

Fig. 2. Landsat MSS false color composite image (bands 7,5,4 in RGB) acquired on April 05, 1986 (Northern part of maritime Syria). (see also page 304)

Figure 3. Landsat MSS false color composite image (bands 7,5,4 in RGB) acquired on March 23, 1993 (Northern part of maritime Syria).
(For color figure, see page 305)

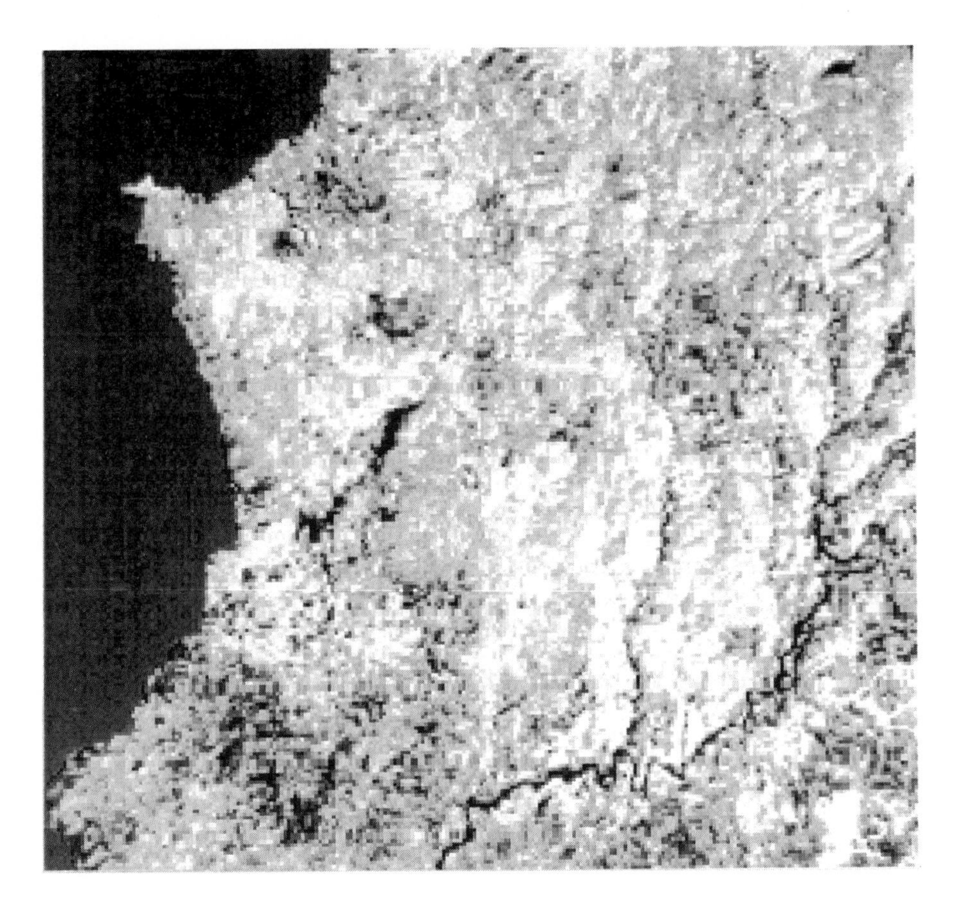

Fig. 4. NDVI image derived from the MSS scene of March 23, 1993. Light grey indicates high NDVI values, dark grey and black low NDVI values. (see also page 306)

This radiometric correction was however not easily feasible with the data produced by ESA, so that we decided to apply only a scene-to-scene radiometric normalization using pseudoinvariant features. This technique is based on the assumption of constant reflectance selected in-scene elements. Differences in the grey-level distributions of these invariant objects are assumed to be a linear function and are corrected statistically to perform the normalization.

The 1986 image was selected as reference to normalize the other two images. An image by image geometric correction was then applied to make them superimposable. The 1977 and 1993 images were corrected relative to the image of 1986. Three false color composite images of the study area were realized (bands MSS 7, 5, 4 in RGB) (Fig. 1, 2, 3). Visual analysis of these images allows one to recognize the main changes that occurred during the 1977-1993 period, such as the urban growth of Lattaquié city, the variations in hydrographic outlines, especially for the Al Kebir stream after a dam building, the decrease in land vegetation cover in the North-Western part of the study area as a consequence of forest fires.

The normalized difference vegetation index was calculated for each date. In order to avoid the negative values of the NDVI, we chose to use the following formula:

255*NIR/NIR+R <=> 255*MSS7/MSS7+MSS5.

Therefore, the values vary between 0 and 255 instead of -1 and +1.
The vegetation index images were visualized by monochromatic display, where NDVI values are in the range (0-255), dark grey indicates low NDVI values whereas light grey indicates high values (Fig. 4). The visual comparison of these three NDVI images can give indications about the biophysical changes for the 1977-1993 period. Subtracting a pixel's NDVI value on one date from the corresponding pixel NDVI on another date constitutes another way by which to quantify and locate NDVI variations (Fig. 5).

But the interpretation of such documents is often difficult since grey levels are hardly perceptible by human vision.

4 THE RGB-NDVI COMPOSITE IMAGE TECHNIQUE

In order to facilitate the interpretation of the changing NDVI values for each date and to study changes that occurred in the 1977-1993 period, a simple and logical technique was applied to visualize changes by combining the NDVI calculation and RGB image display functions. In this study, NDVI of 1977, 1986, 1993 were assigned to red, blue and green respectively (Sader and Winne, 1992).

In the resulting RGB-NDVI color composite image (Fig. 6), white, black and grey indicate no major change in NDVI values between the three dates while colors illustrate change in vegetation cover. For example, magenta is created by a high NDVI value 1977 and 1986 and a low value in 1993 (green biomass reduction); green is created by a low NDVI intensity in 1977 and 1986 and a high value in 1993; yellow represents high NDVI value in 1977 and 1993 and low in 1986. For more details, the reader is referred to Table 1.

Fig. 5. Subtracted NDVI image obtained by subtracting March 1993, MSS data from March 1977 data. Very light tones indicate vegetation decrease while very dark tones indicate vegetation increase in 1993. (For color figure see also page 307)

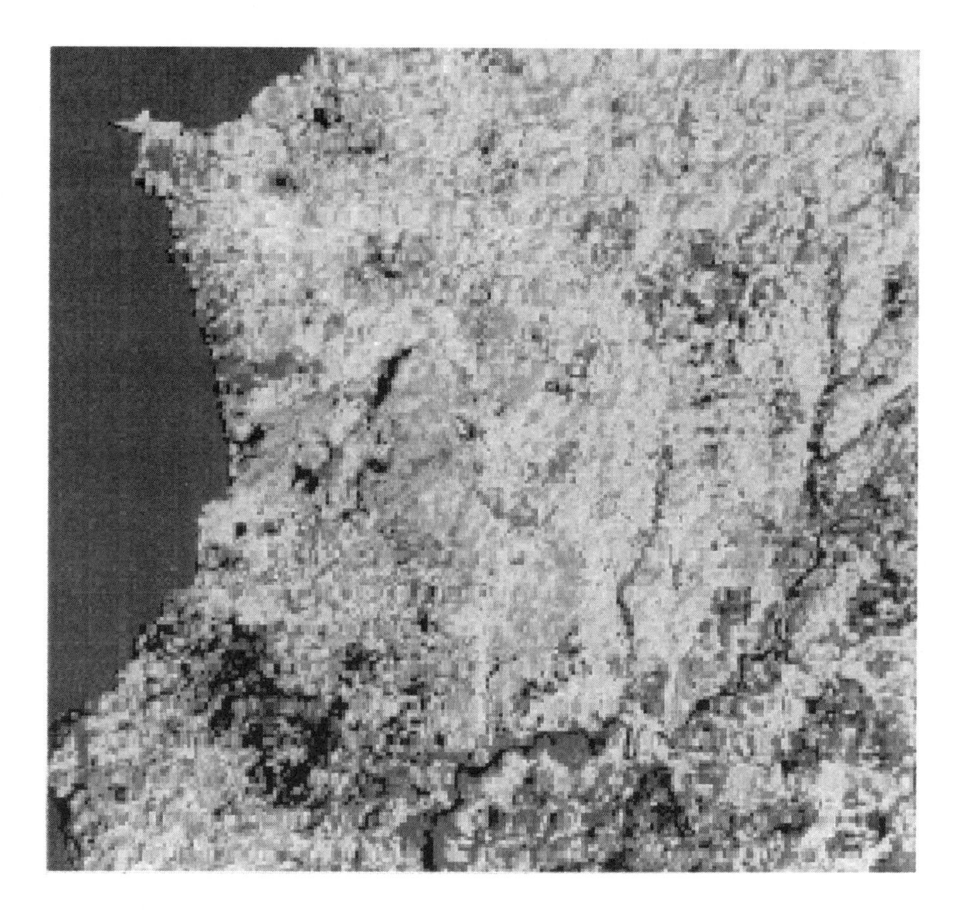

Figure 6. RGB color composite image of the 1977, 1986, 1993 NDVI assigned to RBG. White, black and grey express no major change in NDVI values between the three dates while colors illustrate change in vegetation cover (magenta = decrease in NDVI value in 1993; green = increase in NDVI value focussed in 1993). (For color figure, see page 308)

Figure 7. RGB color composite image of the 1997, 1993 NDVI assigned to R and BG respectively. White, black and grey express no major change in NDVI values between the two dates while light red illustrates low decrease, dark red high decrease in NDVI value in 1993. Light and dark cyan indicate an increase in NDVI value in 1993. (For color figure, see page 309)

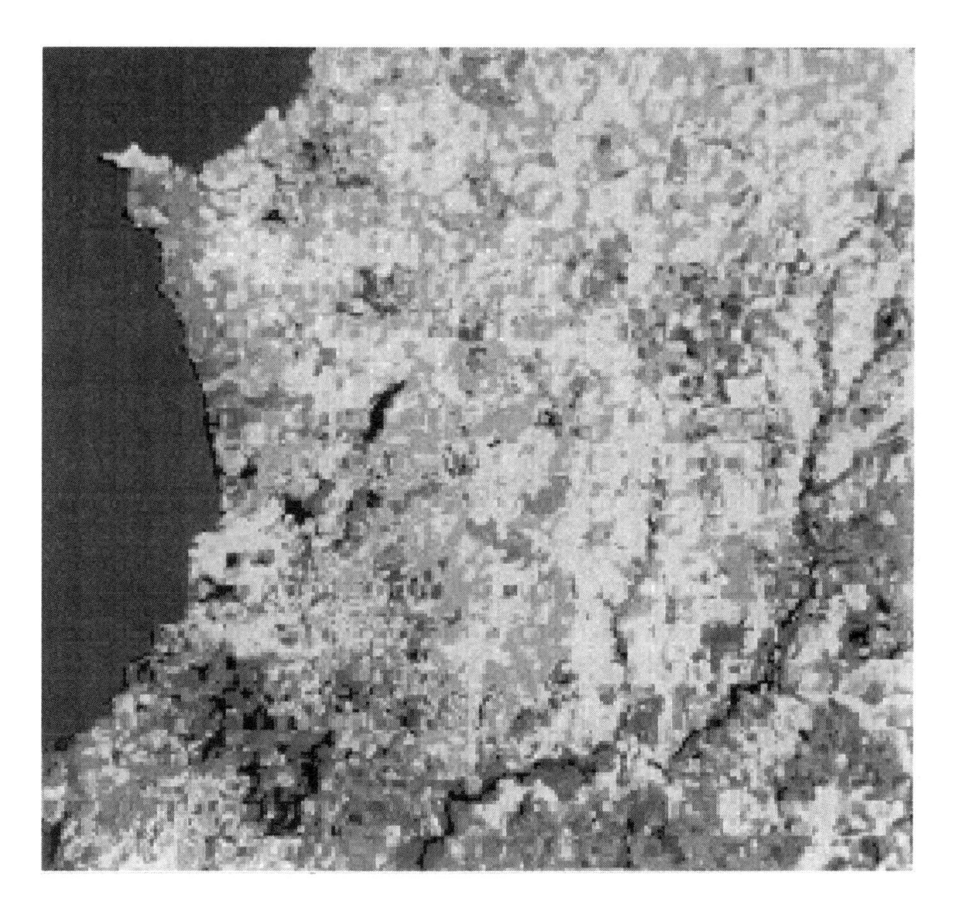

Figure 8. Sixty-four classes RGB-NDVI image. The 1977, 1986, 1993 classified NDVI are assigned to RBG respectively. White, black and grey express no major change in NDVI values between the three dates while colors illustrate change in vegetation cover (magenta = important decrease in NDVI value in 1993; green = important increase in NDVI value in 1993; yellow = important decrease in 1986 and 1993; light blue = increase in 1986 and 1993; orange = important decrease in 1986 and slight increase in 1993). (For color figure, see page 310)

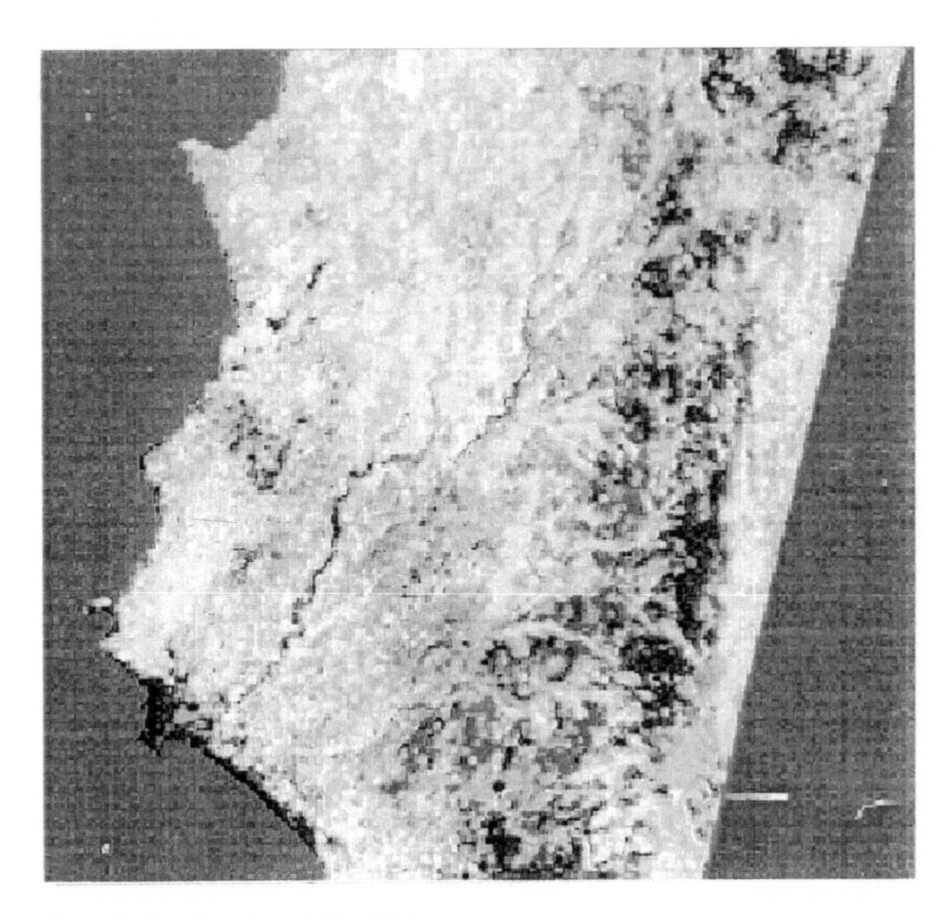

Figure 9. Sixty-four classes RGB-NDVI image: generalization to a larger area. Same color coding as in Fig. 8. (For color figure, see page 311)

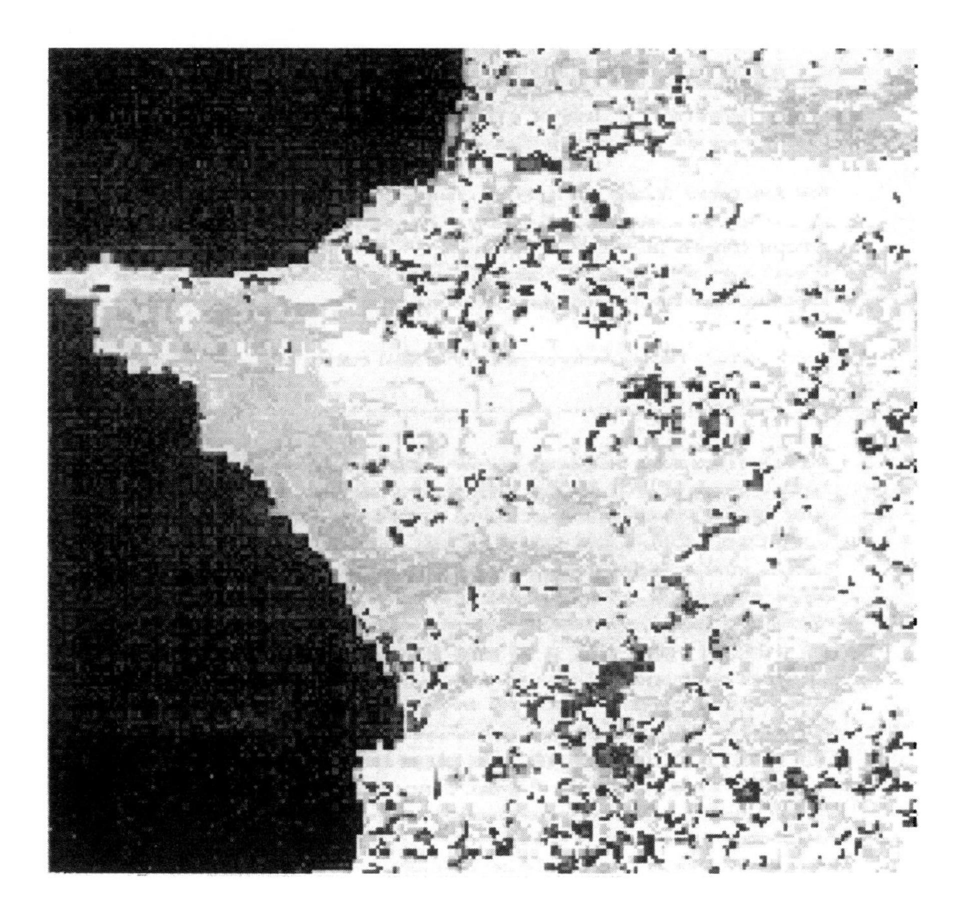

Figure 10. Mapping the magnitued of vegetation changes in maritime Syria. Subtracted NDVI image was obtained by subtracting the 1993 classified vegetation index image from the 1977 classified NDVI image. White indicates no change in NDVI slice units. Magenta, red and yellow represent classes whose NDVI value was lowered by one, two and three slice units respectively. Cyan, blue and black (except sea surface) represent classes whose NDVI value rose by one, two and three slice units respectively. (For color figure, see page 312)

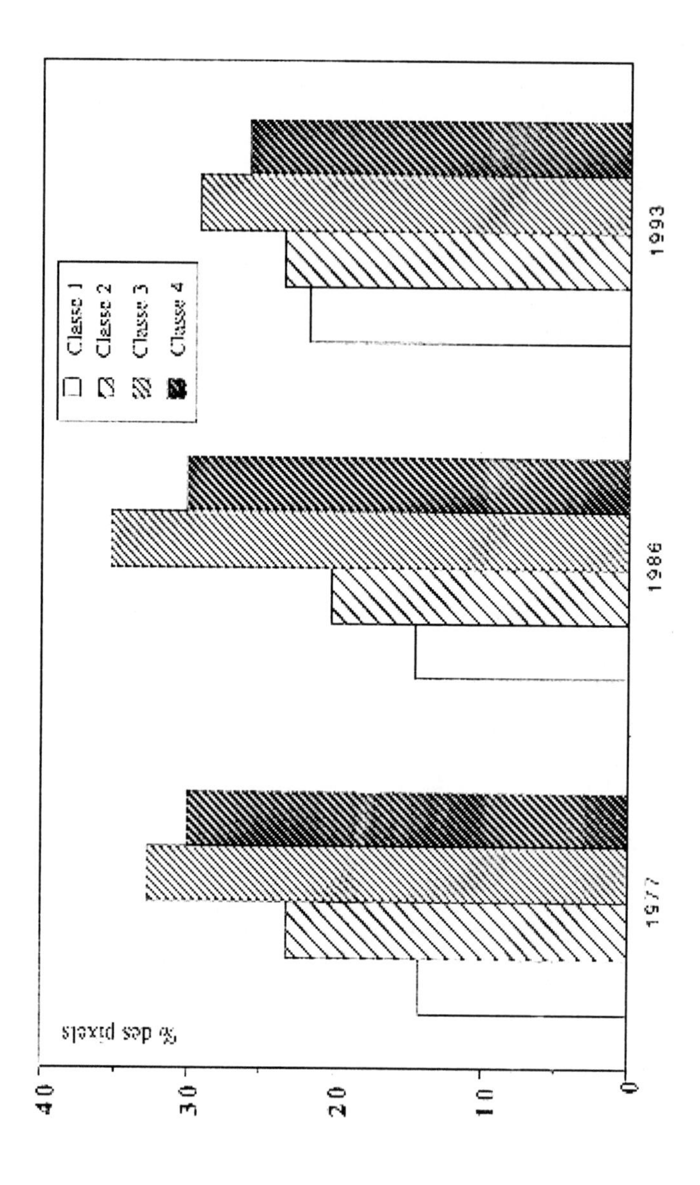

Figure 11. Multi-temporal analysis of the distribution of pixels into four NDVI classes (expressed in % of the total number of pixels in the terrestrial part of the image).

The RGB-NDVI color composite image technique can also be used to compare two NDVI dates. In order to detect changes between 1977-1993 dates, NDVI 1977 and NDVI 1993 were assigned to red and blue+green respectively (Fig.7). Red illustrates a decrease in NDVI values in 1993 and cyan indicates that there was more green biomass in March 1993. Once more, black, white and grey mean no change in vegetation cover. The RGB-NDVI composite image technique allows vegetation cover differences to be mapped and quantified. But, such a method is of limited interest when the topics to be covered include not only the detection but also the nature of change. Moreover, the user must interpret a myriad of colors necessitating laborious tasks. Below we describe methods which facilitate the interpretation of these documents as they qualify the detected changes.

5 Modified RGB-NDVI Composite Image Technique

Each image includes 256 NDVI values. When combining three dates, in the RGB-NDVI composite image method, 256^3 different colors that must be interpreted are produced. A reduction of color space is then preferable, without losing the most important information. The NDVI images were statistically analyzed and related to the main types of land cover in the study area. NDVI values can be divided into the four following classes connected to the density of vegetation:

- Class 1: NDVI=0-165: N; absence of vegetation: water surfaces, urban sites, bare soils;
- Class 2: NDVI=165-180: L low vegetation density: degraded natural vegetation;
- Class 3: NDVI=180-200: M medium vegetation density: maquis, garrigues and cultivation;
- Class 4: NDVI=200-255: H high vegetation density: forests, intense cultivation.

Each NDVI image was classified according to these four density slices by the look-up table transformation method. Then a post-classification filter (3 by 3 window) was applied to smooth the image. This filter removes single, isolated pixels and replaces the pixel with the most common or majority value of surrounding pixels. The resulting images are once more visualized in the RGB system (Fig. 8 and 9). This technique allows one to qualify the nature of change in a much more efficient manner than would be possible from the simple use of the NDVI-RGB composite technique. Table 1 gives examples of such interpretations as good tracers for land cover evolution during the 1977-1993 time period.

6 Change Mapping (1977-1993)

In order to map the change in land cover between 1977 and 1993, we subtracted the classified vegetation index image derived from the 1993 scene from the classified NDVI image obtained from the 1977 data. In order to avoid the negative values of the NDVI class, we add +4. The distinguishing class image resulting from this procedure contains seven classes that reveal two main vectors of change: loss or gain in vegetation cover. Each of these seven classes represents the shift from one unit to another in the land cover which can be higher or lower in the predefined NDVI slices. (see Fig. 9).

7 Discussion

From the analysis of the results, it can be deduced that maritime Syria experienced important changes in its overall land cover types during the 1977-1993 period (Table 2).

Statistical analysis of the differentiation of the two classified NDVI images showed that change occurred in 47% of the total terrestrial part of the image. In the 1977-1993 period, lowering of the NDVI values is registered by 31.08% of the total terrestrial pixels: 24.74% by one, 5.50% by two and 0.84% by three slices respectively. Meanwhile, increase of the NDVI values is recorded by 15.81% of the total terrestrial pixels distributed as: 13.14% by one, 2.30% by two and 0.37% by three slices. The major change is observed for one degree of variation of NDVI classes (-24.74% and +13.14%).

Surface distribution of each class for the two dates is depicted as follows: 14.23% of the terrestrial part of the scene was occupied in 1977 by class 1 (non-vegetable elements) which increased to 21.69% in 1993; class 2 (low vegetation density) registered no significant change (23.07% in 1977 and 23.41% in 1993); class 3 and 4 clearly record an important decrease in vegetation. In fact, these two classes which correspond to a high and medium density of vegetation are affected by the previously noted decrease of NDVI by one slice. The surface of class 3 occupied 32.72% of the total terrestrial surface in 1977 and in 1993 29.15% (78.5 km^2). Class 4 was 29.98% in 1977 and became 25.75% in 1993 (93 km^2). These results have been compared to those relative to the 1986 scene. It is deduced that no major change occurred between 1977 and 1986, i.e. during an eleven year time span. What caused such an important change between 1986 and 1993? The factors responsible for this change can be subsumed under five main themes:

1 - Forest fires in summer 1992 in the North-Western part of the region;
2 - The building of the Al Kebir dam after 1984;
3 - Development of communication structures (roads, railways, yards...);
4 - General urban growth particularly appreciable in Lattaquié city;
5 - Deforestation practices.

However, the real impact of each kind of factor needs to be all the more controlled by field observations as other factors may be involved in change processes.
Additional investigations being conducted include the analysis of seasonal variations especially during and after the dry season when the main degradation processes in vegetation cover were shown to occur in the study region.

8 Conclusion

The multi-temporal NDVI analysis applied to maritime Syria made it possible to detect land cover differences during the 1977-1993 period. The RGB-NDVI composite image technique contributed to visualize and quantify vegetation change but was unable to qualify the nature of change. This procedure was combined with the post-classification comparison method. Classified NDVI values were assigned to four classes referring to vegetation density. The interpretation of the resulting image demonstrates the possibility of evaluating change both quantitatively and qualitatively.

Table 1. Interpretation of additive colors in three date RGB-NDVI image.

Color observed	1977 (Red)	1986 (Blue)	1993 (Green)	Interpretation of change
Black	N	N	N	no change (water, urban)
Dark grey	L	L	L	no change (low biomass)
Light grey	M	M	M	no change (medium biomass)
White	H	H	H	no change (high biomass)
Light pink	H	H	M	* decrease in 1993
Pink	H	H	L	**decrease in 1993
Magenta	H	H	N	***decrease in 1993
Yellow	H	L	H	**decrease in 1986; **increase in 1993
Orange	H	L	M	**decrease in 1986; *increase in 1993
Cyan	L	H	H	**increase in 1986 and 1993
Light blue	M	H	H	*increase in 1986 and 1993
Green	M	L	H	*decrease in 1986; **increase in 1993
Dark green	L	L	M	*increase in 1993
Light Green	M	L	H	*decrease in 1986; **increase in 1993
Dark green	L	L	M	*increase in 1993
Red	H	L	L	**decrease in 1986 and 1993

Legend: NDVI classes: N - No vegetation; L - Low vegetation density;
M - Medium vegetation density; H - High vegetation density;
*, **, ***: degree of change (NDVI slice units).

Table 2. Distribution of NDVI classes over the 1977-1993 period (in % of the total terrestrial pixels of the scene).

	1977	1986	1993	(1993-1977)
Class 1: no vegetation	14.2	14.5	21.7	+ 7.5
Class 2: low vegetation density	23.1	20.2	23.4	+ 0.3
Class 3: medium vegetation density	32.7	35.3	29.1	- 3.6
Class 4: high vegetation density	30.0	30.1	25.8	- 4.2

The statistical comparison of the different land cover classes between 1977 and 1993 shows that maritime Syria experienced a severe decrease in vegetation cover mainly due to increased agricultural and urban activities because of rapidly expanding population. In the 1977-1993 time interval, twelve percent of the forest vegetation (high and medium density) changed class with a trend from a forest-maquis to a maquis-grassland signature, accounted for by physical processes such as fire, lightning, disease. Non-vegetable elements which we define as regrouping urban infrastructures and bare soils show a surface increase in the range of 7% between 1977 and 1993. The multi-temporal NDVI analysis demonstrates the important changes brought about in maritime Syria during the 1977-1993 period. These changes reveal a continuous trend towards degradation of the vegetation cover. This phenomenon can be explained by man's direct or indirect action. Through a mechanism of bio-geophysical retroaction, the increase in bare areas due to man's action can generate significant climatic changes (Hobbs, 1990). This study permits an appreciation of the magnitude of the degradation processes in maritime Syria over the last 16 years. This situation underlines the importance of constant monitoring of the land surface state in Mediterranean regions. Multi-temporal remotely sensed data are likely to make an important contribution to a better understanding of global and regional environmental change.

KEYWORDS

Remote sensing, Mediterranean region, degradation, vegetation cover, satellite image processing, GIS, vegetation index, ecosystems, biodiversity, forest degradation.

REFERENCES

[1] Abi-Saleh, B., Etude phytosociologique, phytodynamique et écologique des peuplements sylvatiques du Liban. Thèse Univ. Aix-Marseille. 185p. (1978).

[2] Graetz, P., Remote sensing of terrestrial ecosystem structure: an ecologist's pragmatic view. In Remote Sensing of biosphere functioning, R.Hobbs and H. Mooney eds. Springer-Verlag, New York, pp 5-30 (1990).

[3] Hobbs, R., Remote sensing of spatial and temporal dynamics of vegetation. In Remote Sensing of biosphere functioning, R. Hobbs and H. Mooney eds. Springer-Verlag, New York, pp 203-219 (1990).

[4] Khreim, J.F., Lacaze, B., Cartographie de la végétation forestière méditerranéenne dans la région du Baer-Bassit (Nord-Ouest de la Syrie) à partir des données Landsat-TM multidates. Bull. Comité Français de Cartographie, 127-128:130-135 (1991).

[5] Quezel, P., Barbero, M., Carte de la végétation potentielle de la région méditerranéenne. Feuille N 1: Méditerranée Orientale. CNRS Ed. France (1985).

[6] Rouse, J.W., Haas, R.H., Shell, J.A., Deering, D.W. and Harlan, J.C., Monitoring the vernal advancement and retrogradation (Greenwave effect) of natural vegetation. NASA/GDFC TYPE III Final Report, Greenbelt, Mol, 371 p. (1974).

[7] Sader, S.A., Winne, J.C., RGB-NDVI color composite for visualizing forest change dynamics. Int. J. Remote Sensing, vol 13, 16:3055-3067 (1992).

[8] Schott, J.R., Salaggio, C., Volcoch, W.J., Radiometric scene normalization using pseudoinvariant features. Remote Sensing of Environment, 26:1-16 (1988).

[9] Singh, A., Digital change detection techniques using remotely sensed data. Int. J. remote Sensing, vol 10, 6:989-1003 (1989).

INDEX OF ENVIRONMENTAL QUALITY FOR LAND MANAGEMENT

Alberto QUAGLINO

Dipartimento di Georisorse e Territorio, Politecnico di Torino, C.so Duca degli Abruzzi, 24 - 10129 Torino, Italy

ABSTRACT

This paper refers to a methodological study whose goals are to provide integrated information to be used as an environmental index. The aim of this study is to contribute to the determination of a methodology which, by the investigation of a few descriptive parameters pertaining to forests, would be able to judge the objective quality independent from consideration based on a single experiment or from the analysis of a data survey.

Having limited the research to a domain of silvicultural ecology, the possibility of determining a series of synthetic indicators, which would be able to describe a system as complex as a forest, was evaluated. Such indicators should be able to obtain complete and readily readable information, for land management users. As a result of this study an index relative to the stability of the forest was developed, by analysis of the forest structure. The numerical index obtained rapidly demonstrated that it is able to discriminate different structural types by evaluating a few parameters.

Index de Qualité Environnementale pour la Planification Foncière

RÉSUMÉ

Cet exposé se réfère à une étude méthodologique dont l'objectif est de fournir une information intégrée que l'on pourra utiliser comme un index environnemental. Le but de cette étude est de contribuer à déterminer une méthodologie qui, par le biais de l'examen de quelques paramètres descriptifs propres aux forêts, serait en mesure d'exprimer un jugement sur la qualité objective, détachée de toute considération basée sur une seule expérience et sur l'analyse d'une étude de données.

En ayant limité la recherche au domaine de l'écologie sylviculturelle, l'opportunité de délimiter une série d'indicateurs synthétiques capables de décrire un système aussi complexe qu'une forêt, a été évaluée. Ces indicateurs pourraient obtenir une information complète et facile à lire à l'usage de l'exploitation des terres. Comme résultat de cette étude, en analysant la structure de la forêt on a obtenu un index relatif à la stabilité de la forêt. L'index numérique a démontré rapidement qu'il pouvait, en évaluant quelques paramètres, distinguer différentes formes de structure.

1 INTRODUCTION

This paper presents some contributions to the silvicultural ecology domain, with the aim of defining a methodology useful for determining an environmental quality index. This index, obtained by a simple and quick survey procedure, is an expression of the quality of the forest and is linked to the forest's ecological, biological and physionomical characteristics. It is particularly useful, when, for economic reasons, more accurate analyses cannot be performed.

The examined parameters represent characteristic qualities of the forest. A synthetic numerical index was processed in order to determine the stability of the forest by analyzing its structure.

2 METHODOLOGY AND RESULTS

Coniferous forests of the Western Alps (Piedmont Region) were analyzed during the first research stage: particular attention was paid to *Pinus sylvestris*, *Abies alba*, *Picea excelsea* and *Larix decidua* forests under different conditions.

Forty-five different sample areas were studied, in a random locality, each 600 square meters each. A total tree inventory was performed in the above mentioned sample areas, by measuring all the trees with the following characteristics: a minimum b.h. diameter of 5 cm , and a height not less than 2 m.
The thus determined structural complexity index determination was implemented in four different steps:

- determination of foliage height (reference height);
- tree component classification divided according to height;
- analysis of the height distribution;
- determination of a numerical index evaluated by using a few parameters (diameter distribution).

2.1 First step: determination of foliage height (reference height)

The average height of the dominant layer was measured by calculating the average value of the first six trees with the largest diameter. The obtained parameter, called reference height (H_2), is the comparison value used to determine the variability of the tree height distribution in the sample area.

2.2 Second step: tree component classification divided according to height

The space occupied by stems and crowns of every sample area was divided using the highest tree height of the same sample: five different layers were obtained and every tree was included in one of these layers. The determined layer thickness is shown in Figure 1.

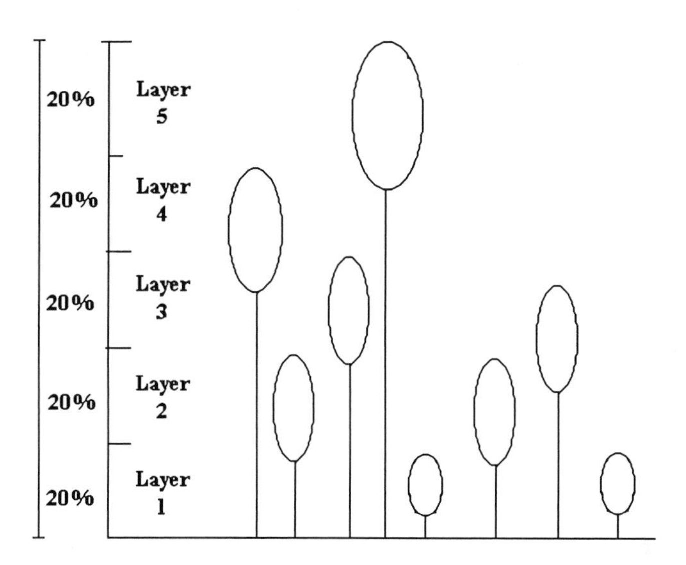

Figure 1. Layer thickness.

2.3 Third step: analysis of height distribution

The sample tree height variability was calculated by using the following equation:

$$Sh_r = \sqrt{\left[\sum \frac{(H_i - H_r)^2}{n}\right]} \tag{3.1}$$

where:

H_r = reference height,
n = number of samples,
H_i = tree height,
Sh_r = reference height mean spread.

The Sh_r value was weighed with reference to the height and expressed in percentage as follows:

$$Sh_r\% = \frac{Sh_r}{h_r} \text{ x } 100 \tag{3.2}$$

One then had to evaluate whether the parameter $SH_r\%$ as determined in (3.2) was able to describe, with sufficient precision, the distribution of tree heights in the different layers.

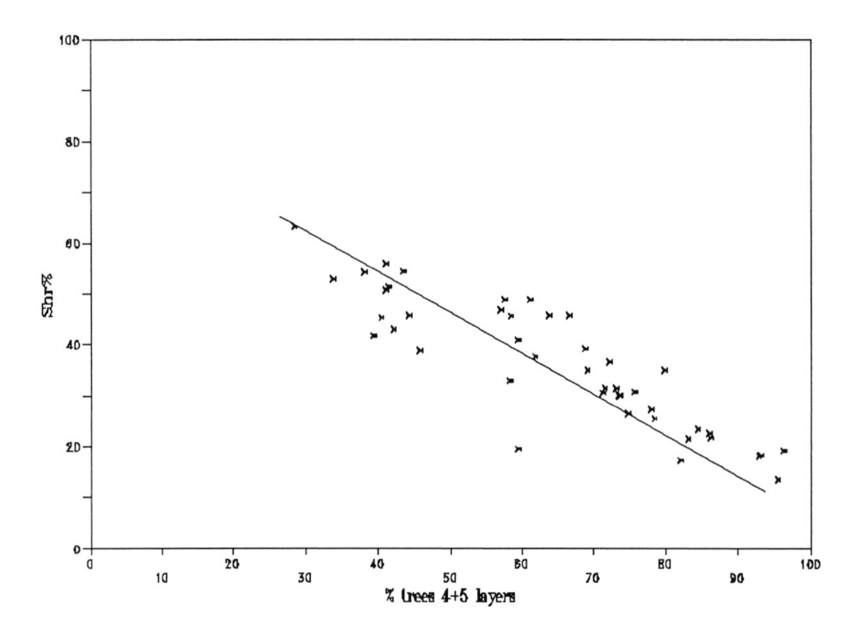

Figure 2. Regression between $SH_r\%$ and the dominant layers.

As far as the previous hypothesis is concerned, higher $SH_r\%$ values represent a greater distribution of tree heights in the dominant layers (4 and 5), where lower values are typical of a uniform distribution in all the different layers. The regression (Fig. 2) calculated by taking the dominant layers and $SH_r\%$ into account, shows (r^2=0.78) that the index obtained is well able to describe tree height variability in the sample area.

2.4 Fourth step: determination of a numerical index evaluated by using a few parameters (diameter distribution).

Having created a structural complexity index representative of the prefixed goals, it was then necessary to simplify the above-mentioned ground surveying methodology by taking a great deal of time to measure the large number of tree heights (this is not easy in a forest).

Another index, i.e. the height distribution frequency was determined by calculating the diameter distribution frequency according to the following equation:

$$Sd_m = \sqrt{\left[\frac{(D_i - \overline{D})^2}{n}\right]} \tag{3.3}$$

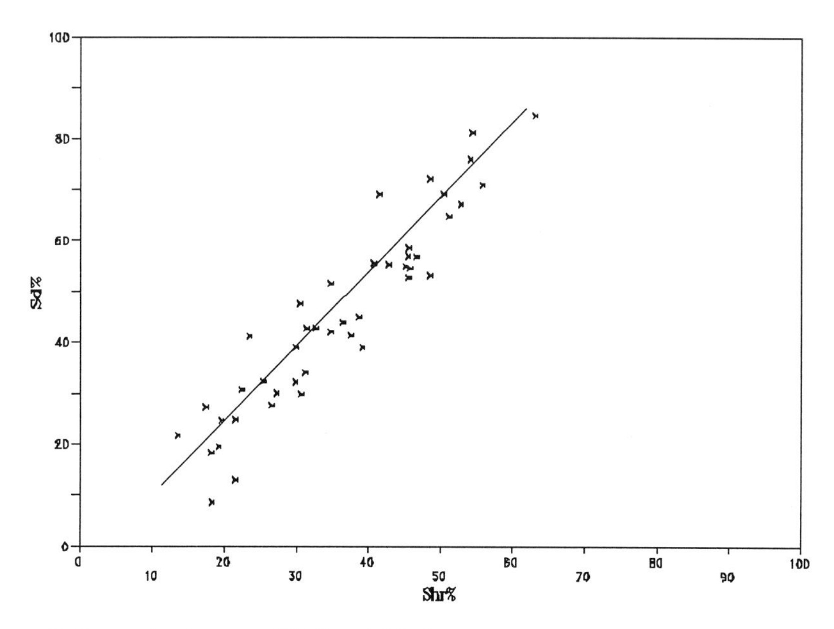

Figure 3. Regression between SH_r% and Sd_m%.

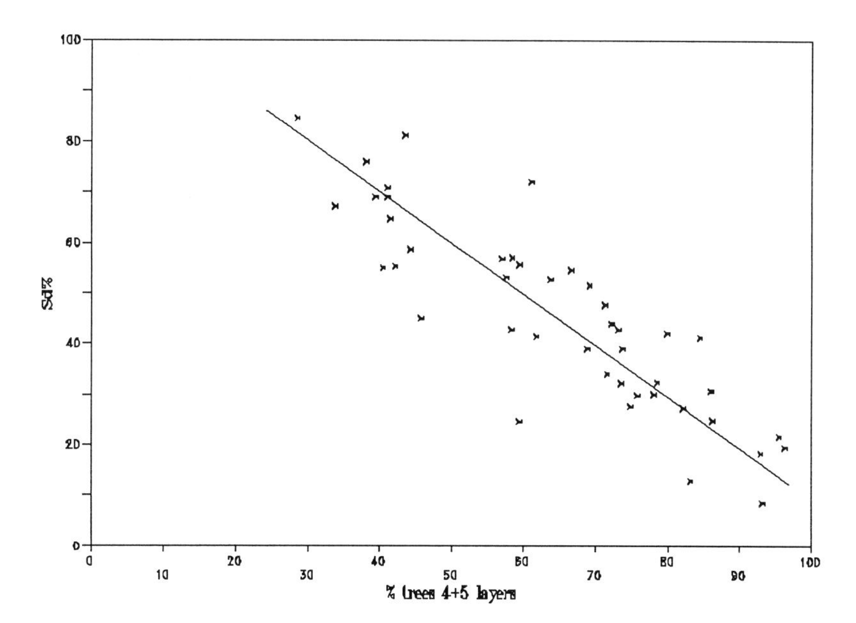

Figure 4. Regression between the Sd_m% and the number of trees in the last two layers.

where:

D_i = tree diameter measured at 1.30 m,
n = number of samples,
= mean diameters,
Sd_m = mean diameter spread.

The Sd_m value was divided into the mean diameter and expressed, in percentage, as follows:

$$Sd_m\% = \frac{Sd_m}{D} \times 100 \qquad (3.4).$$

The correlation (Fig. 3) between the height percentage spread and the last parameter was calculated using the regression procedure (r^2=0.88).

In conclusion, the correlation between the $Sd_m\%$ and the trees in the last two layers was calculated, confirming that it is possible to replace the height index with that of the diameters. The regression obtained in this case (r^2=0.78) is also good (Fig. 4).

3 CONCLUSION

If one takes into account the results obtained in the case of the forest types (Western Alps coniferous forests) examined, the diameter measurements are sufficient to identify the stratification degree of the forest.

As expected, the lower $Sh_r\%$ values obtained describe a mono-stratified community; on the other hand, the spread values proportionally increase with an increase of the forest complexity, allowing one to place each forest in its proper structural type group.

The obtained index (Fig. 5) is included in a theoretical range of 0 to 100, where 0 indicates mono-stratified forest, while 100 indicates a hypothetical maximum variability (multi-stratified forest). A single layer forest shows guide values of between 18 and 35 (this being the case of almost all the *Pinus sylvestris* forests examined); guide values of between 36 and 55 characterize quite an irregular height distribution; the higher values consist of forests where the different height classes are mixed, such as the *Abies alba* more natural forests surveyed.

The environmental quality index, in relation to the forest stability determined by the forest structure analysis, presents several advantages:

- it is surveyed rapidly;
- it can be objectively determined;
- it is separated from considerations based on a single experiment or the analysis of the data surveyor;
- it enables one to obtain complete and easy information for land management use.

Guide value Sdm%	% trees 4+5 layers
10	95
20	87
30	78
40	69
50	61
60	52
70	44
80	35
90	27

Figure 5. The obtained index is included in a theoretical range of 0 to 100.

The application field was limited to coniferous forests, both mono-specific and mixed with other types of conifers. The good results obtained encourage continuing experimental work by adapting this methodology to deciduous forests, to mixed deciduous-coniferous forests and to coppice forests as well.

KEYWORDS

Silviculture, ecosystem, risk analysis, environmental quality index, forest quantification of risk, Alps.

REFERENCES

[1] Giau B. "Considerazioni sul valore pubblico dei boschi e sulle sue procedure di stima - II forum nazionale", " I beni pubblici e privati: tendenze evolutive e criteri di valutazione ", C.N.E.L., Roma (1994).

[2] Leibundgut H. "Die Waldpflege", Paul Haupt, Berne, CH (1984).

[3] Mayer H., Ott E. "Gebirgswaldbau schutzwaldpflege", Gustav Fischer, ST, (1991).

[4] Quaglino A. "Analisi ecologico-selvicolturale del bosco ai fini della pianificazione territoriale", GEAM, **1-2**, 27 -31 (1993).

[5] Susmel L. "Normalizzazione delle foreste alpine", Liviana Editrice, Padova (1980).

[6] Viola F. "Considerazioni ecologiche su alcuni biotopi e cenosi forestali in Comelico", *Atti dell'istituto di ecologia e selvicoltura*, Università degli studi di Padova, **5**, **8**, 257-292 (1988).

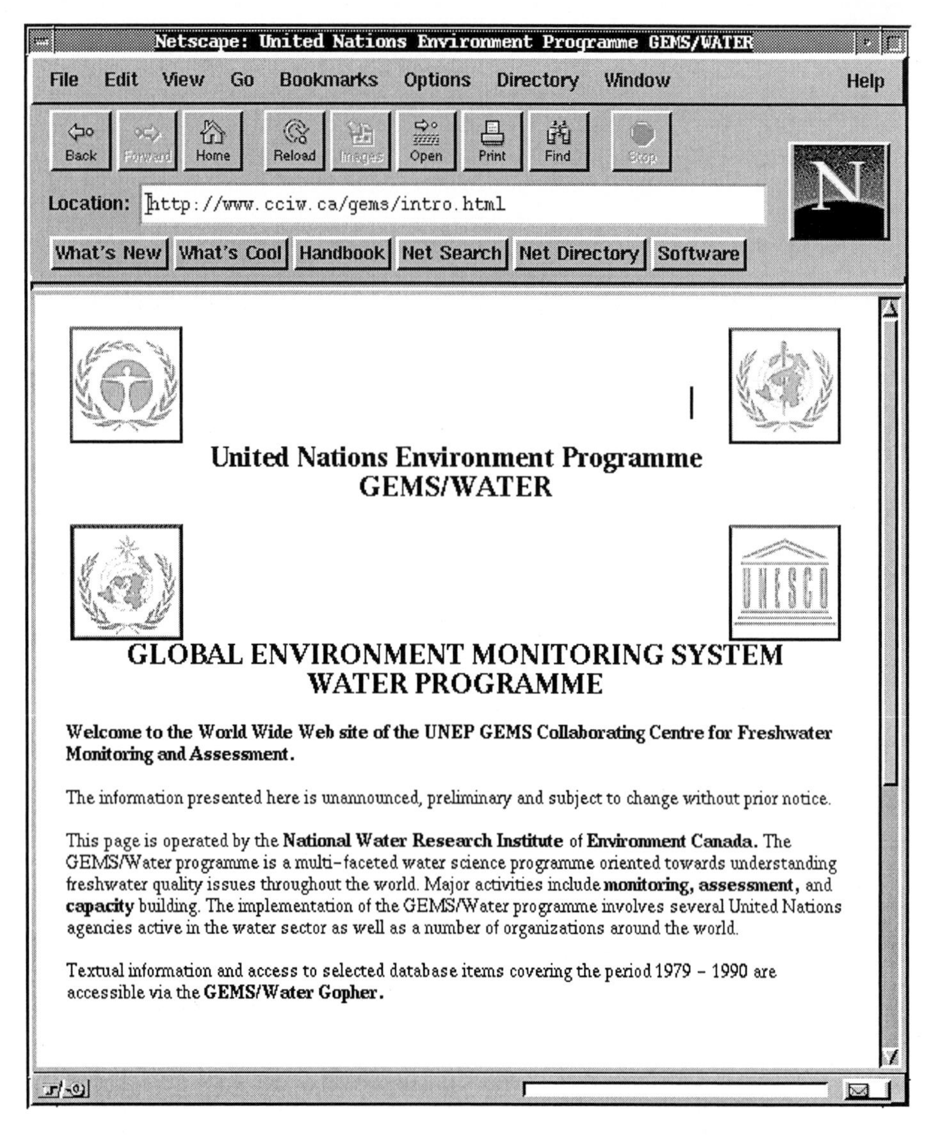

On-line information systems and databases are now available on the Internet. Among others, the Canada Centre for Inland Waters (CCIW) is one of the world's leading centres for water research, generating environmental information and knowledge about ecosystems. These servers host information from the UNEP GEMS Collaborating Centre for Freshwater Monitoring and Assessment Ecological Monitoring and Assessment Network, the Annotated Digital Atlas of Global Water Quality and an electronic document presenting information and findings on 82 major watersheds of the world that selectively summarizes data contributed by countries participating in the GEMS/WATER Programme over the period 1976 - 1990. The servers can be consultated on Internet at location : http://www.cciw.ca

Chapter 5

Accessing Geoenvironmental Data From On-line to Expert Systems

Large Database Access and Usage in the Earth and Space Sciences through Computer Networks

James R. THIEMAN

NASA / Goddard Space Flight Center, National Space Science Data Center, Code 633.2, Greenbelt, MD 20771, USA

Abstract

The development of the worldwide computer network has fostered the creation of and access to many data information systems and databases. These offer the capability for users to identify, access, use, and intercompare data from remote locations. It is not always easy to find the best sources of information for a particular problem. Directories for this purpose have existed for some time, and new methods are being developed for network information location, especially for systems such as the World-Wide Web and other network tools. It is still important, however, that there be consistency and standard approaches to the information content for effective information retrieval.

Accès à Travers les Réseaux et Utilisation des Bases de Données en Sciences de la Terre et de l'Espace

RÉSUMÉ

Le développement du réseau mondial informatique a favorisé la création et l'accès à de nombreux systèmes d'information et de banques de données. Ces dernières donnent l'opportunité aux usagers d'identifier, d'accéder, d'utiliser et de comparer entre elles les données venant de contrées lointaines. Il n'est pas toujours aisé de trouver les meilleures sources d'informations concernant un problème particulier. Pour ce faire, des annuaires ont été développés pendant quelque temps mais de nouvelles méthodes d'accès sont actuellement à l'étude pour localiser l'information pertinente sur un réseau, et en particulier sur le World-Wide Web. Il est cependant important qu'aujourd'hui il puisse y avoir une cohérence et des approches standards quant au contenu de l'information pour une exploitation performante de l'information.

1 INTRODUCTION

Locating needed data and information among and within the many large databases which now exist online is a formidable task. Data directories have existed for a long time to help find useful data in a particular domain. In particular, an International Directory Network (IDN) [1] was established in the Earth and space sciences for this purpose. This system uses keyword searching techniques to access databases where the information is stored in relational database management software. It also provides automated connections to other information systems which have detailed information about the data as well as the data themselves.

Other methods such as Gopher, Wide Area Information Servers (WAIS), World-Wide Web (WWW), MOSAIC, etc., are available which allow free text searching of textual databases. They offer additional capabilities for visualization and hypertext linking which help to get the user to the data. These tools were made to be used across the entire range of information available over the network, both bibliographic and data-oriented, so it is sometimes difficult to narrow the search down to a topic, such as specific interests in the Earth and space sciences.
A hybrid approach which uses both controlled keyword and free text searching is sometimes necessary. The variety of tools available can be daunting to a user. Which will serve the need best? *How many types of systems should a user have to learn to access and use to assure a comprehensive search?* Ideally an untrained user should be able to use any of the systems from the start in a simple and common approach.

Searches often lead to information systems which have existed for a long time and offer very specific information and capabilities for accessing data. Testing is being done to make existing information systems easier to use in conjunction with each other.

One technique, context passing, allows a file of information to be passed to another information system as the user is being transferred to that system. This information file can then be employed to aid the user in searching for useful information.

Remote querying of information systems as well as the return and comparison of results is being tested by the EOS Data and Information System (EOSDIS).

Incorporation of standard software packages to achieve interoperability through uniformity is another approach. Still other techniques are being developed by several groups to provide an interoperable environment within particular disciplines or projects.

The nature of the network-accessible information systems will be highlighted. The progress of making these systems easier to use and an overview of what might be expected for the future will be discussed.

2 ACCESSING DATA

New satellites and the inevitable expansion of the research being done has caused the amount of available data to increase exponentially. Some of the individual databases, especially those associated with satellite imagery, are growing extremely large. Thus, it is becoming impossible to centralize the archives of all related data in a single location. The growth of computer networks has made this an unnecessary task, however.

A virtual archive of related databases can be created through the networks even though the individual databases are widely scattered. Not many of these coordinated virtual archives have been created, though many individual databases have been made available through the networks. The user must try to find what is needed among this ever-growing conglomeration of information and data.

Several solutions to finding useful data were popular in the past. One could collect paper catalogs of data holdings from the various major archives. Phone calls among peers or known experts in a particular field was also an effective method. There was also the approach of writing letters to a variety of places expected to have useful data. The networks have provided several additional efficient methods. Letter writing is usually done by electronic mail now. There are a number of online information systems which were built specially to give information about the locations of specific types of data and metadata. Finally there are the more generic tools which are freely available and can be used throughout the networks for general searches.

The use of networks for data and information search and transfer is growing quickly. Technical advances have improved the bandwidth available at low cost to the general user. Also, network tools are freely available which make network searching as well as putting one's own data and information on the network an easy task.

For example, *File Transfer Protocol* (FTP) is the most popular method of data and information transfer. FTP servers containing data and information are pervasive, but it is not always an easy task to determine what servers have the needed data or information. Many use the "Archie" software to perform free text searches through textual descriptions about the FTP servers which have been identified.

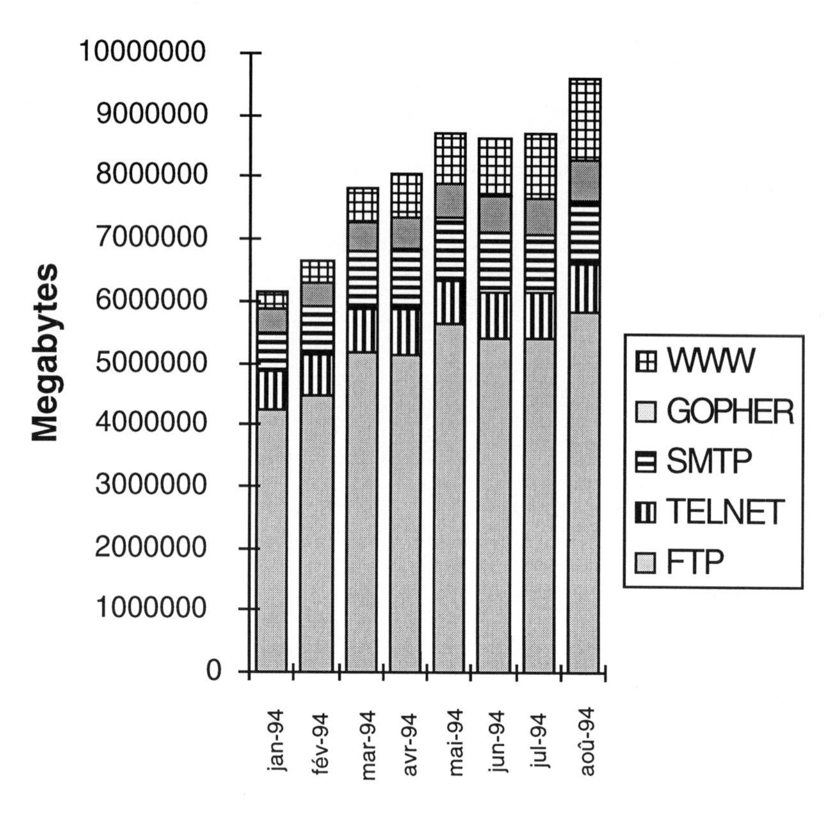

Figure 1. Statistics of the use of the National Science Foundation network (NSFNET).

The "gopher" system has also had great popularity in recent years. Gopher servers provide hierarchical menus to reach needed information. These are interconnected to each other through multiple links and ultimately lead to various services such as data access or image display. Again, there is a problem of knowing which routes to take to reach particular data or information needed.

For this reason the "Veronica" software was developed to allow free text searching through information about the gopher servers registered throughout the network. The information on where to go can be determined quickly if the gopher is registered properly. The Veronica system, however, does not allow searching of the contents of the data or information that one ultimately reaches. For this reason the *Wide Area Information Servers* (WAIS) have gained in popularity.

The WAIS system is composed of a number of servers on which textual documents are contained. All of the documents on each server have been processed so that all of the words in the documents are indexed. The user can then make free text searches matching query words against the indexes of selected servers. Documents are given a numeric ranking based on the abundance, proximity, location, etc. of query words. The user then gets a listing of the applicable documents in ranked order.

This service provides more complete searching capability, but is still somewhat limited to mainly text capabilities.

Most recently, the World-Wide Web (WWW) and the Mosaic interface for using it have become available and have grown phenomenally in popularity. The interface offers "pages" having both text and graphics in a pleasing format which is easy to create.

In addition, the text can be highlighted to indicate a hyperlink which will automatically connect the user to another page on the network upon clicking a mouse on the highlighted text. This in addition to the fact that gateways to FTP, gopher, and WAIS are readily available make the WWW approach comprehensive to needs. There is still, however, the problem of finding which pages have the information needed by the user.

Many groups have created search tools for the WWW. These are often called "spiders" since they crawl around on the Web keeping a textual record of what is found. The spider services then offer the capability of searching their information databases usually through a free text query. Though these services are very useful, they are still prone to missing important items for several reasons. The methods used by these programs to traverse the Web must handle a very complex interconnection setup. The programs are usually offered as freeware to all users. This makes them readily available to all, but it also means that there is no support for keeping a complete, error-free, and continuously supported software system. Nonetheless, the WWW has proven to be a very useful system and this is clear from the growth in its usage.

Figure 1 shows some recent statistics of the use of the National Science Foundation network (NSFNET). The statistics indicate the number of megabytes which have flowed through the network using the protocols associated with the various network tools for the months of January through August of 1994. Five examples are given though many more are available. These represent general FTP statistics, gopher usage, Simple Mail Tranfer Protocol (SMTP) indicative of one type of email flow, and WWW usage. FTP is clearly the most popular method of transferring information on the network at present. FTP usage, gopher, and electronic mail usage have remained about the same or grown slightly during the eight month period. WWW and the Mosaic interface, however, have only been recently introduced, but are showing phenomenal growth. In the eight month period, the number of bytes being transferred using WWW has tripled to the point that it is now second only to FTP in this set of examples. It is for this reason that some call the WWW the "killer application" since there is worry that it will grow to the point of choking the network with an overwhelming volume of information flow.

Despite the many useful aspects of the WWW it is still not completely satisfactory for finding useful data and information. Some experts in information retrieval, such as librarians, use the WWW, but prefer the more established information finding systems. They know that many services, such as the commercial service Dialog, support searches by putting information in a standard format and creating software that makes good use of the format. The commercial services have also existed for a long time and are committed to maintenance of the software and the continuation of service into the future.

This is not to say that WWW is not committed to serving on into the future, but there is not an effort to put information into a consistent format that makes it easier to assure that a search will come up with consistent, comprehensive results. The software is also offered freely to everyone and therefore should not be expected to have a high level of support for fixing errors.

3 THE INTERNATIONAL DIRECTORY NETWORK

There are many non-commercial services as well which offer the capability of searching information which has been placed in a defined format. In the Earth and space sciences the International Directory Network (IDN) was set up specifically for this purpose. Through the cooperation of member agencies of the international Committee on Earth Observation Satellites (CEOS) a number of computer network sites were established which contained directory software designed to provide information for the community about available Earth and space science data sets. The sites gather information in their area about available data sets and write descriptions about the data sets in a standard format called the Directory Interchange Format (DIF) [2]. Anyone interested in advertising the availability of a data set can submit a description of their data in this format to an IDN site. A Directory Interchange Format Manual can be obtained from the author.

The sites have all agreed to exchange this information on a periodic basis. Specific sites do this through the computer networks using automated exchange programs. Three sites, the American coordinating node (the Global Change Master Directory) at Goddard Space Flight Center in Greenbelt, Maryland, the Asian coordinating node in Hatoyama, Japan, and the European coordinating node in Frascati, Italy have agreed to collect all DIF information in their databases and make it freely available to users accessing their directories through networks or dial-in lines. Other sites are more specialized, such as the NASA Master Directory, also at Goddard Space Flight Center which emphasizes NASA-funded data information with special emphasis on the space sciences. Information about accessing or contacting most of the nodes has been compiled in a brochure entitled "The International Directory Network" [3] which is available from the author.

4 INTEROPERABILITY

The unifying factor of the Directory Interchange Format provides a common "look and feel" to the information available through the IDN. On the World-Wide Web, standard interfaces such as the Mosaic interface provide this common "look and feel" as well. The commonality makes it easier for the user to access differing information systems yet know how to operate in each. The ability to use different information systems in a similar way is a measure of the interoperability of these systems. Finding or displaying information in a common way is only half the battle, however. Users must also be able to quickly acquire, understand, and use the data. To make these types of services "interoperable" so that users can quickly understand and use the services is a more difficult task.

The National Space Science Data Center has been working on this problem in various ways. Within the NSSDC many user access, data information, and data distribution services are offered using computer systems that have been built over a period of many years. There are data available in astronomy, space physics, planetary science, solar physics, life science, and microgravity areas and these are distributed by magnetic tapes, printed materials, CD-ROMs, and direct network connections. Other services offered are online newsletters, bulletin boards, and bibliographic information on relevant standards and technologies. NSSDC is presently engaged in a program to make it easier for the user community to access and use these systems in an interoperable manner. Thus, the NASA Master Directory and underlying more detailed information systems are being integrated to make an overall interoperable NSSDC system. We invite users to try the NSSDC information systems at any time as they evolve and provide us with the feedback we need to improve the service.

The Interoperable Systems Office within NSSDC has responsibility for this program. NSSDC shares interoperability techniques with other NASA organizations as well as other federal organizations and world-wide groups. We are always interested in the techniques they have developed as well. Periodic interoperability workshops have been held to encourage this sharing of ideas. Contact the author for further information.

5 CONCLUSION

In conclusion it can be said that it is clear that access to data via network connection is increasing rapidly. Heavy usage is being made of the publicly available interoperable network search tools, but the text searching approach that some of these tools use does not always guarantee satisfactory results in a non-uniform search environment. The online directories and interoperable systems that take a structured, standardized approach to finding data and information are still very worthwhile. The National Space Science Data Center continues to work in this area in order to provide the community with improved tools for data access and management. The true test of these approaches is the amount of usage the community makes of such tools and the feedback acquired as a result.

KEYWORDS

World-Wide Web, internet, network, database, space science.

REFERENCES

[1] Thieman J. R., "The International Directory Network and Catalog Interoperability", *Remote Sensing Reviews*, 241-253 (1994).

[2] "Directory Interchange Format Manual, Version 4.1", *Publication #93-20, National Space Science Data Center*, NASA/GSFC (1993).

[3] "IDN World Guide", *Publication of the Committee on Earth Observation Satellites*, (1994).

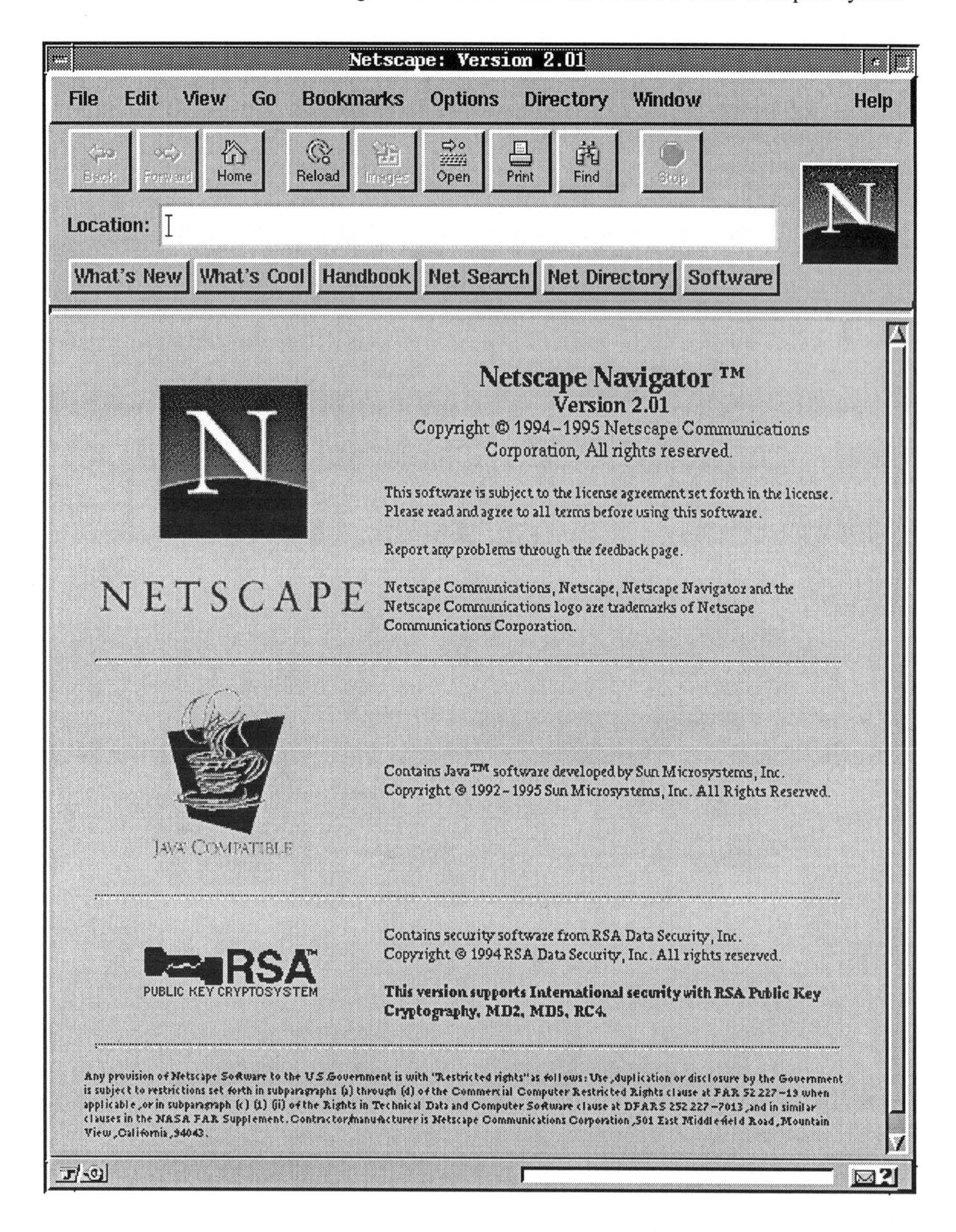

The World-Wide Web (WWW) is a distributed information system in which data, images and documents are located on different computers dispersed around the Internet. The access to this information is facilitated by users oriented programs (called browser). Above the Netscape front page, one of the most used browser around the Internet community (see page 223).

International Directory Network - Russian Branch

Vitaly A. NECHITAILENKO

Geophysical Center, Russian Academy of Sciences, Molodezhnaya Str. 3, Moscow 117296, RUSSIA

ABSTRACT

The prospects of the WDC system development, its increasingly important role in space and earth science and continuing efforts to perfect it in every way are, to a great extent, associated with the initiatives and participation of its creators in setting up globally distributed information-retrieval systems. The CEOS International Directory Network project is at the moment deeply and crucially involved in these activities. The network recently launched an extension to Russia through the Russian IDN Cooperating Node managed by the Geophysical Center of the Russian Academy of Sciences.

Three main fields of activity are considered: (i) improvements of the users-IDN information resources links, (ii) plans for creating a GEONET information-computer network by joint efforts of the RAS Department for Geology, Geophysics, Geochemistry and Mining, which is viewed as an IDN prospective component part, and (iii) improvement of the on-line information system GOLDIS, Version 2, which is now a division of the IDN catering to Russian users of IDN resources.

Réseau International des Répertoires en Russie

RÉSUMÉ

Les perspectives du développement du Système WDC, l'importance croissante de son rôle en matière de sciences de l'espace et de la terre ainsi que les efforts continus pour son perfectionnement, sont en grande partie associés aux initiatives et à la participation de ses créateurs à mettre au point des systèmes largement distribués de recherche documentaire. Le projet du Réseau de l'Annuaire International CEOS est actuellement très fortement impliqué dans ces activités. Le réseau a lancé récemment une extension à la Russie par le "Russian IDN Cooperating Node" géré par le Centre Géophysique de l'Académie Russe des Sciences.

On analyse trois champs principaux d'activité: - l'amélioration des liens en ressources d'information pour les utilisateurs IDN, - les projets pour créer un Réseau d'Information Informatique GEONET au moyen d'efforts conjoints du Département RAS de Géologie, de Géophysique, de Géochimie et des Mines, qui est considéré comme un élément d'avenir IDN, et - l'amélioration du système on-line GOLDIS, version 2, qui est actuellement une division de distribution des services IDN pour les usagers russes.

1 Communications with and within Russia Related to Planetary Geophysics

Most important in this area are initiatives of NASA, ESA and other agencies in expanding Internet to Russia. Russian Space Science Internet (RSSI) managed by Space Research Institute (IKI), Russian Academy of Sciences provides important opportunities to both users and developers of Russian scientific information systems.

The Geophysical Center participates in these activities by installing a dedicated link to the RSSI node at IKI and managing the WDCB domain of Internet.

The WDCB domain includes Sun-3 based Local Area Network of Ethernet type (WDCBnet). WDCBnet serves WDCB and EDNES branch users and is the host for GOLDIS, Version 2. At the moment WDCB net is connected via dedicated links to RSSI (14.4 Kbaud) and to IASNET (9.6 Kbaud) (Fig. 1).

The Institute for Automated Systems (IAS) under agreement with Geophysical Center supports transparent access to WDCBnet from IASNET public telephone concentrators installed in Moscow and several big cities in Russia.

2 DGGGMS Information Network (GEONET)

In June 1993 the Department of Geology, Geophysics, Geochemistry and Mining Sciences of the Russian Academy of Sciences launched an initiative establishing the DGGGMS non-commercial computer-based information network to ensure access of both Russian and foreign users to information of scientific institutions of the Department as well as to international information resources in Earth sciences.

GEONET will include not only academic institutes but institutions of other agencies as well (e.g. Roskomnedra). It is planned that the network will provide access to most important national and international information systems.

It is further planned to obtain the necessary level of standardization of hard- and software used at the network as well as standardization of model description and data and metadata formats, using recommendations of the WDC System and the CEOS Working Group on Data.

Figure 2 shows a structure of the DGGGMS Information Network with main links between institutes participating in the project and in international networks.

One of the first steps of the GEONET project was to create the list of information resources handled by the participating institutes. The list contains over 100 items, each representing factographic or documentary data bases of scientific interest. Unfortunately an essential part of these data are not in computer readable form. DGGGMS now seeks financial support from national and international foundations to digitize the most interesting data and make them easily accessible to the national and international scientific community.

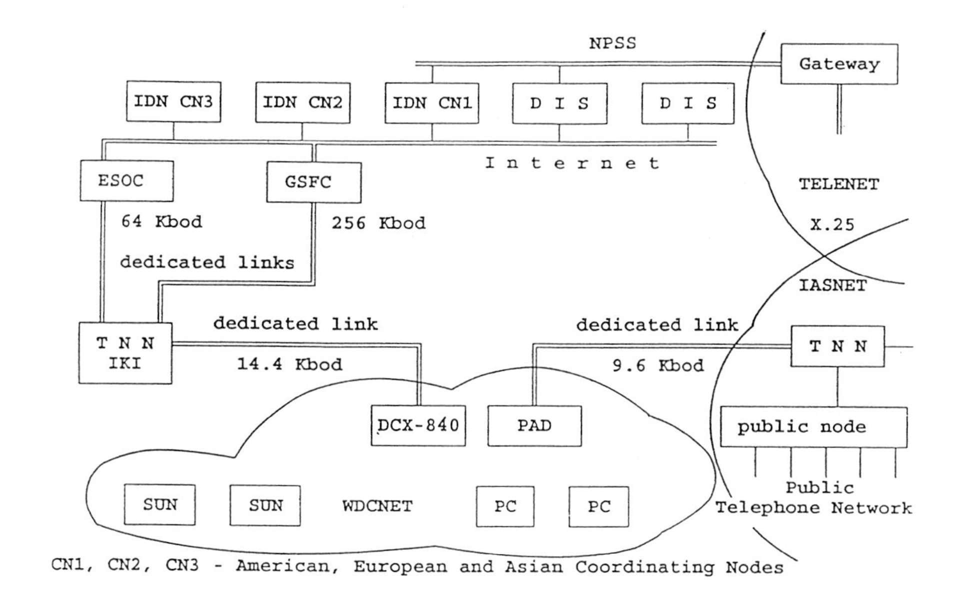

Figure 1. International Directory Network.

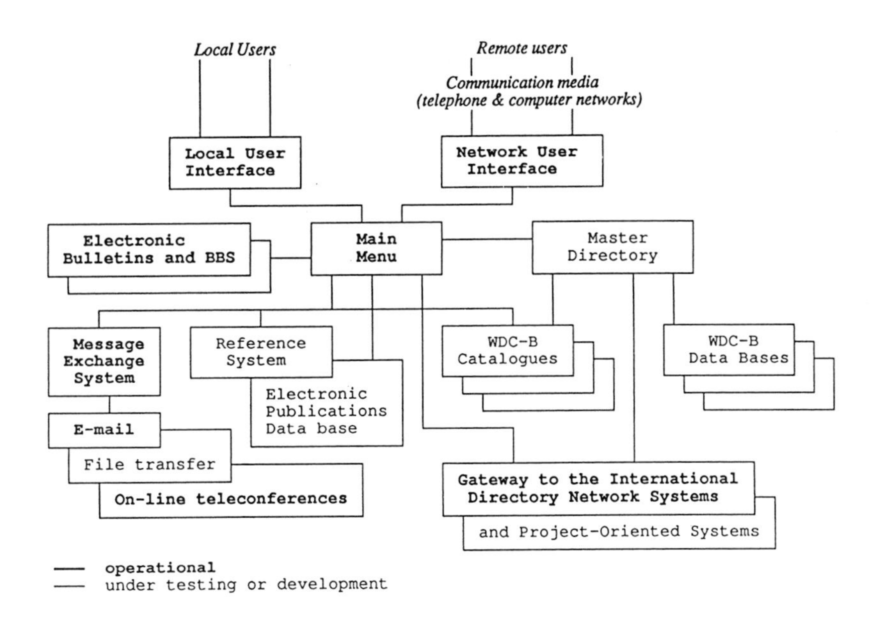

Figure 3. GOLDIS - Geophysics On-Line Data & Information System (Version 2).

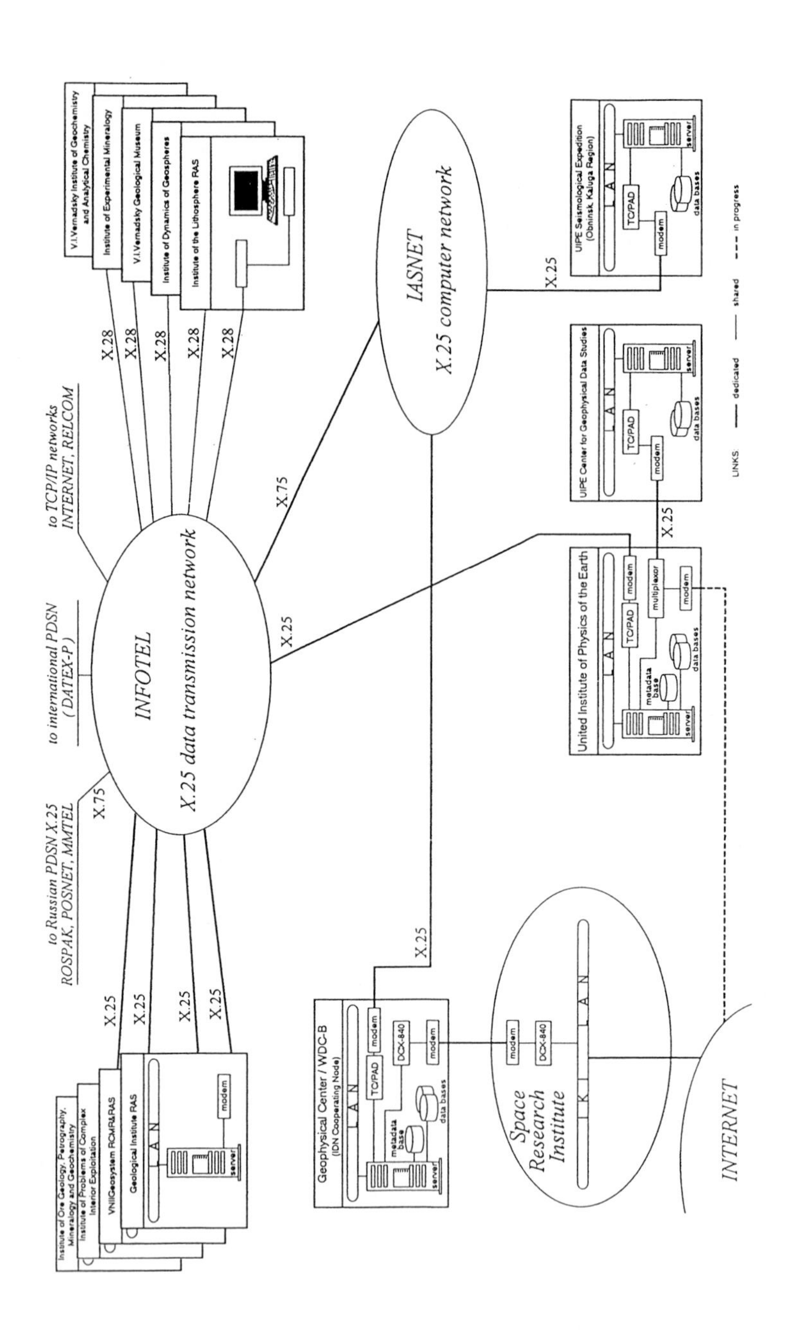

Figure 2. Structure of the DGGGMS RAS Information Network.

3 GOLDIS : THE GEOPHYSICS ON-LINE DATA AND INFORMATION SYSTEM

In the framework of the GEONET project GOLDIS is considered as an upper level information system for managing centralized metadata resources and as a gateway between GEONET users and IDN. GOLDIS is also a component part of the Planetary Geophysics Data and Information System (PGDIS) which is being developed at the Geophysical Center of the Russian Academy of Sciences to bring together numerous on-line and off-line information systems which are in use now at the GC and the WDCB.

The general plan for the development of the Geophysics On-Line Data & Information System (GOLDIS) is to create a system to provide geophysicists with:

- Information on international geophysical programs.
- Directories of international and WDC-B datasets and collections of data prepared under international projects.
- A directory of scientists (working bodies, etc.)
- Transparent access to the Cooperating and other nodes of the International Directory Network (IDN).

Version 2 of GOLDIS is based on Sun-3 platform under UNIX Operating System. This hardware and software environment provides developers with important advantages compared to the previous PC based version, i.e. multiple access to the system for both local and remote users, high flexibility to user's terminal settings, etc.

The main parts of GOLDIS are shown in Figure 3. Some of them are still being developed and tested; the following are operational at the moment :

- Electronic Bulletins. Current version includes National Geophysical Committee Bulletin (in Russian, using 866 ASCII coding table), STEP International (which is submitted by the USSCO) and RUSCO Bulletin, submitted by the Russian STEP Coordination Office.
- Remote Systems & Networks. Current version provides users with 'transparent' access to remote (domestic and international) systems and OPEN ENTRY to X.25 networks. With implementation of full connection to Internet, OPEN ENTRY to Internet will be accessible as well.
  To use these services, user must be registered as Net User (see below). This branch works as a gateway to remote networks and systems, which provides users with automatic log-on to systems of International Directory Network (IDN) and IASNET and Internet based systems.

GOLDIS's RSN system as well as NASA MD or ESA PID provides users with physical and logical access to remote systems and automatic log-on procedures for open systems. It is the user's responsibility to obtain a personal username and password if remote system requests them.

OPEN ENTRY. The current option provides users with a possibility to connect directly from GOLDIS to any X.25 (and Internet in the nearest future) network address (domestic or international) typed by user. System checks correctness of syntax and Decimal Network Identification Code (DNIC).

- Registering system. The current version supports a self-registering procedure, possibility to type and edit user's card, change password and view list of registered users. Inclusion in the User Net list must be authorized by GOLDIS administrator.

KEYWORDS

Databases, network, on-line information system, geophysics, scientific data policy, internet.

G.D.Q.-Expert: An Expert System for Assessing Geotechnical Data Quality Application to the Menard Pressuremeter Test

Lyes MATMATTE[a], Jean Louis FAVRE[a,1] and Reda M. BAKEER[b]

[a]*Ecole Centrale de Paris, Lab. MSS, CNRS URA 850, 92295 Chatenay Malabry Cedex, France*

[b]*Department of Civil and Environmental Engineering, Tulane University, New Orleans, Louisiana, USA*

ABSTRACT

This paper describes an expert system for assessing the quality of geotechnical tests. Data from any geotechnical test must be analysed and assessed for quality before use in any formal design. This task requires an extensive geotechnical experience and background which makes it ideal for expert system type software. We used an object-oriented methodology associated with production rules for representing knowledge. The system was developed using an object-oriented expert system shell, called Level 5, to run under the Microsoft Windows graphical environment.

G.D.Q-Expert: Un Système Expert pour Tester la Qualité des Données Géotechniques. Application aux Essais Pressiométriques de Ménard

RÉSUMÉ

Dans cet article, on présente un système expert qui a pour tâche le contrôle de la qualité des données géotechniques. En effet, les données obtenues à partir des essais géotechniques doivent être vérifiées et analysées avant toute utilisation. On a choisi une méthodologie orientée objet couplée à une technique de règles de production pour la conception du système et la représentation de la connaissance. Ainsi, on a défini les classes, leurs attributs et les procédures et règles qui les manipulent. Ce système est développé sous Level 5 Object et tourne sous Windows.

---

[1] To whom correspondence may be addressed

1 INTRODUCTION

A geotechnical investigation is done by means of drilling, sampling, in-situ and laboratory tests. This investigation provides qualitative and quantitative information regarding specific locations on the site where soil information, borings and/or tests are available. This common practice may only cover a small fraction of the area of the site under investigation, which introduces variable uncertainty in the soil characteristics. But this uncertainty is unavoidable from an investigation-cost point of view. Another important source of uncertainty is the bias in the measuring procedures.

Geotechnical tests involve examining a given soil with a specific device according to a standardized procedure. Response of a given soil to a particular test depends, in ideal execution conditions, on the soil type, as well as other factors including initial state of stress and test type. Unfortunately, the ideal execution conditions may not be well defined and may considerably vary according to the soil type. Data obtained from geotechnical tests have to be analysed and assessed for quality before use in any formal design in order to take into account the effect of the influencing procedural factors.

2 BASIC ASSUMPTIONS IN CHECKING QUALITY PROCESS

Geotechnical tests are divided into two main groups: laboratory tests and in situ tests. Laboratory tests can be divided into two main classes: identification tests performed on disturbed and "assumed" to be representative samples, and mechanical tests performed on what are "assumed" to be undisturbed samples. The basic theoretical assumption made for interpreting the results of a mechanical laboratory test is that the stress-strain field is homogeneous, or in other words, the stresses and strains can be measured at a particular point then assumed to be valid across the whole specimen. Interpretation of mechanical laboratory tests is done with the assumption that the disturbance during sampling, storing, preparing and testing samples has been highly reduced.

In situ tests are performed at the site either by pushing a testing device into the soil, or by pre-drilling a borehole before the introduction of the testing device. In either case, interpretation of the results is made assuming that a minimum amount of soil disturbance occurs during the test. In most cases, interpretation of in-situ tests is done by empirical methods developed and verified through experience.

Other than identification tests which involve simple, direct and systematic procedures, results of mechanical laboratory and in situ tests are highly subjective and depend on the quality of the test performed.

3 DATA QUALITY ANALYSIS

Checking the quality of any geotechnical test data involves verifying the validity of the basic assumptions made for interpreting the test which produces the data. If, in the triaxial test, a localization of deformations on a specific plane occurs, the continuity assumption of the stress-strain field becomes invalid.

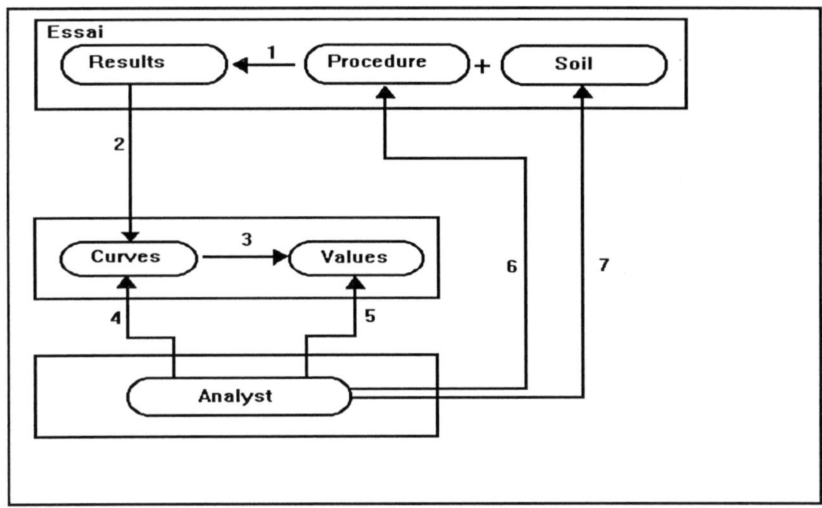

Figure 1. Quality assessment process for a geotechnical test.

Performing a geotechnical test is a sequence of different operations (Fig. 1): execution of the test according to a given procedure, gathering of the row data (1), correcting row data and tracing curves (2); from these curves, extracting values of geotechnical parameters (3). In order to assess the quality of the test results, the analyst checks if there is any noticeable deviation during the execution of the test (6) or on the curve (4) and verifies the coherence of the test results with the soil type (5) (7).

4 METHODOLOGY FOR ASSESSING THE QUALITY OF GEOTECHNICAL DATA

The adopted methodology for assessing the quality of geotechnical data is based on a two level strategy: the first level concerns the site investigation program test examination (checking data quality of each test and verifying coherence between all the tests performed at the site). The second level concerns the comparison between site tests and similar tests on the same soil type currently available in the database.

4.1 A one site verification - This level involve two parts

4.1.1 Particulate verification

This concern each individual test performed at a specific location which involves:

- Procedure verification regarding norms, including drilling or sampling methods and a test procedure.
- Calculated parameter verification, to ensure that they fall within the typical range of values for this type of soil.
- Checking the curve shape with respect to the test type and soil type.
- Obtained value coherence in one test.

4.1.2 General coherence verification

This part concerns the verification of the general coherence between different tests performed at the site by means of established correlations.

4.2 Site independent verification

The second level of data quality assessment concerns site independent comparison between parameters from tests performed at a given site and typical parameters obtained from the same type of tests performed on a similar type of soil at other sites available in the database.

5 APPLICATION TO THE MENARD PRESSUREMETER

There is a large array of in-situ tests currently in use in geotechnical investigations. Some of the tests are performed by pushing a test device directly into the ground while others require the drilling of a borehole before introducing the testing device. The selected test for use in the prototype expert system is the Menard pressuremeter test and it belongs to the latter group. The pressuremeter test is performed by expanding a cylindrical cell into a soil mass by increasing the internal pressure and measuring the corresponding volume change. The quality of a pressuremeter test is highly dependent on the quality of the boring and the quality of the test procedure.

The result of the test is a curve of pressure versus volume from which we can calculate three mechanical parameters: a pressuremeter deformation modulus E, a creep pressure Pf, and a limit pressure Pl.

These three parameters are related by statistically established correlations for different classes of soils depending on the soil nature, its consolidation state and its consistency. The ratio E/Pl for a normally consolidated clay, for example, is within the range of 8-15. The Pl for the same soil is within 0.5-3 MPa if the clay is soft, 3-8 MPa if the clay is plastic and 8-40 MPa if we have a stiff clay. The ratio Pl/Pf is about 1.7 for clay and between 1.7 and 2 for sand.

This knowledge allows us to check pressuremeter results and to detect any deviation. A one site verification can be made by using knowledge about the factors which could affect the quality of the pressuremeter test results:

- The boring operation is very important for the quality of the pressuremeter test and must be checked. The French standard NF P-94 110, has divided boring methods for pressuremeter use into four classes: recommended, tolerated, prohibited and misfit. We must take into account the class of the method used, but the effect of one tolerated method may vary from one soil type to another. Knowledge gathering on the subject is necessary.

- The choice of the type of pressuremeter is important for the precision of the data.

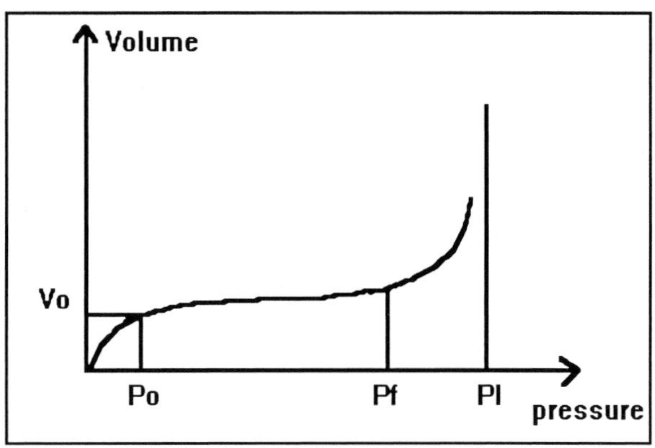

Figure 2. Pressuremeter curve and its typical values.

- E and Pl must be verified to be within typical range according to the soil type.
- The ratios E/Pl and Pl/Pf must be verified, as well as the value of Vo, which is a good indicator of the calibration of the hole.
- The obtained curve shape must be checked, as it can show us any deviations during the test execution.

6 METHODOLOGY AND SYSTEM DESIGN

To build any system in any specific area requires a good comprehension of the problem domain and the system tasks. This understanding should be formalized and explicitly expressed in order to facilitate knowledge exchange between analyst, domain experts, and end users.

Building an "intelligent" system implies representing knowledge. Knowledge representation consists, from an Artificial Intelligence point of view, of finding appropriate data structures to store and manipulate information [6]. Object-oriented methodology satisfies these requirements and allows us to ensure a good level of system stability, objects being the least volatile elements in system specification [5].

6.1 Object definition and object relations

A geotechnical investigation is done by means of borings. For the Menard pressuremeter test, a hole is drilled before performing the test; drilling the hole must respect some rules in order to have a good calibrated hole with the most intact faces. Two objects are defined to represent this knowledge: "the boring" object and "the test" object. The "boring" is a class object, boring *can be* a pressuremeter boring, a penetrometer boring..., which are specializations of "the boring" object. The generalization-specialization relation is represented by a half circle symbol in Figure 3. Any boring involve layers and water (if any); "the layer" object and "the water" object are parts of "the boring" object. In order to express the system tasks in the best way, each test object involves two parts: "results" object and "testing device" object ("probe" in this case).

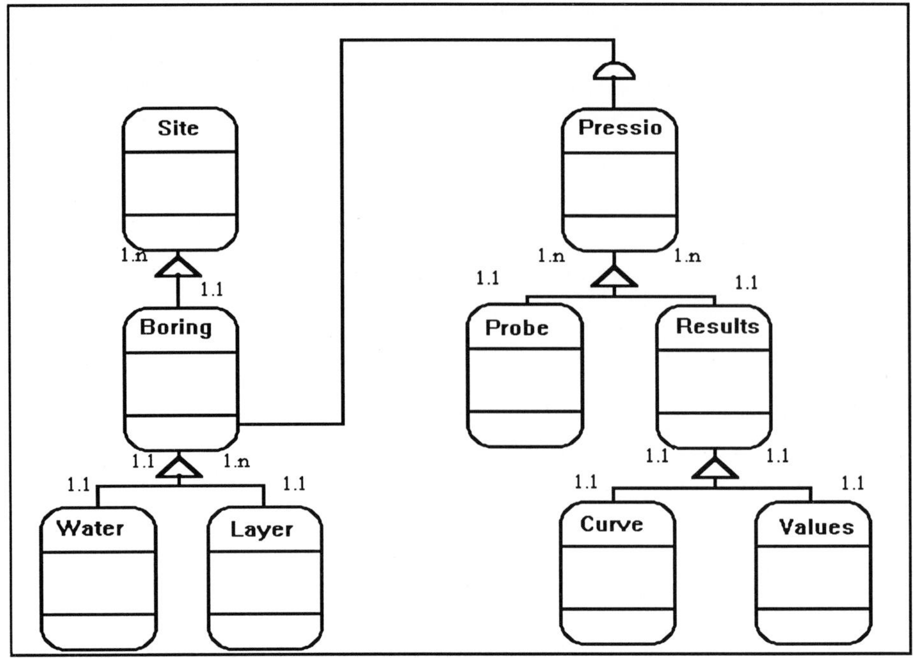

Figure 3. Object structure of the pressuremeter test.

The whole-parts relation is represented by a triangle symbol in Figure 3. This object structure allows separating the tasks of result checking from those of procedure checking. The test gives row data on which we make some corrections. From these corrected values we trace curves. These curves are analysed, interpreted and then mechanical parameters are extracted. Therefore, the "results" object will be decomposed into two parts: the "curve" object and the "values" object for the obtained values. Each of these objects has a specific task. The task of the "curve" object is to check the curve and to determine whether it will be admitted for interpretation. The task of the "values" object is to check the validity of parameters extracted from the curve.

6.2 Propagation of information

To judge the quality of one test, we have to judge the quality of the drilling operation, the quality of the test execution and the resulting data coherence with the soil type. This is a continuous process, in other words, the test can be declared defective at the boring level or at the test object level (Fig. 3). Inheritance from the boring object to test objects allows propagating information from the boring to the pressuremeter objects.

6.3 System development and programming language

The expert system was developed using an object-oriented expert system shell to run on IBM-type personal computers under the Microsoft Windows graphical environment. From a knowledge representation point of view, this shell is hybrid:

it combines the *class* concept with concepts from *frames* systems, particularly, *demons* and *rules*, procedural and declarative *facets*. Thus an object is defined by its class, its attributes, declarative and procedural facets attached to each of these attributes and demons and rules which manipulate attributes and their values. In order to clarify what we have said, we give the "values" object example:

Object "Values":
Attributes introduced by the end user:
Nom_Couche : layer name, the value of this attribute is imported from the layer object.
Etat_Consistance : layer state of consistency, the value of this attribute is imported from the layer object.
Etat_Consolidation : layer state of consolidation, the value of this attribute is imported from the layer object.

Attributes calculated or introduced by the end user:
Module_pressio : pressuremeter modulus value.
Pression_limite : limit pressure value.
Volume_initial : initial volume value before the starting of the test.

Calculated attributes:
Module-sur-limite : Value of the E/Pl ratio.
Fluage-sur-limite : Value of the Pl/Pf ratio

Attributes deduced from the knowledge base:
Couche_E : state of coherence of the pressuremeter modulus value with the soil type.
Couche_Pl : state of coherence of the limit pressure value with the soil type.
E_Pl : state of the E/Pl ratio according to the soil type.
Pl_Pf : state of the Pl/Pf ratio according to the soil type.
Etat_essai : state of the test after examination of obtained values.

These attributes are storage variables for different values of an attribute. Control or computing procedures are attached to these attributes in order to avoid introducing invalid values by the end user or to compute the value of an attribute from those of others. Production rules are also attached to these attributes so as to allow reasoning. Thus, a "When Changed" method is attached to the attribute Module_pressio which is fixed when the attribute value changes. This method prevents the end user from introducing a negative or zero value (the modulus takes positive values only), and sends a message to the end user each time the modulus value is non positive.

```
When Changed Values.Module_pressio :

    IF Values.Module_pressio > 0 THEN
        Values.Module := TRUE
        ELSE
            Message.variable text := "le module pressiométrique"
            ASK Message
        END
    END
```

Examination of the values of the initial volume Vo and the limit pressure can show us a deviation. A production rule called Demon is written for such a task.

```
DEMON 24

        IF Values.Volume_initial > 200
        AND Values.Pression_limite < 1
        THEN Values.Etat_Essai := defective
```

7 CONCLUSION

This expert system project is a response to a current need in the geotechnical engineering domain and follows a tendency in research to reinforce the quality assessment process in information systems. The GDQ-EXPERT system described above provides a tool for design engineers to assess the quality of test data and existing geotechnical databases.

KEYWORDS

Expert system, geotechnical data, quality assessment, property of soils, soil erosion, object oriented language, geotechnics, geomechanics.

RÉFÉRENCES

[1] Cassan M., "Les essais en mécanique des sols", Editions Eyrolles (1988).
[2] Favre J.L., "Milieu continu et milieu discontinu. Mesure statistique indirecte des paramètres rhéologiques et approche probabiliste de la sécurité", Thèse d'état, Paris VI (1980).
[3] Jézéquel J. et al., "Le pressiomètre Ménard, quelques problèmes de mise en oeuvre et leur influence sur les valeurs pressiométriques", Bulletin de liaison des LPC, n° 32 (1968).
[4] Matmatte L., "Système expert pour l'analyse des données géotechniques", Rapport interne, Ecole Centrale Paris (1993).
[5] Coad and Yourdon, "Object Oriented Analysis", Editions Prentice Hall (1991).
[6] Masini et al., "Les langages à objets", Editions InterEditions (1989).
[7] Spigolon S.J. et Bakeer R.M., "GEODREDG, Geotechnical factors in dredging, a prototype knowledge based expert system", Technical report submitted to US Army Corps of Engineers, Waterways Experiment Station, Vicksburg, MS, USA (1991).

THE USE OF G.I.S. IN PETROLEUM INDUSTRY FOR OIL EXPLORATION AND PRODUCTION

François BOUILLÉ

L.I.S.T., Boîte 1000, Université P.M.Curie, 4 Place Jussieu, 75252 Paris cedex 05, FRANCE

ABSTRACT

Processing the relevant knowledge from oil exploration and production mainly concerns the use of a Geographic Information System. At the present time, a GIS must be completely object-oriented, must accept fuzziness, present all capabilities of a very large database and knowledge base, including an expert system, a discrete simulation system, a neural system, accepting genetic algorithms. That is exactly what the HBDS model ensures, as the kernel of an integrated GIS, particularly well-suited to oil data processing; among main themes, it deals with all kinds of pipe networks, as well as data relevant to geophysics, diagraphies, or technological equipment such as off-shore drilling platforms.

We emphasize the need of a tool allowing simultaneously to handle facts, rules, processes and neurons in a complex scientific field such as the oil industry.

Utilisation des S.I.G. dans l'Industrie Pétrolière pour l'Exploitation et la Production de Pétrole

RÉSUMÉ

Le traitement de la connaissance dans l'exploration et de la production pétrolière concerne principalement l'emploi d'un Système d'lnformation Géographique. Actuellement, un SIG doit être intégralement orienté vers l'objet, doit accepter le flou, présenter toutes les possibilités d'une très grande banque de données et banque de connaissances, incluant un système expert, un système de simulation discrète, un système neuronal, et acceptant des algorithmes génétiques. C'est ce que propose le modèle HBDS, en tant que noyau d'un SIG intégré, particulièrement bien adapté aux traitements pétroliers; parmi les thèmes principaux, il traite tous les types de réseaux, de même que la géophysique, les diagraphies, ou les équipements de haute technologie comme les plates-formes offshore.

Nous insistons sur la nécessité d'un outil permettant en même temps de traiter faits, règles, processus et neurones dans un champ d'application scientifique aussi complexe que celui de l'industrie pétrolière.

1 INTRODUCTION

The petroleum industry is largely diversified, and managing the produced data first requires knowledge of many rather different topics, but always centred on two purposes: exploration and production. It needs various studies such as geophysics and diagraphies, together with many other topics dealing with the 2-D or 3-D (and even 4-D) coordinates; in fact, they merely deal with the geography, with metrical properties as well as topological relationships. That is why the GIS (Geographic Information Systems) are so important in oil exploration and production activities.

2 THE MAIN TRENDS IN THE GIS

The concept of GIS, appeared in the 1970s but was really accepted fifteen years later. Many tools have been developed during the last decade, generally based upon concepts designed a long time ago, all stemming from "brainware" (Kitagawa, 1974), with implications on software and hardware. New ways have been introduced for considering geographical problems, new ways of programming them have been developed, new integrated systems have been realized, and new devices have been built, revolutionizing the way of using a computer; and finally, new built-in geographical components have been prepared, introducing the concept of re-usability of geographical types (Bouillé, 1994d).

2.1 New concepts in GIS

These concepts are essential in GIS, where they appeared in 1976. If everybody agrees with multiple inheritance, the use of methods, possible communication through messages, etc..., some other ideas introduced eighteen years ago have not been completely understood by all: -persistency, -graphical representation of types, -fuzziness, ...

The most important aspect concerns *persistency*: any geographical type must be created, updated, used as if it was resident in a core memory, and it must be automatically stored in a very large hidden database, without any human action; the user will find this type the next time he is connected, as if it had never left the core memory. This point is closely linked to O.-O. databases. Working with geographical types requires persistency, because it provides the most simplicity for the user and the greatest efficiency for the system.

Some authors trying to describe a phenomenon with a programming language speak about "*data representation*". It is a very poor representation; for a complex phenomenon requiring 3000 abstract data types, 1500 pages are sometimes necessary.

Associating graphical representation with types, independent of any programming language, is a powerful tool for communication between human beings, even if they come from several different countries, because it is a common language; it is a way too of avoiding classical programming in most basic operations, and it explains the solution to a problem, no matter how complex, by a path inside the graphical knowledge structure. The next step, more recent, is the use of 3-D graphical representation with images and texts appearing inside this structure, by clicking on any component, using the new vizualization devices. Most problems can be solved this way, combined with other tools (expert system, neural models, etc....).

Largely developed in all fields of industry, particularly studied in Japan, *fuzziness* is present everywhere in geographical phenomena; by approximation, in coordinates for instance; by uncertainty, for instance concerning the possible superimposition or intersection of networks; by subjectivity because of the opinion that any geographer may express about the types he has captured and/or processed; by multiplicity, because of the existence of several data instead of one (where we expect only one), among which we cannot choose. Considering a nuclear factory, we know that the effective places of components do not correspond exactly to those mentioned on the plans; all information, even in such a matter, are fuzzy, combining the four kinds of fuzziness: approximation, uncertainty, subjectivity, multiplicity. Geography, like all activities, is deeply involved in mastering fuzziness, even if some geographers have not realized that the time has come to deal with this.

Though closely linked to the preceding topics, the quality is rather different, and is limited or surrounded by fuzziness; geographers must evaluate the quality of the data captured and stored; cartographic activity for instance may be compared to a complete path inside a car factory; the concept of a flexible workshop may be applied, as well as methods such as the KANBAN, nevertheless, with a difference: the geographer uses the existing world and does not build it in a factory; but the products he builds, such as maps, can be controlled by the same efficient methods. Data quality evaluation is only just starting in GIS but will very soon find the same place as in all other industries.

One of the most important features of the O.-O. approach is self-extensibility; new types may be defined owing to the basic types, by elementary schemes; it is then very easy to build prototypes which can be used as a toy ("lego" for instance) for making copies, for assembling them, combining each prototype partially or entirely; using prototypes such as "graph", "network", "isoline", "polyhedron", etc... immediately provides schemes composing any geographical structure, whatever the theme (electricity, sewer, metroliner, geology, remote-sensing, etc...). It is the first step to re-usability. Designing a geographical structure no longer requires several months but only two or three days. Compared to the work needed for building a relational model, such a method appears nearly instantaneous.

2.2 GIS: the features of the system

O.-O. databases are unlimited, shareable, protected, they can be built for storing persistent abstract data types, and they can store simultaneously all kinds of information: digital, textual, graphical, image, sound, etc.... The distinction between metrical databases and topological databases appears particularly out-of-date: using an O.-O. structure shows that a class carries both its topology and its metrics, and the term "database" looks obsolete. There are not several databases in an O.-O. persistent context; there is only the automated storage of all persistent types, about which we need not know what device, what file, what table, etc... are used for ensuring this storage. An O.-O. database supports all aspects of fuzziness with no problem.

An Expert System must be integrated in a GIS kernel, thus requiring a knowledge base (Bouillé, 1984, 1988). While a database is supposed to store "data", a knowledge base stores the knowledge made of facts and rules; it does not imply an opposition between both terms; it might be considered as an extension, a knowledge base simply using a database for ensuring its storage; in an O.-O. context, a database stores abstract data types, whatever they semantically represent; if a class is a class of facts or a class of rules, it is stored likewise; but the main difference does not come from the knowledge base itself; taken independently, it is very easily implemented; the difference comes from the way the rules are used, launched, chained, interpreted and first of all compiled; it is the task of the inference engine, making this knowledge base "intelligent". The term deductive database (or deductive knowledge base) is more adequate. As the reader may see, the "intelligence", is not in the "base", itself but in the engine connected to it. As a typical example, the toponymy must be stored with each geographical object; but the way this toponymy is drawn on an output map follows some complex rules which must be also stored in the same system. At run-time, the inference engine will generate the most well-suited coordinates for each toponymical element.

The Expert System of a GIS must not be based on the stupid tree-structure, must not use the obsolete pattern matching and back-tracking, and must work on fuzzy facts and rules. In the near future, most GIS will include an O.-O. expert system, working on a partitioned knowledge base, with classes of facts and rules, the classes of rules being nothing but particular facts, and the expert system using several cooperating inference engines. For more than ten years, we can build such expert systems whose speed does not depend on the number of facts and rules stored, because they no longer use pattern matching. They use the pre-existing relationships recognized at the structuring step. Such methods have been successfully applied since 1987 for toponymy repositioning on maps (Tite 89).

A GIS must be able to allow decision-makers to test some hypotheses; a geographic dynamic model requires thousands of processes, considered as persistent objects implicitly stored in the knowledge base of the GIS. This tool is one of the best ways for testing phenomena in which components move in interaction: for instance, car flows in an urban area, schedules of transportation networks, pollution propagation in pipes, plant disease expansion in a forest, etc.... But sometimes, this discrete simulation must be connected to continuous processes, for instance when the trajectories of some mobile components must be displayed on a digital/topological terrain model; or when you build a very large geological model... Simulation models will be one of the main tools in environmental studies during the next decade.

Complementary tools are provided by the connection approach. Classical neural models are essentially based upon matrices, on which various filters are applied. On the other hand, O.-O. neurons are objects of classes of neurons, and their behavior is implicitly launched when an external action is performed on one of their attributes or links previously specified as a launcher. The behavior is able to perform any task on any other persistent type(s). Of course if these other types are launchers of other neurons, the reader may imagine the propagation which can be built very easily. While the matrices of the classical connection approach are not extensible, the persistent O.-O. neurons are unlimited because of the hidden knowledge base, and new ones can be generated at any time. The behavior is algorithmic or heuristic, as complex as required. Moreover, the neural engine allows automated learning.

In the area of GIS, neural models are particularly convenient for very large models dealing for instance with pollution propagation, resource distribution, information transmission, and more generally wherever an automated learning is needed. Everything modeled with processes can be done with neurons; but both tools are not strictly equivalent. Processes are more convenient when times have to be managed, whereas neurons are better suited to the automated searching of one solution (not simply solution...) (see Bouillé, 1993b).

The new (not so new...) fashion is constraint programming. This consists of expressing the constraints between several types; a very simple example will present the problem. For instance, in a first thermodynamic simplified approach, P, V, and T are linked by a simple relation which is $P*V/T$ = constant. Try to explain that to your database ! Give P, no problem; give V, no problem; but now T is fixed... You could have first given T, etc.... If you define rules with an expert system, you need several rules and it will be neither elegant nor efficient. To write a procedure and call it every time will go back to prehistoric times. You need just express the formula as a constraint. Of course, constraints in an O.-O. context, are objects of classes of constraints managed by a special engine...

In the GIS area, many constraints can be expressed; to use again the example of the toponymy automated positioning, the use of constraints could largely improve the method presently based on rules (this being still very fast). The constraints are particularly interesting when dealing with superimpositions, intersections and interconnections of networks, mainly in urban areas.

2.3 The devices for GIS: a revolution in use

The era of multimedia has changed most uses, and mainly minds; a simple example in this brief paragraph: considering a complex urban network, you have to deal with urban furniture; when pointing to an object of a class of furniture, as a telephone cabin, you may have the image of this cabin, or you may have some associated plans. In an environmental application, concerning the distribution of birds, when pointing to an object of a class of birds, you may hear its song, and simultaneously you may see its image, or for expressing the way it flies, you may see a brief movie.

Nowadays, cartographic robots, or exploring robots in hostile environments, are very useful and very numerous; thousands of robots work for us, generally on materialized networks, for instance, in oil exploration and production industry, for cleaning pipes; or for controlling the position of pipes under the sea; or for digging under the sea, welding pipes, burying pipelines, etc.... A GIS must be able to be connected to such applications.

Among other real-time operations, requiring high speed, a GIS can be used for cars wandering through all streets of a town and emitting radio signals which launch local transmitters, providing immediate and automated gas-meter readings without stopping cars and without entering houses. Many new applications like this will be launched in the next decade, including the remote controlled vehicles.

Two wonderful tools are modifying our way of looking at geographical models: stereoscopic TV, as it is currently used in Japan, for the public at large. Adapted to GIS applications, it completely transforms our sight of a terrain model, for instance in a flight simulator, or our impression of the town for an urbanist.

Virtual reality, rather different, works at another scale, possibly putting the geographer inside his model, providing moreover an interaction with this one..

3 BRIEF REMINDER OF O.-O. TOOLS

A model named HBDS was developed in 1976, together with a complete system, and many applications devoted specifically to GIS and oil exploration and production, both these areas being deeply interconnected (Bouillé, 1994a).

3.1 The "HBDS" model

HBDS is based on mathematical concepts: set theory, graph theory including hypergraph, theory of categories, fuzzy set theory. It uses abstract data types (a.d.t.), with six basic ones, and six extensions, respectively named:

- class - attribute of class - links between classes, composing the skeleton,
- object - attribute of object - links between objects, composing the realizations,
- hyperclass - hyperattribute -hyperlink, allowing group components,
- embryo, - prototype, - structure, allowing the design of new families of components.

We shall not detail these aspects which were published before in various papers.

3.2 The "HBDS" system kernel

The HBDS system is built on a multilayered architecture, composing the kernel, and successively providing:

- an indexed-sequential file system, supporting any level of indexes, the only limit being the available disk space,
- a virtual memory mechanism providing the system with the impress that all data are resident in core memory, taking into account the turn-over, and optimizing exchanges, possibly with a look-ahead approach,
- an object-oriented database, based upon persistent abstract data types, (protected, simultaneously shareable and distributed), accepting fuzzy data
- a list processing supporting an unlimited number of unlimited lists, whose elements are abstract data types,

- a functional engine using functional algebra (considerably more efficient and user-friendly than relational algebra), whose efficiency is based on swappers and rollers,

- an expert system (Bouillé, 1984) working in a fuzzy context (Bouillé, 1994c), with several classes of inference engines which are able to manage an unlimited number of facts and rules, considered as objects of classes, (without using the obsolete concepts of "pattern matching" and of "back-tracking",

- a simulation system, based upon a discrete simulation engine, managing an unlimited number of interacting discrete processes, possibly mixed with continuous processes,

- an automated learning system based upon a neural engine, managing an unlimited set of neurons, considered as objects of various classes of neurons,

- a 3-D graphical stereoscopic animated interface, allowing, among other capabilities, to manage dialogues with users, to handle graphical representation of HBDS data types, and to ensure the graphical outputs of various applications, for instance in the field of GIS.

All these mechanisms are integrated in the kernel, not as a simple juxtaposition of different tools. That is why we can say that HBDS is an integrated object-oriented platform for GIS.

For managing the HBDS types, a very high-level object-oriented programming language has been designed, named ADT'81.

This language performs scientific computation, inference processing, discrete and continuous simulation, neural processing, graphical handling and fuzziness processing.

The HBDS system is associated with an ADT'81 compiler together with its optimizer, its decompiler, its executer, and a compiler-compiler. These tools themselves are based upon HBDS and directly programmed in the ADT'81 programming language.

All preceding tools could be considered separately but if used simultaneously, very great difficulties would occur, such as the development of interfaces between non compatible products. That is why it is of the most importance that all the tools be designed on the same basis, in order to be directly integrated into only one system, thus composing an integrated platform as the kernel of an O.-O. GIS.

For the designers of the future GIS, "integration" will be the system keystone.

Rules, processes and neurons are not equivalent, and a simple comparison shows they must be used in cooperation, present in the same integrated system (Bouillé, 1993b).

4 "GSC": THE GEOGRAPHICAL STRUCTURE CATALOGUE

Most geographical phenomena have their structure completely designed and now belong to a catalogue, the GSC; they thus can be re-used for any new application, partially or as a whole; two structures, or more, can be assembled to provide a new one. Time and brain are then minimal for any new problem. We briefly give here a non-exhaustive list of the only urban phenomena whose structures belong to the GSC:

- ways (segments, crossing, blocks, and block-faces), composing the urban texture, pavement, sidewalks, gutter, administrative boundaries,
- electrical networks (very high-, high-, medium-, low-voltage),
- water networks (domestic water, industrial water, fire network), heating (hot water and vapour), conditioning (cold water), compressed air,
- gas, pipelines, various other gases such as nitrogen, argon, etc...., grease,
- various telecommunication networks, lightning, traffic regulation and video control,
- transportation (trains, metroliner, bus, taxi, tramways, etc.),
- sewers,
- etc.

Some essential factories have their structures too (generally using networks), such as:

- railway and train stations,
- airports,
- harbors,
- gas-works and oil-refineries,
- damps, electricity production (hydro-, thermal, or nuclear),
- etc.

For that purpose, prototypes (Bouillé, 1993a) have been developed, such as "GRAPH", "NETWORK", "POLYHEDRON", etc... allowing a fast and easy design of any new geographical theme. Many other networks have been processed, concerning land planning, and not only urban. The most important structures for a country have also been designed, most of them using the prototype "network":

- geology, mining and hydrocarbon production and distribution,
- topography, hydrology,
- pedology, vegetation, agriculture,
- health, education,
- defence...

Some specific studies have more deeply considered some networks (Jung, 1987) or the cartographic aspects (Pasquier, 1987), or the application to the environment (Bouillé, 1994b). Moreover, a particular effort has been made for managing interconnections, intersections, superimpositions, embedments of these various elements belonging to several networks and representing more than 30 000 cases. People develop structures which were previously developed by other people for other applications; for instance for urban areas, somebody develops a knowledge structure dealing with road networks; but somebody had previously developed approximately the same structure for studying the impact of the roads on the environment in forest areas, and another team is developing a third version for preparing the connection between cities. They speak about the same road concept, they probably will reach the same set of components, but three teams have to develop and do the same work; moreover, if they try to connect their applications later, they will have to interface them, and it will not be very easy. Why waste brain, time, effort and money, when it is possible to choose in a catalogue the corresponding structure, generally made of more than 500 classes, 2000 or 3000 attributes, probably the same number of links ? Then, they can reuse this structure, entirely or partially, to compose it with other sub-structures, thus defining a new application. Only a few hours are necessary, even one or two days, instead of months.

The concept of "geographical structure kit" will provide many benefits, and improve the quality of the structure used, because it is developed by specialists and used by all others ensuring all future connections. For geographical purposes, it is one of the most important practical tools in the near future. This one is not relevant to software, but to the brainware. A knowledge structure is not a set of programming statements, as some elementary O.-O. systems have tried to apply in GIS.

5 Structuring the Main Themes of the Oil Industry

An important effort was launched a few years ago by the petroleum industry, named the POSC project in order to develop tools for an object-oriented approach of oil applications. Nevertheless, this project initiated by US companies is completely obsolete because tools were developed a long time ago, far in advance compared to the trends of the POSC tools. These at present do not exist and people are re-inventing what was designed more than ten years ago for modeling all oil themes, such as drilling, geophysics, diagraphies, etc.... HBDS may be efficiently used (Sinsoilliez-Clauser, 1992).

The GSC presents all the basic themes necessary for implementing more specific topics:

- the topography, orography, hydrography and cadastral themes, composed of more than 300 classes, considered as the basis for knowing the environment of all expedition, mission, exploration and station positioning,

- in order to dispatch some oil equipment at the earth's surface, some other items are absolutely necessary, such as road networks (300 classes approximately), railways (likewise around 300 classes), electrical networks (very high -, high -, medium - and low voltage), telecommunication networks, etc...composing the infrastructure needed by any industry.

- then geology, including not only boundaries, surface and sub-surface information, but also all underground information; this includes things not directly relevant to the GIS, such as stratigraphy, paleontology, tectonics, representing approximately 500 classes, and completed by information issued from the mining industry (more than 300 classes).

- connected to the preceding field, there are themes dealing with gravimetry, magnetism, electrical methods, but the most important topic is the seismic, including marine exploration; the detailed structure has more than 300 classes, with links to the reservoir that we will discuss later.

- drilling is an important activity, for which a complete structure has been designed, with a second one devoted to off-shore platforms (around 500 classes), taking into account the different models of platforms, according to their size (from 50 000 tons to 500 000 tons), nature (concrete, iron, mixed), purpose (exploration, production, storage, housing), structure (piles, jacket, etc....), and location (Arabia, North Sea, Asia, etc....). Drilling is one of the topics where *discrete simulation techniques* are particularly interesting, and the use of *object-oriented processes,* working in quasi-parallelism, in an unlimited number, is one of the most convenient approaches.

- connected to the drilling activity, an important topic is the diagraphical with two major themes according to whether the diagraphy is obtained during the drilling or when stopped. The number of potential diagraphies which can be obtained, deriving from the simplest ones, can reach 1000 and the reader understands that the number of classes is not restricted on such a topic. Diagraphies for a given hole must be correlated together; performed at different places, they must be reconnected, and performed at different times, their evolution must be interpreted for a better understanding of the reserves for instance. Diagraphic studies is one of the topics in which an *object-oriented expert system* is particularly convenient, representing a powerful tool for a better mastering of too bulky knowledge.

- the preceding activity has implications for another, reservoir study requiring approximately 500 classes. We must particularly emphasize the impact of the connections approach on reservoir modeling; the use of *object-oriented neurons* is particularly interesting in trying to obtain a dynamic model, and to provide *automated learning* (Bouillé, 1994e). The use of *genetic algorithms,* based upon genes belonging to classes, in cooperation with neurons, may prove promising. At the present time, it has not given any fascinating results, but studies must not be stopped.

- the pipeline network, with all the basic equipment, connected to large centers such as terminals or refineries (requiring approximately 500 classes). Pipelines are easily modeled by neurons, in order to provide a more dynamic view of oil and gas flows; this theme is connected to specific components allowing a real-time interaction.
  The pipeline structure directly derives from the object-oriented prototype "NETWORK" and is enriched by specific components.

6 CONCLUSION

The basis of the whole structuring activity in oil exploration and production is a graphical object-oriented model managing implicit persistent abstract data types, which can be grouped into prototypes and structures and can be re-used, thus avoiding an important waste of brain, effort, time and money.

This model, named HBDS, integrates facts, rules, processes, neurons and constraints, all objects of classes, possibly considered in a fuzzy context; it provides the kernel of a GIS which is used as a basis for an object-oriented system dealing with all the activities of oil exploration and production, at the different steps of the studies, and including all the common tools of the field (geophysics, diagraphies, platforms, piping, etc.).

The same model, used for databanking, problem solving, simulation, automated learning, provides a convenient tool for decision-makers, and most of all, a common tool providing easier communication between the different activities of a very large industry.

KEYWORDS

GIS, petroleum industry, oil exploration, object oriented database, expert system, neural system.

REFERENCES

[1] Bouillé, F., "A structured expert system for geodata banking", 9th Int. CODATA Conf., Jerusalem, June 24-27, in "the Role of Data in Scientific Progress", Elsevier, 417-420 (1984).

[2] Bouillé, F., "Developing strategies in GIS, by problem-solving methods based on a structured expert system", Int. Conf. EUROCARTO Vll, Enschede, Sept.19-22, Proceed. 12p. in "Environmental Applications of Digital Mapping", ITC.Pub., N-8, 42-50 (1988).

[3] Bouillé, F., "Prototyping in GIS: methodology, uses and benefits", EUROCARTO Xl Int. Conf. Kiruna, Dec. 8-11, Proceed. 48-66 (1993a).

[4] Bouillé, F., "Comparing processes, rules and neurons, for dynamical phenomenon modeling in GIS", EUROCARTO Xl Int. Conf. Kiruna, Dec.8-11, Proceed. 177-194 (1993b).

[5] Bouillé, F., "Methodology of building a class-based integrated platform for GIS design and dcvelopment", EGIS/MAR1'94 Int. Conf., Paris, March 29-April, Proceed., Vol. 1, 909-918 (1994a).

[6] Bouillé, F., "Object-oriented methodology in GIS environmental studies", Intercarto Int. Conf. on GIS for Environmental Studies and Mapping, Moscow, May 23-25, Proceed., 39-46 (1994b).

[7] Bouillé, F., "Fuzziness structuring and processing in an object-oriented GIS", AGIT'94, Salzburg, Proceed. Salzburger Geographische Materialien, Heft 21, 113-122 (1994c).

[8] Bouillé, F., "Towards 2000: the actual main trends in future GIS", GIS Int.Conf. "Europe in transition", Brno, Aug. 28-31, Proceed., chap. K, 13-27 (1994d).

[9] Bouillé, F., "Principles of automated learning in a GIS, using an illimited set of object-oriented persistent neurons", ISPRS - III - SSIDPCVN, Munchen, Sept. 6-9, Proceed. Ebner-Heipke-Eder Editors, Vol. 30, Part 3/1, 69-76 (1994e).

[10] Jung, F., "La cartographie assistee par ordinateur du reseau de transport d'Electricite de France", Thèse de Doctorat, Univ. Paris, 180 (1987).

[11] Kitagawa, T., "Brainware concept in intelligent and integrated systems of information", Research Inst. Fund. lnf. Sci., Kyushu Univ., Research Report 11-39 (1974).

[12] Pasquier, B., "Cartographie numérique des domaines. Structuration, modèlisation, algorithmique", Thèse de Doctorat ès Sciences Math.lnfo., Univ. Paris, 350 p. (1987).

[13] Sinsoilliez-Clauser, C., "Analyse de systèmes d'information. Etude des méthodes classiques et conception d'un prototype intégrant le modèle HBDS. Application aux activités de développement et de production de TOTAL", Thèse de Doctorat, Univ. Paris, 350 p. (1992).

[14] Titeux, P., "Automatisation de problèmes de positionnement sous contraintes: une méthode intégrant des techniques de systèmes experts, de compilation et de bases de données. Application en cartographie", Thèse de Doctorat Math. lnfo., Univ. Paris, 414 p. (1989).

Chapter 6

ECOTOXICOLOGICAL ISSUES

CLEANING OF GAS FLOW USING CHEMICAL REACTION DATABASE

E. A. PHILIMONOVA[a], *D. A. PHILIMONOV*[b] *and M. B. ZHELEZNIAK*[b,1]

[a] *Institute for High Temperatures, Russian Academy of Sciences, Russia*
[b] *National Research Center for Biologically Active Compounds, 23 Kirova Str., Staraya Kupavna, Moscow Region, 142450, Russia*

ABSTRACT

To protect the environment it is necessary to clean waste gases, reducing the emission of toxic impurities. During past years considerable efforts have been made to develop plasma methods for removal of pollutant components. The source of non-equilibrium plasma operates in a waste gas flow and produces chemically active species. These species initiate the chain of reactions in which the pollutant molecules participate, and as a result gaseous or condensed non-toxic components arise. The investigation of the time evolution of the reactive gas systems is very important for these purposes. Software applications for solving the problems of removing toxic compounds in the gas flow have been developed.

---

[1] To whom correspondence may be addressed

These include a chemical reaction database and programs for serving it, program for initial preparation of data, a solver unit, a program for tracing and controlling the set of operating equations, a program of reduced kinetics, and tools for representing results. The integrated user environment allows one to operate with all these parts of the system.

Purification des Rejets Gazeux: utilisation d'une Base de Données de Réactions Chimiques

RÉSUMÉ

Pour protéger l'environnement, il est nécessaire de se débarrasser des gaz rejetés dans l'atmosphère et de réduire ainsi l'émission de substances toxiques. Durant ces dernières années un effort considérable a été fourni pour développer les méthodes à torche plasma pour éliminer les polluants. La source de plasma en déséquilibre agit sur les rejets gazeux et produit des composés chimiquement actifs. Ces composés provoquent des réactions en chaîne auxquelles participent les molécules polluantes qui sont alors transformées en condensés et gaz non toxiques. L'évolution dans le temps de la réactivité du système gazeux est très importante pour ce type d'application. Un logiciel pour résoudre le problème de la dégradation des toxiques contenus dans un flux gazeux a été développé. Celui-ci comprend une base de données des réactions chimiques et un programme d'interface, un programme de préparation des données, un module de résolution, un programme pour tracer et contrôler l'ensemble des équations concernées, un programme de cinétique et un outil de représentation des résultats. Un environnement utilisateur permet l'utilisation de chacun de ces modules de manière intégrée.

1 DATABASE

The database contains about 2000 reactions for nearly 200 organic and inorganic substances. These substances are as follows: neutral components, electrons, negative and positive ions, excited atoms and molecules, clusters, (which can be formed from the elements N, O, H, C, S), temperature dependent rate coefficients and equilibrium constants are represented in the Arrhenius form for the most part. Each reaction is given a reference and a remark if it is necessary.

The main sources of the data are [1, 2, 3, 4]. There are server programs for working with the database: a program for inputting new data, a check program for conservation of charge and number of elements in each reaction, programs for ordering reactions and forming a list of all species which participate in them. These programs use the ASCII file form of the database.

The database in internal format is created from a set of such ASCII files. The internal Database contains only unique reactions. If an identical reaction is already contained in it then the reaction will not be included. The information on all reactions not included is saved in a special ASCII file.

2 Data Preparation

The data preparation program forms the set of chemical reactions for each specific problem. In the developed applications software dialogue windows are used for this purpose. The main window contains two lists with symbols of species and reactions. At the beginning the symbols of chemical components with non-zero initial concentrations are selected in the species list. The reaction system is supplemented with all possible reactions and reagents arising from these initial substances by a search in the internal Database. As a result complete lists of species and reactions are generated. The user can delete some components and/or reactions from lists and correct the reaction systems.

The consequence of this dialogue is the creation of the reaction system file. An editor window is used to set values of the initial concentrations of components and other parameters.

3 Solver

The standard Gear algorithm is used as a solver. All necessary improvements have been made to adapt this program to solving the set of operating equations.

4 Program of Tracing and Control

During calculation the tracing and control program creates the set of operating equations. At each calculation step the program works out criteria to bring a component into operation or to eliminate it from the set of equations. The criteria depend on the concentration of the component and its derivative.
In such a manner the reduced lists of equations and reactions are formed on the basis of the complete lists. Reducing the possible number of solving equations decreases the computation time. In addition, the solving becomes more stable.

5 Program of Reduced Kinetics

This program is executed after operation of the main program. In line with given criteria, the program creates reduced lists of components and chemical reactions. The necessary information is selected from the tracing and control program during solving.

One of the obvious criteria for including the component in the reduced list is if its concentration exceeds any reference value. The criteria for reactions are not so simple, and they contain the analysis of forward and backward processes, both in the reactions and in the total derivatives.

In addition, the reduced kinetics depends on the list of species with non-zero initial concentration and the physical time of the problem.

```
        Component under consideration   NO           Pulse number 20
Concentration                        0.63E+16  0.62E+16  0.57E+16  0.57E+16  0.61E+16  0.63E+16
-----------------------------------!---------!---------!---------!---------!---------!
474  NO         <= diffusion       !  7-6-5--43-2-1---------------------------       !
383 N(2D) + CO2 => NO + CO         !67                                               ! 0.253E+12
212 HNO + OH => H2O + NO           !           8------76                             ! 0.858E+12
158 N + O2 => NO + O               !76---5-6--------7---65                           ! 0.210E+13
159 N(2D) + O2 => NO + O           !23-45                                            ! 0.276E+13
161 O2 + HNO => NO + HO2           !          7---6--42---------                     ! 0.287E+13
 97 NO2 + O => NO + O2             !5---4-----5-------3-4--3                         ! 0.545E+13
 39 N + NO2 => NO + NO             !4--3-------4-----2--3--4                         ! 0.676E+13
388 OH + N => H + NO               !32-----------3----45-                            ! 0.143E+14
293 H + NO2 => OH + NO             !1-------------2-4568                             ! 0.616E+14
---Sign dn/dt-----------------------------------------------+++++++++++++++++++++++++----- -.50E+14
292 NO2 + M1 <= NO + O + M1        !1------------------2-3----4-----                 ! -.495E+15
323 NO + OH + M1 => HNO2 + M1      !3---------------2---1----2----------------------1! -.286E+15
190 N + NO => N2 + O               !2---------------3----4---5                       ! -.239E+15
317 HO2 + NO => OH + NO2           ! 6----5---4----------32--1---------------------- ! -.117E+15
324 H + NO + M1 => HNO + M1        !54--------5-----                                 ! -.411E+13
 99 NO + O3 => NO2 + O2            !                5--4-3--------------------3! -.186E+13
214 -O2(H2O) + NO => -NO3 + H2O    !45----6                                          ! -.147E+13
191 N(2D) + NO => N2 + O           !67                                               ! -.191E+12
-----------------------------------!---------!---------!---------!---------!---------!
                                  -7.       -6.       -5.       -4.       -3.       -2.
                                                   log10(t)
```

Figure 1. Chart of the leading reactions for NO production

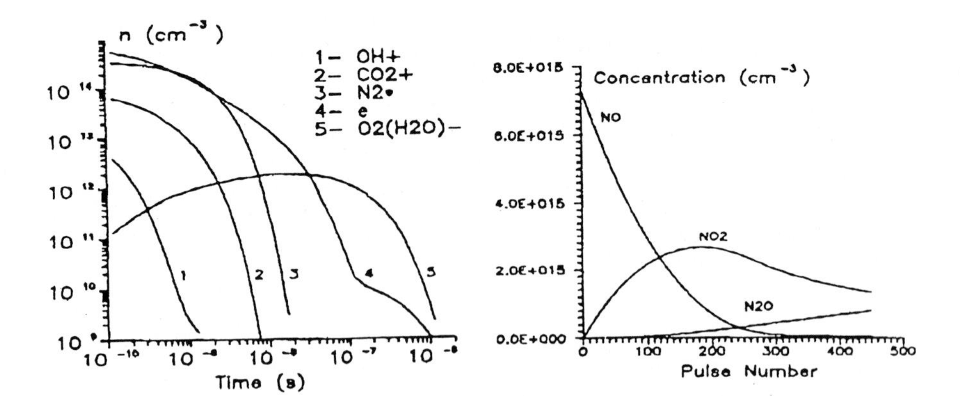

Figure 2. Change of concentration in the pulse number N=60.

Figure 3. NO_X removal as a function of the pulse number.

Our experience has shown that the total lists of components and reactions can be reduced by more than a factor of two. The results obtained with the reduced kinetics are practically identical to those for the total set of equations. This allows users to solve problems faster and with reasonable accuracy. For example, we studied the evolution of a system that was described with 160 species and 1250 reactions. For 1-10 sec of physical time the solution required 4-5 min computing time on a *DELL PowerLine PENTIUM-60*. This time was lessened by a factor of 1.5 when the reduced kinetics were used.

6 TOOLS FOR REPRESENTING RESULTS

The various information in convenient form is given by the tools representing results. The user can find the operating time intervals for each component when it is included in the operating set of equations. This depends on development of the chemical processes. More detailed information about each component with leading reactions is given as a diagram (an example is shown below).

The interactive graphic representation is used for the concentrations of components versus time dependence. The user can select (unselect) any component in the species list and its time-dependent concentration will be drawn on the diagram marked in its own colour. It is possible to change the scale of axes, their type (linear or logarithmic), and to move the diagram along the axes. The diagram can be printed in the graphic mode.

7 USER INTEGRATED ENVIRONMENT

The user integrated environment was developed using Borland Pascal with Objects 7.0. It operates under DOS on IBM compatible PCs. We can incorporate it into Windows.

The described applications software has been used for simulating the cleaning processes of possible flue gases. CO_2, H_2O, O_2 and impurities NO, SO_2, CO are considered. It is assumed that the flow field (velocity, density) is known and a high voltage pulsed electrical discharge of the corona type operates in the flow. Usually discharges of this type propagate in gases as many thin filaments (streamers) in which diverse active species are produced. Usually the discharge slightly disturbs the main flow and the initial volume of the streamers is small. In such a manner the non-uniform reactive gas flow takes into consideration the streamer interactions due to their diffusion expansion. The diffusion model was described in [5]. After execution of the program involving reduced kinetics the operating files contain 84 reagents and 474 reactions.

The numerical results show that the main process leading to removal of toxic components is oxidation of SO_2 and NO. The diagram for NO is presented in Figure 1. On the left side there is a list of leading reactions. Integer numbers in the main part of the chart correspond to the intensity of the reaction in which a chosen chemical component is produced by each reaction. Examples of concentration variation vs. time is shown in Figure 2. After operation of 450 pulses the total degree of cleaning NO_x reached nearly 80% as viewed in Figure 3. This result is in good agreement with experiments. The reactions NO_2 and SO_2 with radicals result in acid formation that, together with the added ammonia, form non-toxic condensed particles. The role of other active components in removal process is defined as well.

ACKNOWLEDGEMENTS

The authors are grateful to IVTAN Analytical and Numerical Research Association for financial support of this work.

KEYWORDS

Database, chemical reaction, gas pollution, kinetics, data, chemistry, cleaning, remediation

REFERENCES

[1] Matzing H. *"Chemical kinetics of flue gas cleaning by electron beam"*, Tech. Report KfK 4494, Kerfirschunzentrum, Karlsruhe (1989).

[2] Baulch D.L., Cobos D.J., Cox R.A. et al." Evaluated kinetic data for combustion modeling", *J. Phys. Chem. Ref. Data*, **2** (**3**) (1992).

[3] Bochkov N.B., Lovachev L.A., Chetverushkin B.N. "Chemical kinetics of NO_x formation in combustion products of methane.", *Math. Modelirovanie*, **4**, **3**, in Rus. (1992).

[4] Bychkov V.L., Yurovsky V.A. "Modeling beam plasma of water vapor", *Teplophys. Vysokih Temp.*, **31**, **8**, in Rus. (1993).

[5] Philimonova E.A., Zhelezniak M.B. "Removal of toxic components in gas flow using plasma sources of chemical active molecules", in Proceedings: *Int. Symp. On Heat and Mass Transfer in Chemical Processes Industry Accidents*, Rome, Italy, (September 1994).

PHOTO-INDUCED TOXICITY OF PAHS: QSAR MODELS AND ENVIRONMENTAL SIGNIFICANCE

Ovanes G. MEKENYAN[a,b], Gerald T. ANKLEY[c], Gilman D. VEITH[c] and Daniel J. CALL[b,1]

[a]Bourgas University of Technology, Department of Physical Chemistry 8010 Bourgas, Bulgaria

[b]Lake Superior Research Institute, University of Wisconsin-Superior Superior, WI 54880 USA

[c]U.S. Environmental Protection Agency, Environmental Research Laboratory - Duluth, 6201 Congdon Boulevard, Duluth, MN 55804 USA

ABSTRACT

Research with a variety of aquatic species has shown that while polycyclic aromatic hydrocarbons (PAHs) are generally not acutely toxic in conventional laboratory tests, many are extremely toxic in the presence of sunlight. A QSAR model was developed for predicting phototoxicity of PAHs, which incorporated the competing effects of stability and light absorbance of the PAHs as well as irradiation parameters. Interaction of these factors produces a complex, multilinear (bell-shaped) relationship between toxicity and chemical structure.

The molecular descriptor best describing both light absorbency and stability of ground state PAHs was found to be the energy of HOMO-LUMO gap, which could be computed from structure rather than measured empirically. Thresholds of HOMO-LUMO gap values were established, classifying PAHs according to their photo-induced toxic properties. The substitution of PAHs, in general, reduced their HOMO-LUMO gap, which is consistent with the Principle of Absolute Hardness. Alkane and hydroxy substituents did not significantly reduce the HOMO-LUMO gap of PAHs. This held less well for nitro, alkene and chloro substituents.

Interactions (p-π and π-π) between the electronic structures of these substituents and PAHs provided hyperconjugation effects and eventually "red-shifted" the parent compounds. It was concluded, however, that the photo-induced properties of PAHs were mainly due to the electronic structure of the parent chemicals.

[1] To whom correspondence may be addressed

Phototoxicité induite des HAP: Le Modèle QSAR et sa Signification pour l'Environnement

RÉSUMÉ

Alors que les hydrocarbures aromatiques polycycliques (PAH) ne sont pas fortement toxiques au cours de tests conventionnels en laboratoires, la recherche entreprise dans le domaine des espèces aquatiques a montré que de nombreux PAH peuvent devenir extrêmement toxiques lorsqu'ils sont exposés à la lumière solaire. Un modèle QSAR a été conçu pour prédire la phototoxicité des PAH. Celui-ci prend en compte les effets opposés de stabilité et d'absorption de la lumière par les PAH, ainsi que les paramètres d'irradiation. L'interaction de ces paramètres produit une relation multilinéaire complexe (en forme de cloche) entre la toxicité et la structure chimique. Le descripteur moléculaire qui décrit le mieux à la fois l'absorption de la lumière et la stabilité des PAH, est la différence d'énergie de l'intervalle HOMO-LUMO. Ce paramètre peut être calculé à partir de la structure moléculaire plutôt que mesuré empiriquement.

Une classification des PAH en fonction de leurs propriétés toxiques photo-induite est proposée à partir des valeurs de l'intervalle HOMO-LUMO. La substitution des PAH réduit en général l'intervalle HOMO-LUMO, ce qui est en accord avec le Principe de la Dureté Absolue. Les substituts alcalins et hydroxydes ne réduisent pas de manière significative l'intervalle HOMO-LUMO des PAH. Ceci est moins net pour les substituts nitraté, chloré et alcalin. Les interactions (p-π and π-π) entre les structures électroniques de ces substituts provoquent des effets d'hyperconjugaison et parfois de "déplacement dans le rouge" par rapport aux composés primaires. En conclusion, il apparaît que les propriétés de photo-induction des PAH sont essentiellement provoquées par la structure électronique des produits chimiques parents.

1 INTRODUCTION

Polycyclic aromatic hydrocarbons (PAHs) are ubiquitous contaminants of aquatic ecosystems. The predominant sources of PAHs in the environment are linked to human activities such as oil spills and the burning of fossil fuels. PAHs include chemicals which generally have two to six fused benzene rings, with occasional incorporation of heterocyclopentene rings containing nitrogen or sulphur. A wide variation of alkyl substituents gives rise to thousands of different possible PAHs, and many have been identified in environmental samples.

Although some PAHs are biodegraded under aerobic conditions, many are extremely stable, especially in the anaerobic conditions of aquatic sediments. In short-term assays, PAHs are predicted to be, and have been found to be, relatively innocuous [1], exhibiting little or no acute toxicity to aquatic organisms even at aqueous solubility. The traditional toxicological concern has focused on metabolic activation of a number of PAHs to metabolites far more toxic than the parent compounds [2].
However, another danger to aquatic ecosystems may result from photo-induced toxicity caused by PAH excited states which are formed in the presence of ultraviolet (UV) radiation in sunlight.

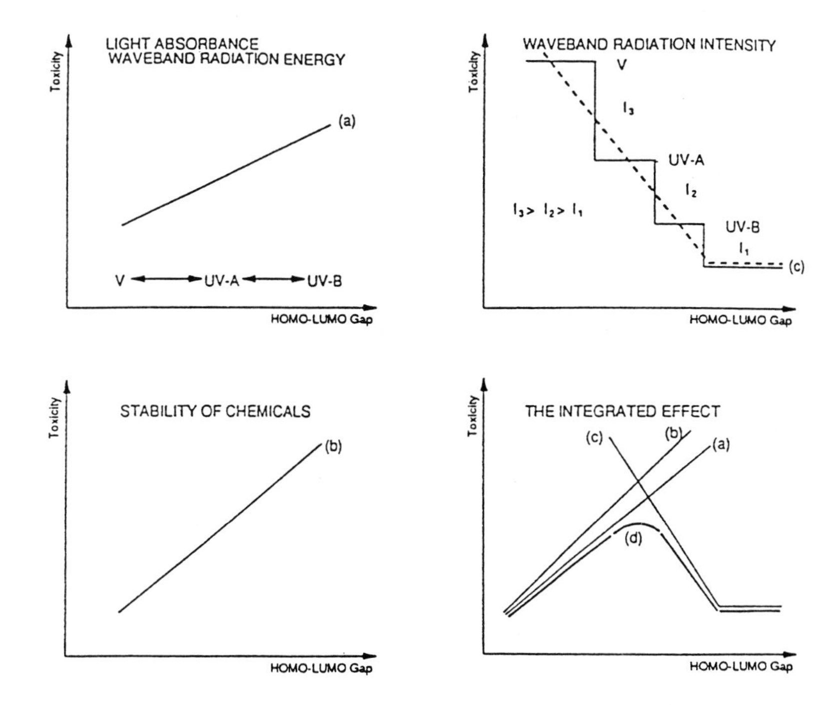

Figure 1. The internal and external factors conditioning the phototoxicity: (a) light absorbency; (b) stability of chemicals; (c) irradiation intensity; (d) integrated photo-induced toxicity.

Studies with a variety of aquatic species including fishes, invertebrates, and plants have shown that PAHs such as anthracene can be orders of magnitude more toxic in the presence of UV light than under typical laboratory lighting [3], [7].

In an earlier study, Newsted and Giesy [4] found the median adjusted lethal time (normalized to a constant concentration) of PAHs for *Daphnia magna* varied with the energy of the triplet molecular states (obtained from fluorescence and phosphorescence emission spectra [8]) according to a parabolic relationship. The result was consistent with the fact that the triplet excited state is more persistent than the single and is more likely to "keep" energy long enough to exert toxicity. However, the parabolic nature of the triplet state-toxicity relationship for a wide variety of PAHs is not readily explained without a formal mechanistic analysis. Using the toxicity data from this study, we have developed quantitative structure activity relationships (QSAR) based on the electronic structure of ground and excited molecular states to predict both the occurrence and potency of photo-induced toxicity for individual PAHs. The effect of chemical substituents on the predicted photo-induced toxicity of these compounds was also studied.

Table 1. Names of PAHs used in model development and in Figures 2 and 3.[a]

1. Anthracene	15. Carbazole
2. Benzo[a]pyrene	16. Triphenylene
3. Dibenzo[a,h]anthracene	17. Chrysene
4. Fluoranthene	18. Perylene
5. Pyrene	19. Benzo[b]anthracene
6. Benzo[a]anthracene	20. Benzo[g,h,i]perylene
7. Benzo[e]pyrene	21. Coronene
8. Benzo[a]fluorene	22. Dibenzo[a,j]anthracene
9. Benzo[b]fluorene	23. Benzo[b]chrysene
10. Acridine	24. Benzo[b]triphenylene
11. Benzo[k]fluoranthene	25. Benzene
12. Benzanthrone	26. Naphthalene
13. Phenanthrene	27. Dibenzo[b,i]anthracene
14. Fluorene	28. Benzo[a]chrysene

[a] Chemicals 1-20 were tested for toxicity to *Daphnia magna* by Newsted and Giesy [4], while the toxicities of chemicals 21-28 were predicted from QSAR models developed in this study.

2 Photo-toxicity of PAHs and Their Ground State Electronic Structure

Recently [9], we proposed that photo-induced toxicity is a cumulative result of several factors including internal, or molecular structure factors, such as light absorbency and chemical stability, and external factors, such as the energy and intensity of irradiation. The HOMO-LUMO gap was proposed as a ground state index of the electronic structure of PAHs which relates to energy absorbency and molecular stability factors.

The rationale of this choice was based on the principle of absolute hardness. The latter was defined [10], [11] as half the HOMO-LUMO gap and regarded as a measure of energy stabilization in chemical systems, whereby chemical structures tend to be most stable at the largest HOMO-LUMO gap. In sunlight or other constant light conditions, the HOMO-LUMO gap was presumed to be an indicator of the wavelengths absorbed and therefore the energy of the intermediate excited states.

Ultimately, photo-induced toxicity was assumed to be a result of competing processes which interact to produce a complex, multilinear relationship between toxicity and chemical structure. The composite, parabolic curve (Fig. 1) resulted from the superposition of individual factors as functions of the HOMO-LUMO gap. The individual chemicals used in the QSAR model are presented in Table 1.

A plot of observed phototoxicities from Newsted and Giesy [4] and electronic gaps of PAHs is presented in Figure 2. As can be seen, the "observed" parabolic relationship corresponds to the "theoretical" one.

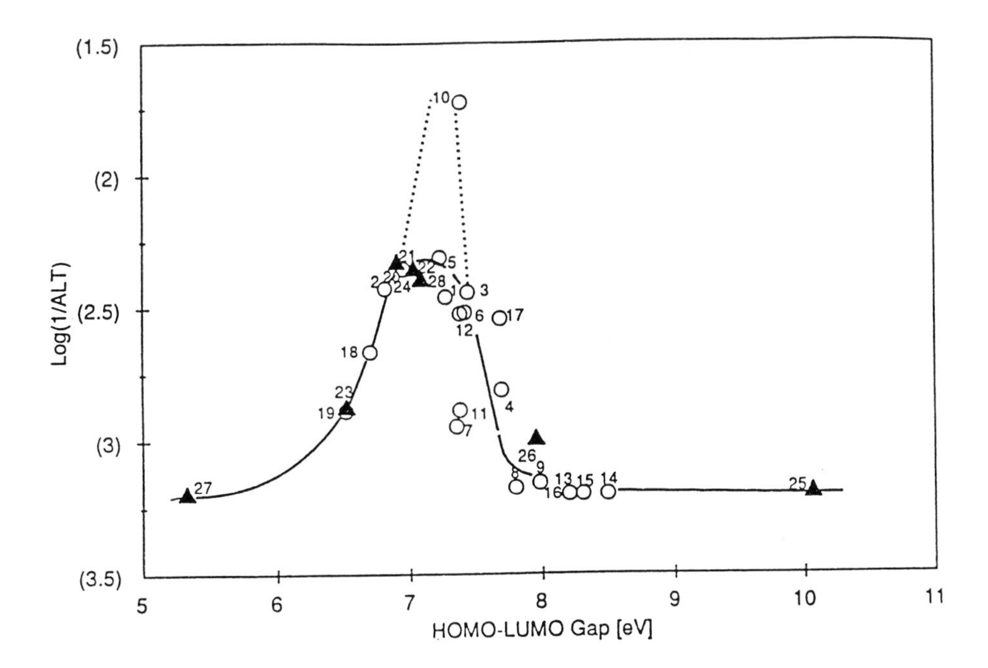

Figure 2. Variation of phototoxicity (log(1/ALT); ALT-Adjusted Lethal Time) with HOMO-LUMO gap [1]. The open circles represent PAHs tested for their toxicity to *Daphnia magna* by Newsted and Giesy (1987). The solid triangles denote the predicted toxicity values for some as yet untested PAHs.

The threshold HOMO-LUMO gap of a PAH to induce phototoxicity in sunlight is about 8.1 eV (as determined by both PM3 and AM1 methods). The most phototoxic PAHs have HOMO-LUMO gaps within the window 7.2 ± 0.4 eV and include anthracene (#1), benzo[a]pyrene (#2), dibenzo[a,h]anthracene (#3), pyrene (#5), benzo[a]anthracene (#6), benzo[e]pyrene (#7), acridine (#10), benzo[k]fluoranthene (#11), benzanthrone (#12), perylene (#18) and benzo[g,h,i]perylene (#20).

Chemicals on the right side of the curve near the baseline have a large HOMO-LUMO gap which prevents light absorption from wavelengths found in sunlight and increases stability. Chemicals having mid-level $E_{HOMO-LUMO}$ absorb more energy from sunlight, and toxicity increases with decreasing HOMO-LUMO gap because of greater intensity at the lower energies of the sunlight irradiation. The toxicity of chemicals with small HOMO-LUMO gaps is less because the absorbed light energy decreases and the stability of chemicals with gaps below 7 eV decreases rapidly. In fact, it is likely that chemicals with extremely small HOMO-LUMO gaps, such as pentacene, would not be stable enough in water even to conduct a meaningful toxicity test.

These results can be explained by the assumption that the frontier HOMO and LUMO orbital condition the lowest excited states of the molecule [11]. Larger energy gaps result in larger energy differences between the ground state and excited states of the same microstate multiplicity. This explains the correlations found [9] between the HOMO-LUMO gap and spectroscopic parameters of the excited states, and justified our assumption that the ground state HOMO-LUMO gap could be an electronic index of absorbed energy.

The advantage of ground state calculations is that of computing speed, because excited state modeling is often prohibitively computer intensive.

3 Photo-toxicity of PAHs and Their Excited State Electronic Structure

The relation of the HOMO-LUMO gap of the PAHs to the energies of the excited states is due to the fact that the quantum mechanical theory of response to a perturbation in a chemical system depends on this gap. The perturbations of electronic structure are usually modeled by mixing excited-state wave functions with the ground-state functions. The coefficient weighing these mixed wave functions is inversely proportional to the energy difference between the states. Therefore, according to the principle of maximum hardness, hard molecules (with larger HOMO-LUMO gap) should less readily undergo changes in electronic structure than soft molecules.

We found [12] the energy convergence procedures of MOPAC6 for optimizing geometries of the excited states to be much too time-consuming (each structure requiring many hours of CPU-time on the EPA Cray). On the other hand, 1SCF calculations without optimized structures for individual microstates yielded energy values for excited states which did not correlate with the observed spectroscopic data.

As a computing compromise, we chose the first MOPAC6 option to perform only geometry optimization of the excited states. To mimic the energy of the excited states, we assumed that the optimized geometry of the excited states described an average molecular geometry of all microstates taken into account in the calculations. Finally, the 1SCF calculations using these optimized structures did provide electronic indices which mimicked the energies of the excited states.

Although the HOMO-LUMO gap does not have physical meaning for the excited states because of the multiplicity of the microstates, the principle of maximum hardness [13] suggests that the computed "HOMO-LUMO gap" [denoted as (HOMO-LUMO)*] of these hypothetical structures might be an appropriate stability estimate, assuming the optimized geometry of excited states is an average geometry of all microstates.

This hypothesis was supported [12] by much better correlations found between (HOMO-LUMO)* gap of the excited states and their observed energies from spectroscopic data than the QSARs found for ground state structures. Further, the models for the triplet state energies were significantly better than those for the singlet states, probably because of the greater stability of the former.

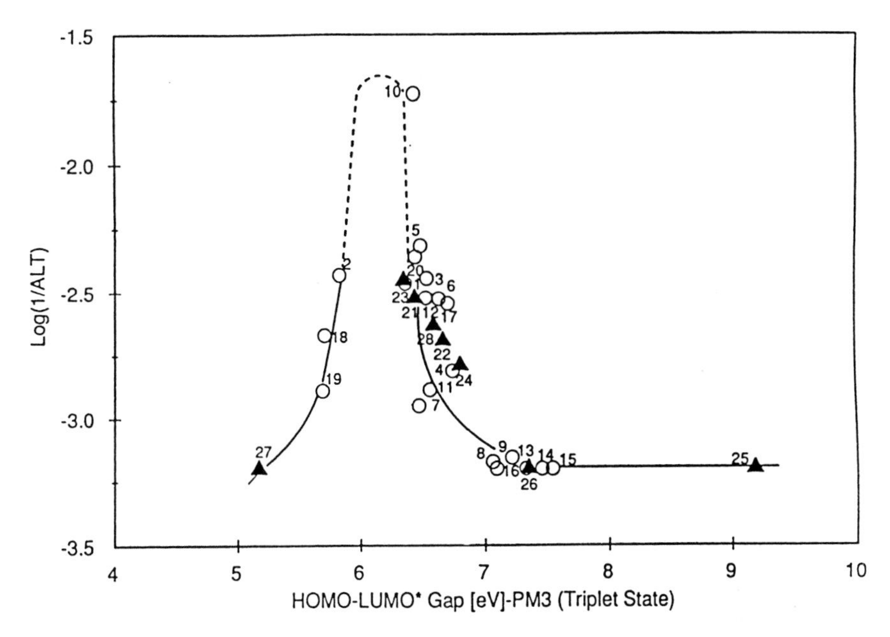

Figure 3. Variation of toxicity (log(1/ALT) with stability parameter of the triplet excited state, (HOMO-LUMO)* gap. The solid triangles denote the predicted toxicity values for some as yet untested PAHs.

The greater energies found [12] to correspond to the more stable excited states means that greater energies are necessary in order to elevate PAHs to the respective excited states. The energetically rich excited states, however, will release a larger portion of energy when going back to the ground state, which ultimately will cause more toxicity to organisms. Thus, one can assume that the linear relationship between the energies of the singlet and triplet states of PAHs and the (HOMO-LUMO)* gap of the excited states correspond to the integrated effects of both stability and light absorbency factors of photo-induced toxicity. If one superimposes this linear relationship on the distribution profile for irradiation, a bell-shaped phototoxicity curve (Fig. 3) similar to that in Figure 2 is obtained.

The only difference from the "ground state relationship" is the clear separation of both sides of the parabolic curve, based on excited state molecular structures.

4 Substituent Effects on Photo-Induced Acute Toxicity of PAHs

PAHs that are phototoxic have HOMO-LUMO gap energies within a fairly narrow range of about 0.8 eV. Rather than calculating this parameter for thousands of chemicals, we systematically examined the effect of substitution on the HOMO-LUMO gap [14]. It was found that alkyl and hydroxy substituents do not significantly change the HOMO-LUMO gap of PAHs. Consequently, alkylated aromatics and phenols are likely to be phototoxic only if the parent aromatic structure is phototoxic. It is also likely that the principal influence of these substituents on toxicity will be through a change in bioaccumulation of the chemicals. Nitro, alkene and chloro substituents have a more pronounced effect on HOMO-LUMO gap. The p-π and π-π interactions permit hyperconjugation between the electronic structures and aromatic rings which causes a decrease or "red-shift" to gap energies. The maximum reduction of HOMO-LUMO gap for monosubstituted derivatives was approximately 0.42 eV, and that for disubstituted derivatives was approximately 0.74 eV. While the magnitude of substituent effects is not large, derivatives of parent aromatic chemicals which have a gap energy between 7.5 and 8.0 eV could be found to be phototoxic after substitution by nitro, alkene and chloro substituents.

5 CONCLUSION

Photo-induced toxicity of PAHs was found to be a result of competing internal (structural) and external (light characteristics) factors. The molecular descriptor of the energy difference between the highest occupied molecular orbital and the lowest unoccupied molecular orbital (HOMO-LUMO gap) discriminated between chemicals which either were or were not phototoxic to *Daphnia magna* in the presence of UV radiation. A bilinear (bell-shaped) curve described the relationship, and QSAR models that predict toxicity were developed for both the ascending and descending components of this curve. The same type of relationship between HOMO-LUMO gap and toxicity was observed with PAHs in both the ground and triplet excited states.

Most phototoxic PAHs have HOMO-LUMO gap energies of 7.2 ± 0.4 eV. The additions of alkane or hydroxy substituents to parent PAHs had negligible effects upon phototoxicity. Phototoxicity in such cases is likely only if the parent compounds are already phototoxic. However, substitutions of nitro, alkene and chloro substituents can reduce the HOMO-LUMO gaps of parent PAHs. This could result in a non-phototoxic compound becoming phototoxic with the addition of these substituents.

KEYWORDS

Toxicity, PAH toxicity, global change, solar photo-induced toxicity, hydrocarbons, chemistry alkene, orbital energy, polycyclic aromatic hydrocarbon

REFERENCES

[1] Veith, G.D., Call, D.J. and Brooke, L.T. "Structure-toxicity relationships for the fathead minnow", *Pimephales promelas: Narcotic industrial chemicals*, Can. J. Fish. Aqua. Sci. **40**, 743-748 (1983).

[2] Thakker, D.R., Yagi, H., Levin, W., Wood, A.W., Conney, A.H., and Jerina, D.M. "Polycyclic aromatic hydrocarbons: metabolic activation to ultimate carcinogens", In, Bioactivation of Foreign Compounds, M.W. Anders, Ed., Academic Press, NY pp. 177-242 (1985).

[3] Allred, P.M. and Giesy, J.P. "Solar radiation-induced toxicity of anthracene to *Daphnia pulex*", *Environ. Toxicol. Chem.* **4**, 219-226 (1985).

[4] Newsted, J.L. and Giesy, J.P. "Predictive models for photoinduced acute toxicity of polycyclic aromatic hydrocarbons to *Daphnia magna*", Strauss (Cladocera, Crustacea), *Environ. Toxicol. Chem.* **6**, 445-461 (1987).

[5] Holst, L.L. and Giesy. J.P. "Chronic effects of the photoenhanced toxicity of anthracene on *Daphnia magna* reproduction", *Environ. Toxicol. Chem.* **8**, 933-942 (1989).

[6] Gala, W.R. and Giesy, J.P. "Photo-induced toxicity of anthracene to the green alga, *Selenastrum capricornutum*", *Arch. Env. Contam. Toxicol.* **23**, 316-323 (1992).

[7] Huang, X.D., Dixon, D.G., and Greenberg, B.M. "Impacts of UV radiation and photomodification on the toxicity of PAHs to the higher plant *Lemna gibba*", Duckweed, *Environ. Toxicol. Chem.* **12**, 1067-1077 (1993).

[8] Morgan, D., Warshawsky, D., and Atkinson, T. "The relationship between carcinogenic activities of polycyclic aromatic hydrocarbons and their singlet, triplet, and singlet-triplet splitting energies and phosphorescence lifetimes", *Photochemistry and Photobiol.* **25**, 31-38 (1977).

[9] Mekenyan, O.G., Ankley, G.T., Veith, G.D., and Call, D.J. "QSARs for photo-induced toxicity: I. acute lethality of polycyclic aromatic hydrocarbons to Daphnia Magna", *Chemosphere* , **28**, 567-582 (1994).

[10] Parr, R.G. and Zhou, Z. "Absolute Hardness: Unifying Concept for Identifying Shells and Subshells in Nuclei, Atoms, Molecules, and Metallic Clusters", *Acc. Chem. Res.* **26**, 256-258 (1993).

[11] Pearson, R.G. "The Principle of Maximum Hardness", *Acc. Chem. Res.* **26**, 250-255 (1993).

[12] Mekenyan, O.G., Ankley, G.T, Veith, G.V., and Call, D.J. "QSAR estimates of excited states and photo-induced acute toxicity of PAHs", *SAR and QSAR Environ. Research*, **2**, 237-247, (1994).

[13] Bader, R.F.W. *Can. J. Chem.* **40**, 1164-1168 (1962).

[14] Veith, G.D., Mekenyan, O.G., Ankley, G.T., and Call, D.J. "A QSAR analysis of substituent effects on the photo-induced acute toxicity of PAHs", *Chemosphere*, **30**, 2129-2142, (1994).

[15] Mekenyan, O.G., Ankley, G.T., Veith, G.D., and Call, D.J. "QSAR evaluation of α-terthienyl phototoxicity", Environ. Sci. Technol. , **29**, 1267-1272, (1995).

[16] Marles, R.J., Compadre, R.L., Compadre, C.M., Soucy-Breau, C., Redmond, R.W., Duval, F., Mehta, B., Momand, P., Scaiano, J.C., and Arnason, J.T. "Thiophenes as mosquito larvicides: structure-toxicity relationship analysis", *Pesticide Biochemistry and Physiology*, **41**, 89-100 (1991).

[17] Davenport, R., Johnson, R.R., Schaeffer, D.J., and Balbach, H. "Phototoxicology. I. Light-enhanced toxicity of TNT and some related compounds to Daphnia magna and Lytechinus variegatus Embryos", *Ecotoxicol. Environment. Safety.* **27**, 14-22 (1994).

[18] Mekenyan, O.G., Ivanov, J.M., Veith, G.D., and Bradbury, S.P. "DYNAMIC QSAR: A methodology for investigating structure-activity relationships", *Quant. Str.-Act. Relat.*, **13**, 303-307, (1994).

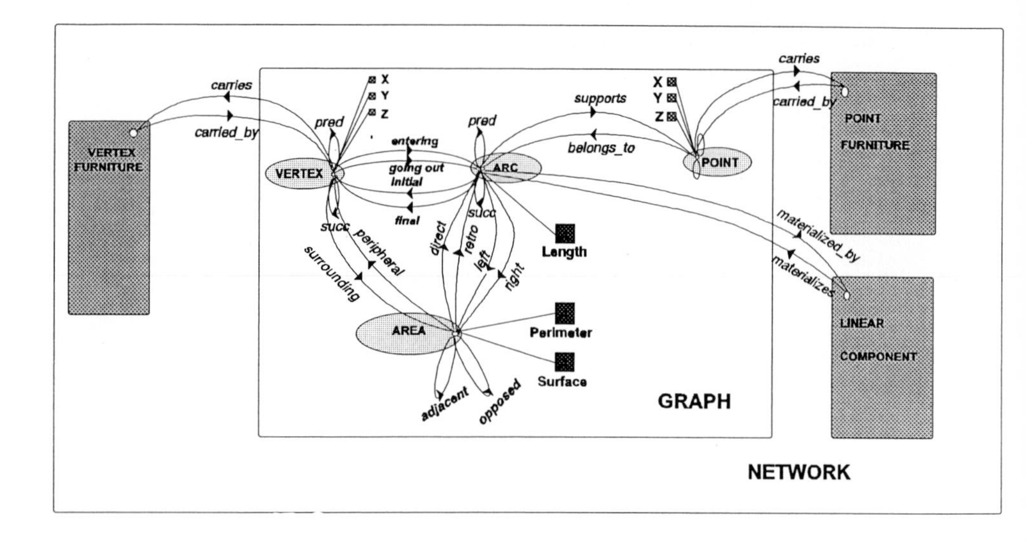

(a) Flow chart explaining the "NETWORK" prototype

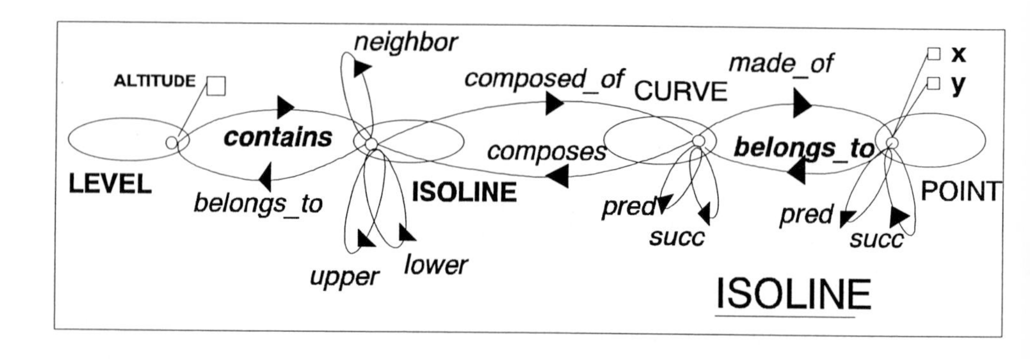

(b) Flow chart explaining the "ISOLINE" prototype

Object Oriented Expert System in G.I.S. - (a) The "NETWORK" prototype is able to model the networks of all the geographical phenomena, the only differences come from the content of the three large hyperclasses respectively named "Vertex_furniture", "Point_furniture" and "Linear_component"; (b) The "ISOLINE" Prototype is a general structure which is able to model any phenomenon represented by prototypes in Geophysics as well as in Topography or Meteorology. (After Bouillé, see page 245).

Cautionary Comments Concerning the Use of Profiles of Polycyclic Aromatic Hydrocarbons (PAH) for Source Apportionment

Arthur GREENBERG

Department of Chemistry, University of North Carolina at Charlotte, Charlotte, NC 28 223 USA

ABSTRACT

Although there are some indications that relative concentrations or profiles of polycyclic aromatic hydrocarbons (PAH) may be useful in source apportionment for environmental receptor modelling, there are reasons for skepticism and caution in their application. Some putative PAH indicators such as cyclopenta[cd]pyrene are extremely reactive under ambient environmental conditions. Furthermore, this specific compound is more widespread than previously thought. The suggestion by some authors that PAH source apportionment may be applicable in winter is not very useful since, in the USA at least, woodsmoke is the dominant winter contributor and PAH emissions from wood combustion are extremely variable in composition.

Précautions à Prendre sur l'Utilisation des Hydrocarbures Aromatiques Polycycliques (PAH) comme Indicateur des Sources de Pollution

RÉSUMÉ

Bien que quelques travaux aient montré que les concentrations relatives ou les profils des hydrocarbures polycycliques aromatisés (HPA) peuvent être utilisés comme indicateur des sources de pollution, il subsiste quelques raisons pour rester sceptique et prudent quant à leurs applications. Les indicateurs PAH hypothétiques, tels que le pyrène cyclopental (cd) se montrent réactifs dans les conditions environnementales ambiantes. D'autre part, ce composé spécifique est bien plus répandu que ce que l'on pouvait croire autrefois. La suggestion faite par certains auteurs selon laquelle la répartition des sources PAH est applicable ne semble pas très utile, car, du moins aux États Unis, la fumée de bois reste le facteur dominant en hiver et les émissions de PAH à partir de la combustion du bois sont extrêmement variables dans leur composition.

1 INTRODUCTION

Polycyclic aromatic hydrocarbons (PAH) comprise a highly diverse class of widespread environmental pollutants ([1], [2], [3]) They are formed as trace by-products of combustion and pyrolysis of organic materials including fossil fuels (gas, oil, coal), wood, plastics, food, municipal waste, etc. Figure 1 depicts molecular structures for a small number of the PAH with particular attention devoted to the structural diversity of this class.

A number of these compounds (*e.g.* benzo[a]pyrene, **18**) are strong carcinogens, while other closely-related species (*e.g.* benzo[e]pyrene, **19**) are non-carcinogens. Source apportionment for receptor modelling is a traditional discipline of environmental research. Following analysis of the complex mixture of PAH (>>100 compounds) in air particulate collected at a given location, its task is to apportion the contributions arising from diesel-powered vehicles, automobiles, space heating (gas, oil, wood), smoking, etc.

Obviously, the most straightforward scenario would be one in which each source emitted a unique "marker" PAH at a constant (ng/BTU) rate or a consistent PAH "signature" and that these marker PAH would have similar physical properties (*e.g.* nonvolatile and thus totally associated with particulate matter) and would be chemically unreactive under ambient environmental conditions.
Unfortunately, these conditions are never met: PAH mixtures are formed from different sources with substantial overlaps in composition and many of the individual PAH are highly reactive under ambient conditions. Indeed, as Grimmer has noted [4] :

"Irrespective of the type of material to be burned, surprisingly similar profiles are formed at a defined temperature. For example, thermal decomposition of pit coal, cellulose, tobacco, and also of polyethylene, polyvinyl chloride and other plastic materials which is carried out at 1000 ^{o}C (in air or nitrogen), yields the same PAH in very similar ratios... In a model installation, under comparable conditions, similar PAH profiles are formed from different organic materials. Consequently, PAH profiles seem to depend predominantly on temperature."

Nevertheless, many researchers have attempted to apply statistical techniques to source apportionment modelling. An ambitious paper, published almost a decade ago, attempted a critical evaluation of the utility of particulate-phase organic compounds for receptor source apportionment [5]. Although the paper is appropriately cautious, it concludes optimistically concerning the prospective use of PAH for receptor source apportionment:

- "PAH profiles of sources which have been repeatedly sampled by the *same researcher* (italics added) appear to be fairly reproducible."
- "The PAH compounds, in particular, appear most promising. Their relative proportions in emissions from a given source type frequently vary over several orders of magnitude which enhances their potential usefulness for source discrimination."
- "There also appear to be several unique or almost unique organic tracers which may be useful such as levoglucosan, C-14 and cyclopenta (c,d) pyrene."

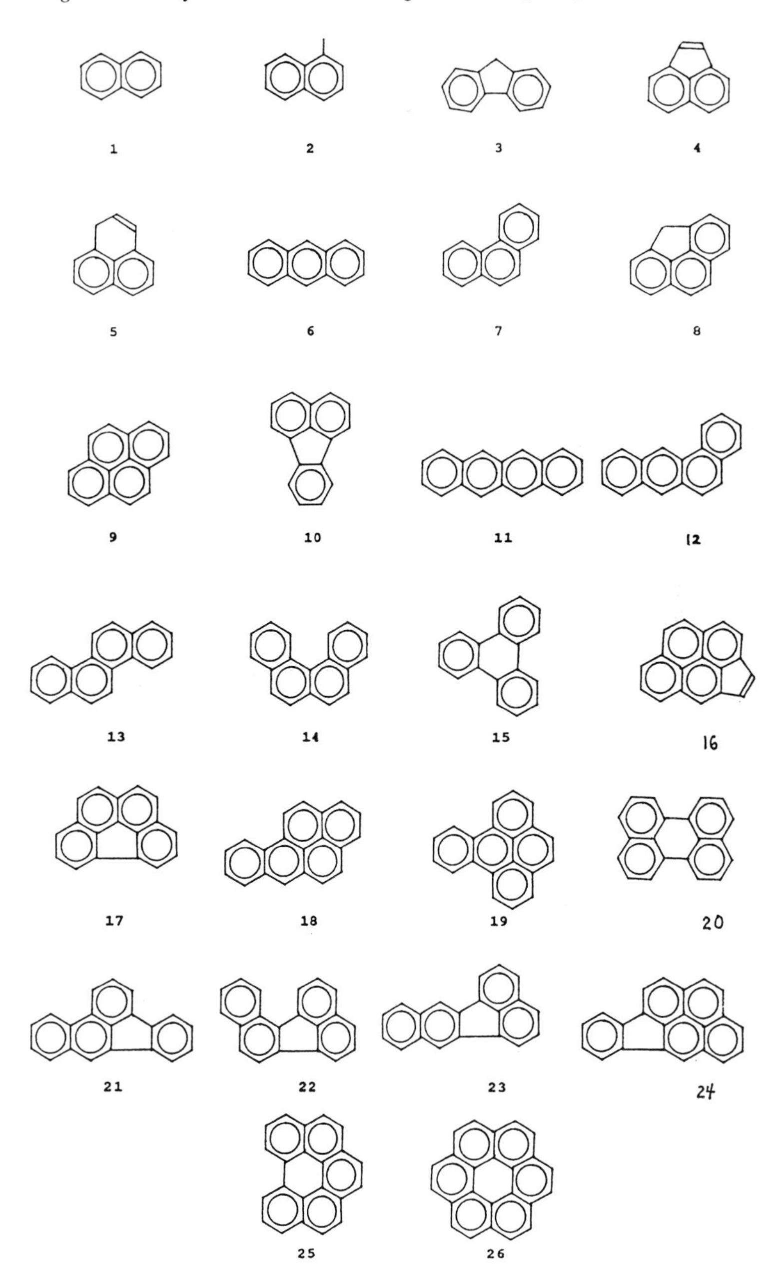

Figure 1. Description of the chemical structures of PAH

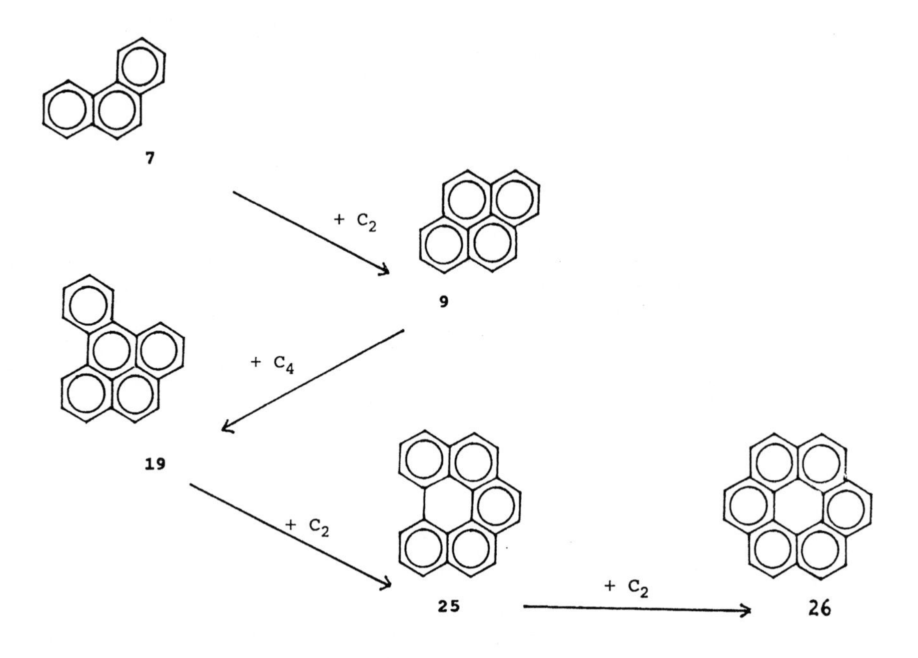

Figure 2. Generation mechanism of PAH

However, in considering the known reactivity of PAH with atmospheric species which have their maximum concentrations during summer periods (*e.g.* photons, OH·, O_3), the authors significantly qualified their conclusions:

- "In view of what is currently known about PAH reactivity, it is advisable at present to attempt to use PAH profiles for source identification only under wintertime conditions."

2 GENERATION OF PAH

Although the specific mechanistic details of PAH formation are not yet known [6, 9], their formation appears to involve stepwise additions to phenyl and phenyl-like radicals of two-carbon species such as acetylene (**27**) and four-carbon species such as diacetylene (**28**) and vinylacetylene (**29**). The growth pattern can be represented in the zig-zag manner of Figure 2 ([8], [9]).

HC≡CH	HC≡C-C≡CH	HC≡C-CH=CH_2
27 ("C_2")	**28** ("C_4")	**29** ("C_4")

Table 1. Relative concentrations (benzo[e]pyrene assumed equal to 1.0) of six $C_{20}H_{12}$ PAH in fresh automobile exhaust [11], fresh oil burner emissions [11] and their comparison with experimental and/or calculated standard gas-phase enthalpies of formation, first ionization potentials and radical localization indices [12].

	Relative Conc.			Properties		
	Auto A	Exhaust B	Oil Burner	$\Delta H_f^o(g)$ kcal/mol	IP_1 (eV)	L_r
Benzo[b]fluoranthene (**20**)[a]	0.8	0.5	0.2	85.4	7.63	2.11
Benzo[k]fluoranthene (**21**)	0.3	0.2	0.04	85.4	7.22	2.13
Benzo[j]fluoranthene (**22**)	0.5	0.3	----	87.4	7.41	2.18
Benzo[a]pyrene (**18**)	1.4	1.4	0.08	69.1	7.12	1.96
Benzo[e]pyrene (**19**)	1.0[b]	1.0[b]	1.0[b]	67.0	7.43	2.23
Perylene (**20**)	0.2	0.2	0.2	73.7	7.00	2.14

[a]. IUPAC Name: Benz[e]acephenanthrylene

[b]. Assumed equal to 1.0

Stein[10] has provided thermodynamic data used for estimating equilibrium concentrations of different benzenoid PAH at high temperature. It is evident from this work that the PAH profile changes as a function of temperature, independent of the fuel.

The molecules depicted in Figure 2 are totally benzenoid and are, therefore, alternant hydrocarbons (every other vertex carbon can be "starred" without two "starred" positions being adjacent). High temperature conditions also produce non-alternant PAH including acenaphthylene (**4**), fluoranthene (**10**), cyclopenta[cd]pyrene (**16**, more properly cyclopenteno[cd]pyrene), benzo[ghi]-fluoranthene (**17**), the benzofluoranthenes **21-23** and indeno[1,2,3-cd]pyrene (**24**).

Although thermodynamics must play a role in the formation of the PAH mixture, it is clearly not a dominant one. Kinetics and mass transport will lead to the formation of kinetically-stable PAH which are not necessarily the most thermodynamically stable. This can be illustrated through comparisons of the concentrations of six isomeric $C_{20}H_{12}$ PAH in fresh automobile exhaust [11] and fresh oil burner emissions [11] with their experimental or calculated gas-phase enthalpies of formation [$H_f^o(g)$] [11] (see Table 1). Based upon $\Delta H_f(g)$ data, and ignoring entropic effects which will not be large, the equilibrium concentrations of the benzofluoranthenes **21-23** should be negligible relative to the benzenoid isomers **18-20**. The compounds benzo[b]fluoranthene (**21**) and benzo[e]pyrene (**19**) are both known to be relatively unreactive [12] (high first ionization potentials (IP_1) and higher radical localization indices (L_r) (see Table 1) and, thus, differences in ambient reactivity are not the origins of this non-equilibrium distribution.

Interestingly, the levels of benzo[e]pyrene found in the cited oil burner emission study [11] are quite high leading to low PAH/B[e]P ratios (Table 1). Thus, it might satisfy some of the criteria as a PAH marker for oil burner emissions. Indeed, in an earlier study we found seemingly elevated levels of B[e]P at three urban New Jersey sites in Winter, 1983 [13].

Table 2. Ratios of selected $C_{20}H_{12}$ PAH to benzo[e]pyrene (B[e]P) as well as to benzo[b]fluoranthene (B[b]F) for Summer, 1982 and Winter, 1983 studies in Newark, New Jersey.[13]

	Summer, 1982		**Winter, 1983**	
	Rel. B[e]P	Rel. B[b]F	Rel. B[e]P	Rel. B[b]F
Benzo[b]fluoranthene	2.1	1.0	0.6	1.0
Benzo[k]fluoranthene	1.0	0.5	0.3	0.5
Benzo[j]fluoranthene	1.3	0.6	0.6	1.0
Benzo[e]pyrene	1.0	0.5	1.0	1.7
Benzo[a]pyrene	1.3	0.6	0.6	1.0
Perylene	0.06	0.03	0.05	0.08

Table 2 lists two sets of ratios (relative to B[e]P and relative to benzo[b]fluoranthene (B[b]F)) for $C_{20}H_{12}$ isomers monitored during Summer, 1982 and Winter, 1983 [13]. The ratios relative to B[e]P look anomalously low for Winter, 1983 compared to Summer, 1982 for the Newark, New Jersey site. However, the variation in ratios relative to B[b]F are much smaller when comparing winter to summer (Table 2). This suggests that winter levels of B[e]P are anomalously high, consistent with high residential oil combustion and the potential utility of B[e]P as a marker (see Table 1) or that there was an experimental artifact in B[e]P determination. This question has not been resolved and we are unaware of any significant use of B[e]P as a marker of residential oil combustion.

3 PAH "Markers" as Artifacts of Unusual Environments

In general, there are sometimes gross differences in PAH composition as a function of source as well as generating conditions. Thus, wood combustion produces high molecular weight unsubstituted (parent) PAH while coal combustion produces alkylated PAH[6]. However, alkylated PAH are very reactive under atmospheric conditions and disappear rapidly with aging [1-3]. One of the most interesting cases for the possible use of a single PAH as marker derives from the work of Grimmer *et al* [14] who found extremely high levels of cyclopenta[cd]pyrene (C[cd]P, **16**) in freshly-collected automobile exhaust and in automobile tunnels. Specifically, the C[cd]P/B[e]P ratio in the tunnel was found to be 5.3 and in fresh automobile exhaust particulate the ratio was reported at 13.6. This compound is known to be highly reactive and Nielsen [15] and our group [16] have demonstrated high loss rates under ambient conditions. This may be demonstrated by the observed winter/summer ratios for selected PAH relative to benzo[b]fluoranthene which are listed in Table 3. Pyrene losses in summer are due to volatility. However, cyclopenta[cd]pyrene is much less volatile than pyrene and its losses should be chemical in nature. Similarly, summertime losses of benzo[a]pyrene exceed those of B[b]F and B[k]F which are about as volatile.

Table 3. Winter/Summer ratios for selected PAH relative to benzo[b]fluoranthene.[16]

PAH	W/S PAH/B[b]F
Pyrene (**9**)	2.75
Benz[a]anthracene (**12**)	1.64
Cyclopenta[cd]pyrene (**16**)	3.35
Benzo[k]fluoranthene (**23**)	0.90
Benzo[a]pyrene (**18**)	1.26
Indeno[1,2,3-cd]pyrene (**24**)	0.90

Table 4. Ratios of selected PAH to B[e]P from residential wood combustion (adapted from ref. 5).

	Seasoned Oak				**Seasoned Pine**	
	Side Draft	Up Draft	Down Draft	High Turbulence	Side Draft	Fire Box
Pyrene (**9**)	3.0	1.7	69	10.8	1.3	3.7
B[b]F (**21**)						
B[j]F (**22**)	4.0	4.3	6.9	1.9	3.0	---
B[k]F (**23**)						
B[e]P (**19**)	1.0	1.0	1.0	1.0	1.0	1.0
B[a]P (**18**)	1.4	3.0	0.49	1.0	1.8	0.9
Perylene (**20**)	0.16	0.19	0.33	1.0	0.26	0.11

Another factor which diminishes the utility of C[cd]P as a marker is that it is generated in many combustion/pyrolysis scenarios. For example, we have found substantial levels of C[cd]P in charbroiled hamburger [17] further confirming that gasoline combustion is not a unique source.
Deep-sea thermal vents have been shown to generate petroleum *de novo* as well as accompanying PAH [18]. One of the most interesting associated findings is the high relative concentration of perylene (**20**) in these mixtures [18]. Perylene is a highly reactive PAH and this is, in part, reflected in its relatively low concentrations in ambient air (Table 2). Its high levels in these deep-sea sediments reflect a probable biotic source (a biological pigment perhaps) but perhaps as important, the reducing nature of the unique environment which thus effectively "preserves" perylene much as cyclopenta[cd]pyrene may be "preserved" in a tunnel which minimizes atmospheric photochemistry.

Since summer time atmospheric chemistry and photochemistry decompose PAH so as to "lose the memory" of the source profile, what of the earlier-cited suggestion to apply PAH profiles to source apportionment in winter only? Here we note the irony that, in New Jersey at least, residential wood combustion is estimated to account for about 6100 kg of the estimated 6300 kg of B[a]P emitted in the State during heating season, [19] and that the aforementioned review [5] notes that the PAH profiles derived from wood combustion are far more variable than those derived from combustion of gasoline, oil or coal (see Table 4). Thus, the ability to use PAH for source apportionment in winter is extremely limited.

4 CONCLUSION

The utility of PAH profiles for source apportionment is far from established. This agrees with the results derived from recent studies employing Multivariate Data Analysis, Principal Component Analysis and Partial Least Squares Regression [20]. The problems include variability of emission from a given source as a function of temperature as well as differences in ambient reactivity. The possibility of using source apportionment in winter [5] is rendered less likely by the overwhelming contribution arising from wood combustion and the high variability in PAH profiles derived from wood combustion [5].

KEYWORDS

PAH production, global change, environmental impact, toxicity, polycyclic aromatic hydrocarbon, combustion emission, automobile exhaust.

REFERENCES

[1] National Academy of Sciences, *Polycyclic Organic Matter*, National Academy Press, Washington, D.C. (1972).
[2] National Academy of Sciences, *Polycyclic Aromatic Hydrocarbons*, National Academy Press, Washington, D.C. (1983).
[3] Harvey R.G., *Polycyclic Aromatic Hydrocarbons: Chemistry and Carcinogenicity*, Cambridge University Press, Cambridge (UK) (1991).
[4] Grimmer G., *Environmental Carcinogens: Polycyclic Aromatic Hydrocarbons*, CRC Press, Boca Raton, p. 29 (1983).
[5] Daisey J.M., Cheney J.L. and Lioy P.J., *J. Air Poll. Control Assoc.*, **36**, 17 (1986).
[6] Lee M.L., Novotny M.V. and Bartle K.D., *Anal. Chem.*, **48**, 1556 (1976).
[7] Howard J.B. and Longwell J.P., In *Polynuclear Aromatic Hydrocarbons: Formation, Metabolism and Measurement*, Cooke, M., Dennis, A.J. (eds), Battelle Press, Columbus, 27-62 (1983).
[8] Fetzer J.C. and Biggs W.R., *Polycycl. Arom. Cmpds.*, **5**, 193 (1994).
[9] Sullivan R.F., Boduszynski M.M. and Fetzer J.C., *Energy Fuels*, **3**, 603 (1989).
[10] Stein S.E., *J. Phys. Chem.*, **82**, 566 (1978).
[11] Grimmer G., Bohnke H. and Glaser A., *Zbl. Bakt. Hyg., I Abt. Orig. B*, **164**, 218 (1977).
[12] Greenberg A. and Darack F.B., In *Molecular Structure and Energetics*, Vol. **4**, Liebman J.F. and Greenberg A. (eds), VCH Pub., New York, 1-47 (1987).
[13] Greenberg A., Darack F., Harkov R., Lioy P. and Daisey J., *Atmos. Environ.*, **19**, 1325 (1985).
[14] Grimmer G., Naujack K.-W. and Schneider D., In *Polynuclear Aromatic Hydrocarbons: Chemistry and Biological Effects*, Bjorseth A. and Dennis A.J. (eds), Battelle Press, Columbus, 107-125 (1979).
[15] Nielsen T., *Atmos. Environ.*, **22**, 2249 (1988).
[16] Greenberg A., *Atmos. Environ.*, **23**, 2797 (1989).
[17] Greenberg A., Hsu C.H., Rothman N. and Strickland P.T., *Polycl. Arom. Cmpds.*, **3**, 101 (1993).
[18] Simonheit B., *Biol. Soc. Wash. Bull.*, No. **6**, 49 (1985).
[19] Harkov R. and Greenberg A., *J. Air Poll. Control Assoc.*, **35**, 238 (1985).
[20] Librando V., *Polycycl. Arom. Cmpds.*, **5**, 175 (1994).

DNA MODIFICATIONS DUE TO OXIDATIVE DAMAGE

J. CADET[1], M. BERGER, I. GIRAULT, M.-F. INCARDONA, D. MOLKO, B. MORIN, M. POLVERELLI, S. RAOUL and J.-L. RAVANAT

CEA/Département de Recherche Fondamentale sur la Matière Condensée, SCIB/LAN, F-38054 Grenoble Cedex 9, France

ABSTRACT

Modified purine and pyrimidine bases which constitute one of the major classes of oxidative DNA damage are likely to be involved in mutagenesis and carcinogenesis and possibly in aging. In the present survey, the currently available information on the structural aspects of oxidized DNA bases and their mechanism of formation are critically reviewed. A survey of the main approaches (HPLC separations associated with various spectroscopic detections, gas chromatography-mass spectrometry, postlabeling techniques, immunoassays ...) involving either initial acid hydrolysis or enzymatic digestion of DNA which were recently developed for monitoring the formation of oxidative DNA base damage in cells, tissues and biological fluids is also presented. The measurement of the above compounds in biological fluids such as urine may be used for assessing oxidative damage to DNA [5]. Other important sources of oxidation processes are provided by physical agents (ionizing radiation, near-ultraviolet/visible light [6,7]). In the latter case, oxidation of DNA would require the presence of endogenous or exogenous photosensitizers including flavins and porphyrins. In this short survey emphasis has been placed on the oxidation reactions of the guanine moiety of DNA and model compounds as the result of one-electron processes and exposure to oxidizing agents including hydroxyl radicals and singlet oxygen. However, it should be mentioned that relevant information on oxidation reactions of adenine, cytosine and thymine, the other DNA bases is available [see for example, 6-10].

Modifications de l'ADN par Oxydation

RÉSUMÉ

L'oxydation des bases puriques et pyrimidiques, principaux composants de l'ADN joue un rôle dans la mutagenèse, la carcinogenèse et peut-être le vieillissement. Dans ce travail, une revue des connaissances actuelles sur les aspects structuraux des bases oxydées de l'ADN et sur les mécanismes de formation est discutée d'un point de vue critique. Les principales approches analytiques sont présentées (séparation par CLHP associée à différentes détections spectroscopiques,

---

[1] To whom correspondence may be addressed

chromatographie gazeuse - spectrométrie de masse, technique de post marquage, immunologie, ...) avec soit une hydrolyse acide initiale, soit une digestion enzymatique de l'ADN. Ces prétraitements ont été récemment développés pour suivre l'oxydation de l'ADN dans les cellules, les tissus, et les liquides biologiques. L'analyse de ces composés dans les milieux biologiques tels que l'urine, peut être utilisée pour estimer les dégâts de l'oxydation sur l'ADN [5]. Le processus oxydatif de l'ADN peut être également étudié au moyen d'agents physiques (radiations ionisantes, lumière dans l'ultra violet proche/visible [6,7]). Dans ce dernier cas, la réaction d'oxydation de l'ADN nécessitera la présence d'agents de photosensibilisation endogènes ou exogènes dont les flavines et les porphyrines. Ce court exposé s'intéresse principalement aux réactions d'oxydation de la guanine de l'ADN et de composés modèles telles que celles résultant de processus à un électron et de l'exposition à des agents oxydants (radicaux hydroxyles et oxygène singulet). Des informations pertinentes sur les réactions d'oxydation de l'adénine, la cytosine et la thymine, autres constituants de l'ADN, peuvent être également trouvées dans la littérature [voir 6-10].

1 RESULTS AND DISCUSSION

The main reactive oxygen species are listed in Table 1. Their reactivity towards DNA components and the chemical reactions of base radical cations that are produced by one-electron reactions are also mentioned. It clearly appears that OH radical and related reactive species including ferryl ions, both of them possibly generated through the Haber-Weiss reaction within cells, are the most reactive agents. Base lesions, abasic sites, DNA strand breaks and DNA-protein crosslinks represent the four main classes of oxidized base damage. The complex nature of the oxidation reactions of the base moieties of DNA is illustrated by considering guanine, the nucleobase which presents the lowest ionization potential among DNA components.

1.1 Oxidation reactions of the DNA guanine moiety

1.1.1 Radical processes involving OH radical and one-electron oxidation

The two overwhelming oxidation products of the purine moiety of 2'-deoxyguanosine (**1**) as resulting from both the reaction OH. radical and the transformation of the guanine radical cation (one-electron oxidation) have been isolated and identified as :

2,2-diamino-4-[(2-deoxy-ß-D-*erythro*-pentofuranosyl)amino]-5(2*H*)-oxazolone (**6**)

and its precursor [10,11] :

2-amino-5-[(2-deoxy-ß-D-*erythro*-pentofuranosyl)amino]-4*H*-imidazol-4-one (**5**)

The mechanism of their production as depicted in Figure 1 may be rationalized in terms of transient formation of the oxidizing guanilyl radical **3** which may arise either from dehydration of the OH. adduct at C-4 **4** or deprotonation of the guanine radical cation **2**.

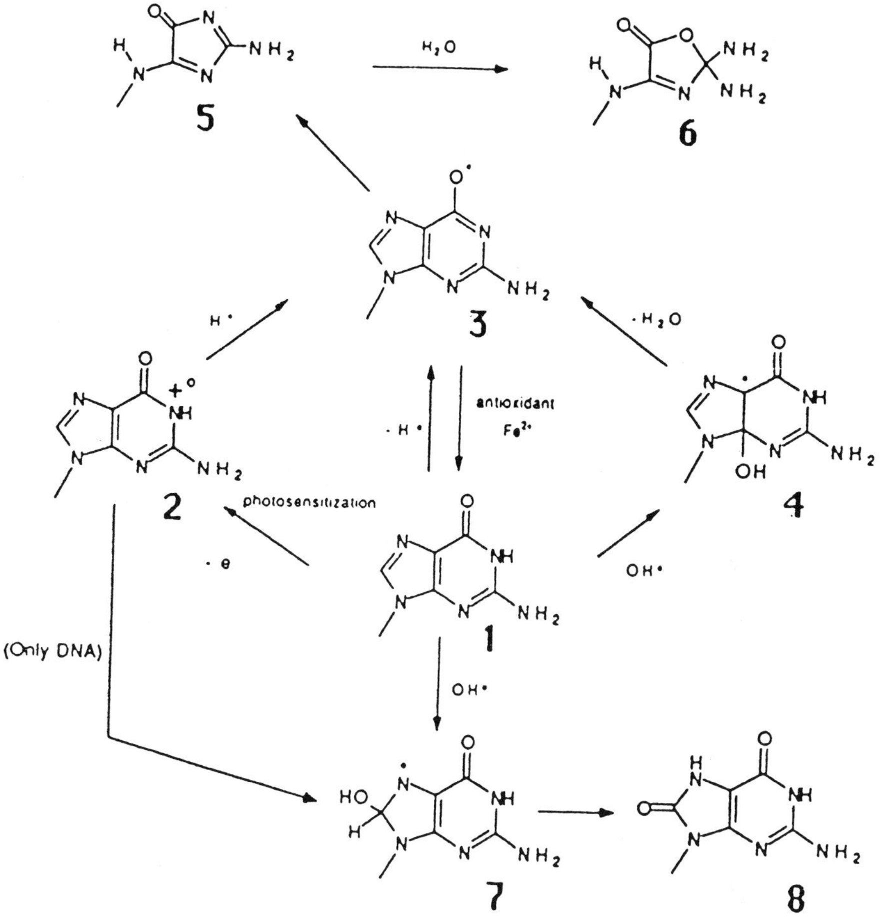

Figure 1. Main radical oxidation pathways of the guanine moiety within nucleosides and DNA.

Then, the resulting neutral radical **3** which may exist in several tautomeric forms is implicated in a rather complicated decomposition pathway. This involves the opening of the pyrimidine ring at the C5-C6 bond, followed by the transient formation of a peroxyl radical arising from the addition of molecular oxygen to tautomeric C(5) carbon centred radical and subsequent nucleophilic addition of a water molecule across the 7,8-ethylenic bond.

This leads after rearrangement to the formation of the unstable imidazolone **5** (half-life = 10 h in aqueous solution at 20 °C) which is then quantitatively converted into the oxazolone **6** [10, 11].

Table 1. Reactive species and radicals involved in oxidative stress

Species - radicals	**reactivity with DNA**
Superoxide radical ($O_2^{\cdot -}$)	not detectable (reduction of ROO·)
Hydroperoxide radical ($HO_2\cdot$)	not detectable
Hydroxyl radical (OH·)	oxidizes bases & sugar moieties.
Ferryl ion	oxidizes bases & sugar moieties
Peroxinitrite ($ONOO^-$)	oxidizes bases & sugar moieties
Singlet oxygen ($^1O_2$)	oxidizes guanine
Ozone (O_3)	oxidizes pyrimidine & purine bases
Hydrogen peroxide (H_2O_2)	oxidizes adenine
Oxyl (RO·) & peroxyl (ROO·) radicals	oxidizes the sugar moieties
Purine & pyrimidine radical cations	hydration & deprotonation

It should be noted that the nucleophilic addition of a water molecule involved in the formation of **6** represents an interesting model system for investigating the mechanism of the generation of DNA-protein cross-links under radical oxidative conditions [11]. The formation of :

8-oxo-7,8-dihydro-2'-deoxyguanosine (8-oxodG) (**8**)

is a minor process when 2'-deoxyguanosine (**1**) is exposed to OH radicals in aqueous aerated solution.

However, the yield of formation of 8-oxodG (**8**) increases at the expense of the oxazolone derivative (**6**) in double-stranded DNA [13]. This is even more striking when considering the transformation reactions of the guanine radical cation within native DNA. Under the latter conditions a significant hydration reaction which was not observed within the free nucleoside **1** and short oligonucleotides was found to give rise to the formation of 8-oxodG (**8**) through the transient [13]:

8-hydroxy-7,8-dihydro-2'-deoxyguanosyl-7-yl radical (**7**)

This illustrates the complexity of the radical oxidation reactions of the guanine moiety of DNA and also shows the similarity in the decomposition pathways mediated by hydroxyl radical and one-electron process.

1.1.2 Oxidation of 2'-deoxyguanosine by Fenton reagents

The use of Fenton reagents to oxidize **1** leads to a complete change in the quantitative aspects of the product distribution since the formation of 8-oxo-7,8-dihydro-2'-deoxyguanosine (**8**) was considerably enhanced, at the expense of those of **5** and **6** which was almost completely abolished.

Table 2. Methodologies for measuring oxidative DNA damage [adapted from [15]]

Methods	DNA	Sensitivity[a]	Amount of DNA (μg)
HPLC/electrochemistry	Hydrolyzed	1×10^{-5}	25 - 50
HPLC/MS (thermospray)	Hydrolyzed	10^{-4} - 10^{-5}	30 - 60
HPLC-fluorescence	Hydrolyzed	5×10^{-5}	4 - 8
CG/SIM-MS	Hydrolyzed	1×10^{-5}	50 - 100
HPLC-^{32}P-postlabeling	Hydrolyzed	1×10^{-6}	1 - 5
Immunology (RIA-ELISA)	Intact	10^{-5} - 10^{-6}	2 - 10
DNA-glycosylases (alkaline elution)	Intact	5×10^{-7}	10

[a] Sensitivity is indicated with respect to normal bases

This may be explained in terms of competition between the reduction of the oxidizing guanylyl radical **3** and the reaction of the latter radical with molecular oxygen [14, 15].

1.1.3 Singlet oxygen oxidation of the guanine moiety

The two main products of $^1O_2$ oxidation of 2'-deoxyguanosine (**1**) were identified as the 4R* and 4S* diastereoisomers of :

8-oxo-4,8-dihydro-4-hydroxy-2'-deoxyguanosine (**9**).

A reasonable mechanism for the formation of the latter specific oxidation products involves [4 + 2] cycloaddition of $^1O_2$ to the guanine moiety according to a Diels-Alder mechanism producing unstable 4,8-endoperoxides [16,17]. It should be added that 8-oxodG (**8**) is also produced by the reaction of $^1O_2$ with 2'-deoxyguanosine (**1**), but in about only 1:7 ratio of **9**. However, in double-stranded DNA, the formation of 8-oxodG (**8**) becomes predominant at the expense of 8-oxo-4,8-dihydro-4-hydroxy-2'-deoxyguanosine (**9**).

2 Methods for Measuring Oxidative DNA Base Damage

The measurement of oxidative base damage in tissue and cellular DNA remains a challenging analytical problem [18]. This may be due, at least partly, to the high level of sensitivity that is required (the threshold of detection is close to one single lesion per 10^5 - 10^6 normal bases in a sample size of DNA lower than 30 μg) together with the multiplicity and, sometimes, the lability of the lesion.

In addition, one of the main limiting factors is the occurrence of autoxidation reactions during the work-up of DNA which may induce a significant level of oxidized base damage. This is particularly true for 8-oxodG (**8**), whose formation is significantly enhanced under Fenton reaction conditions (see above). In addition to the measurement of damage within isolated cells, two main approaches are currently developed for singling out specific oxidative DNA damage.

The available assays for monitoring oxidative base damage within whole DNA or after hydrolysis of the biopolymer have been critically reviewed in two recent surveys [18, 19]. The main methods which in most cases require either enzymatic digestion or chemical hydrolysis of DNA are presented in Table 2.

The most sensitive methods of detection currently available involve radioactive postlabeling of nucleoside 3'-monophosphates and dinucleoside monophosphates which are susbtrates for polynucleotide kinases.

One example is illustrated by the HPLC-^{32}P-postlabeling assay that was recently devised for the measurement of adenine N-1 oxide in cellular DNA exposed to hydrogen peroxide and UVA radiation [20]. The sensitivity of the method is of one modification per 10^7 normal bases in a sample size of 1 µg of DNA.

Similar assays have been developed for the detection of 5-hydroxymethyluracil, 8-oxo-7,8-dihydro-guanine, 5,6-dihydroxy-5,6-dihydrothymine, phosphoglycolate residues and abasic sites [for a recent review, see [21]].

It should be added that chemical postderivatization involving radioactive and fluorescence detections are also available [21]. The most widely used assay deals with the measurement of 8-oxodG **8** by electrochemical detection after high performance liquid separation (HPLC).

Another interesting assay involves the combined use of gas chromatography (GC) with mass spectrometry. However, the method despite its high intrinsic sensitivity, suffers from the fact that about 200 µg of DNA are required [18].

3 Non Invasive Assays

Evaluation of the effects of oxidative stress on DNA in humans requires the development of non-invasive assays. Attempts are currently being made to use the release of oxidized bases and nucleosides in urine as an index of DNA damage. For this purpose, HPLC-EC and GC-MS assays have been applied to the measurement of several oxidized DNA compounds including mostly 8-oxo-7,8-dihydroguanine, 5-hydroxymethyluracil and their corresponding nucleosides [22, 23].

A significant increase in the release of 8-oxodG (**8**) was observed in human urine and leukocytes of cigarette smokers [for a review, see [15]].

These few examples illustrate the potentiality of using oxidized bases and related nucleosides as indicators for epidemiological studies aimed at correlating dietary and life habits with cancer risk.

4 CONCLUSION

This review clearly shows the complexity of the oxidation reactions of DNA. There is still an important need for development of accurate and sensitive methods of detection of oxidative base damage in cellular and tissue DNA. Non-invasive methods are particularly relevant for epidemiological studies. Further developments would involve both molecular biology techniques (polymerase chain reaction) and effective analytical methods such as capillary electrophoresis associated with the sensitive and versatile electrospray mass spectrometry detection technique. Efforts are currently made to determine the biological role of oxidized base damage. One major aspect deals with repair studies which have already shown the major role played by glycosylase proteins in the enzymatic removal of damaged bases [for recent papers and reviews, see [24-26]].

The mutagenic assessment of oxidized DNA bases is based on site specific incorporation of a single modification in sequence defined oligonucleotides [27]. Then, the resulting oligonucleotides can either be used as templates for DNA replication investigations or further elongated prior to incorporation in eukaryotic cells for mutagenic evaluation [for a recent review, see [28]].

KEYWORDS

Oxidative DNA damage, mutagenesis, genetic database, cancer risk, DNA database, toxicology, biological effect, hydroxyl radical, guanine, epidemiological study.

REFERENCES

[1] Ames B.N., Gold L.S. "Endogenous mutagens and the cause of aging and cancer", *Mutat. Res.,* **250**, 3-16 (1991).

[2] Sies H., ed. Oxidative Stress, *"Oxidants and Antioxidants"*, New York, Academic Press Inc., (1991).

[3] Lindahl T. "Instability and decay of the primary structure of DNA", *Nature,* **362**, 709-715 (1993).

[4] Srinivasan S., Glauert H.P. "Formation of 5-hydroxymethyl-2'-deoxyuridine in hepatic DNA of rats treated with gamma-irradiation, diethylnitrosamine, 2-acetylaminofluorene or the peroxisome proliferator ciprofibrate", *Carcinogenesis,* **11**, 2012-2024 (1990).

[5] Shinenaga M.K., Gimeno C.L., Ames B.N. "Urinary 8-hydroxy-2'-deoxyguanosine as a biological marker of in vivo oxidative DNA damage", *Proc. Natl. Acad. Sci. USA*, **86**, 9697-9701 (1989).

[6] von Sonntag C. ed. *"The Chemical Basis of Radiation Biology"*, London: Taylor Francis, (1987).

[7] Cadet J., Vigny P. "The photochemistry of nucleic acids", In: Morrison H, ed. *Bioorganic Photochemistry,* Vol. **1**, New York: Wiley and Sons, 1-272 (1990).

[8] Wagner J.R., van Lier J.E., Decarroz C., Berger M., Cadet J. "Photodynamic methods for oxy radical-induced DNA damage", *Methods Enzym,* **186**, 502-511 (1990).

[9] Wagner J.R., van Lier J.E., Berger M., Cadet J. "Thymidine hydroperoxides: Structural assignment, conformational features, and thermal decomposition in water", *J. Am. Chem. Soc.,* **116**, 2235-2242 (1994).

[10] Cadet J. "DNA damage caused by oxidation, deamination, ultraviolet radiation and photoexcited psoralens", In: Hemminki K et al, eds. *DNA adducts: Identification and biological significance*, Lyon: International Agency for Research on Cancer, IARC Scientific Publications, **125**, 245-276 (1994).

[11] Cadet J., Berger M., Buchko G.W., Joshi P.C., Raoul S., Ravanat J.L. "2,2-Diamino-4-[(3,5-di-O-acetyl-2-deoxy-ß-D-*erythro*-pentofuranosyl)amino]-5-(2*H*)-oxazolone: A novel and predominant radical oxidation product of 3',5'-di-O-acetyl-2'-deoxyguanosine", *J. Am. Chem. Soc.*, **116**, 7403-7404 (1994).

[12] Morin B., Cadet J. "Benzophenone photosensitisation of 2'-deoxyguanosine: Characterization of the 2R and 2S diastereoisomers of 1-(2-deoxy-ß-D-*erythro*-pentofuranosyl) -2-methoxy-4, 5-imidazolidinedione. A model system for the investigation of photosensitized formation of DNA-protein crosslinks", *Photochem. Photobiol* ., **60**, 102-109 (1994).

[13] Kasai H., Yamaizumi Z., Berger M., Cadet J. "Photosensitized formation of 7,8-dihydro-8-oxo-2'-deoxyguanosine (8-hydroxy-2'-deoxyguanosine) in DNA by riboflavin: a non singlet oxygen mediated reaction", *J. Am. Chem. Soc.*, **114**, 9692-9694 (1992).

[14] Berger M., de Hazen M., Nejjari A., Fournier J., Guignard J., Pezerat H., Cadet J. "Radical oxidation reactions of the purine moiety of 2'-deoxyribo-nucleosides and DNA by iron-containing minerals", *Carcinogenesis*, **14**, 41-46 (1993).

[15] Cadet J., Berger M., Buchko G.W., Incardona M.F., Morin B., Raoul S., Ravanat J.L. "DNA oxidation: Characterization of the damage and mechanistic aspects", In: R. Paoletti et al. *Oxidative Processes and Antioxidants*, New York: Raven Press, 97-115 (1994).

[16] Buchko G.W., Cadet J., Berger M., Ravanat J.L. "Photooxidation of d(TpG) by phthalocyanines and riboflavin. Isolation and characterization of dinucleoside monophosphates containing the 4R* and the 4S* diastereoisomers of 4,8-dihydro-4-hydroxy-8-oxo-2'-deoxyguanosine", *Nucleic. Acids Res.*, **20**, 4847-4851 (1992).

[17] Cadet J., Ravanat J.L., Buchko G.W., Yeo H.C., Ames B.N. "Singlet oxygen DNA damage: Chromatographic and mass spectrometric analysis of damage products", *Methods Enzym,* **234**, 79-88 (1994).

[18] Cadet J., Weinfeld M. "Detecting DNA damage", *Anal Chem*, **65**, 675A-682A (1993).

[19] Frenkel K, Klein C.B. "Methods used for analyses of environmentally damaged nucleic acids, *J. Chromatogr.*, **618**, 289-314 (1993).

[20] Mouret J.F., Odin F., Polverelli M., Cadet J. "^{32}P-Postlabeling measurement of adenine N-1-oxide in cellular DNA exposed to hydrogen peroxide", *Chem. Res. Toxicol.*, **3**, 102-110 (1990).

[21] Cadet J., Odin F., Mouret J.F., Polverelli M., Audic A., Giacomoni P., Favier A., Richard M.J. "Chemical and biochemical postlabeling methods for singling out specific oxidative DNA lesions", *Mutat. Res.*, **275**, 343-354 (1992).

[22] Shigenaga M.K., Aboujaoude E.N., Chen Q., Ames B.N. "Assays of oxidative DNA damage biomarkers 8-oxo-2'-deoxyguanosine and 8-oxoguanine in nuclear DNA and biological fluids by high performance liquid chromatography with electrochemical detection", *Methods Enzym.*, **234**, 16-33 (1994).

[23] Faure H., Incardona M.F., Boujet C., Cadet J., Ducros V., Favier A. "Gas chromatographic-mass spectrometric determination of 5-hydroxymethyluracil in human urine by stable isotope dilution", *J. Chromatogr. Biom. Appl.*, **616**, 1-7 (1993).

[24] Boiteux S. "Properties and biological functions of the NTH and FPG proteins of *Escherichia* coli: two glycosylases that repair oxidative damage in DNA", *J. Photochem. Photobiol. B*: *Biol.*, **19**, 87-96 (1993).

[25] Hatahet Z., Kow Y.W., Purmal A.A., Cunningham R.P., Wallace S.S. "New substrates from old enzymes", *J. Biol. Chem.*, **269**, 11814-18820 (1994).

[26] Tchou J., Bodepudi V., Shibutani S., Antoshechkin I., Miller J., Grollman A.P., Johnson F. "Substrate specificity of Fpg protein", *J. Biol. Chem.*, **269**, 15318-15324 (1994).

[27] Guy A., Duplaa A.M., Ulrich J., Téoule R. "Incorporation by chemical synthesis and characterization of deoxyribosylformylamine into DNA", *Nucleic. Acids Res.*, **19**, 5815-5820 (1991).

[28] Grollman A.P., Moriya M. "Mutagenesis by 8-oxoguanine: an enemy within", *Trends Gene*t, **9**, 246-249 (1993).

INDEX OF AUTHORS

K - N

O - Q

R

S - T

U - Z

INDEX

A

B - C

D

E

F

G

H

I

K

L

M

N

O

P

Q

R

S

T

U

V

W

Color Plates

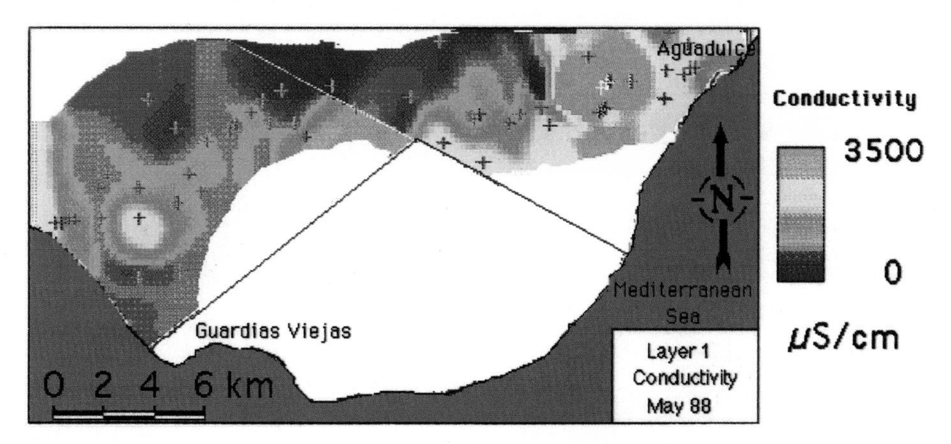

Fig. 6. Map of conductivity in Layer 1, May 1988. (see also page 68)

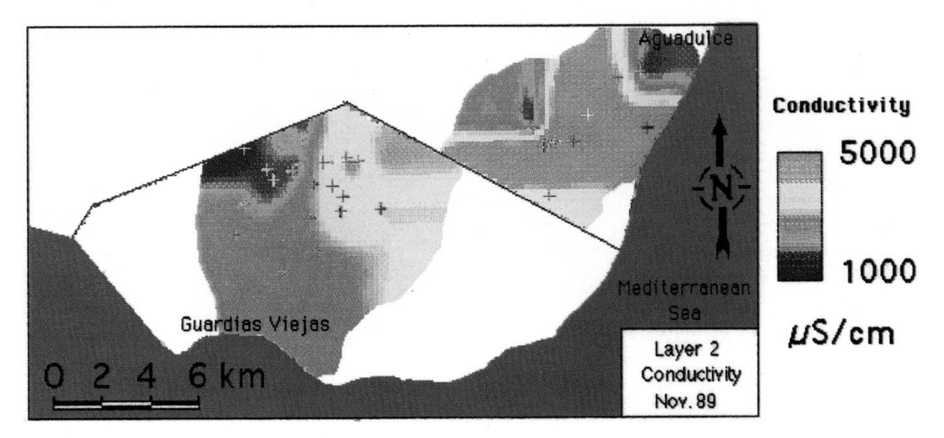

Fig. 7. Map of conductivity in Layer 2, November 1989. (see also page 68)

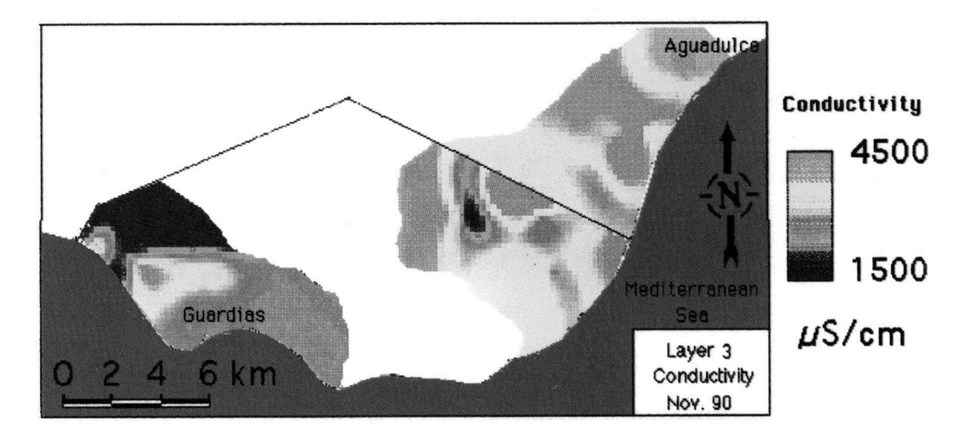

Fig. 8. Map of conductivity in Layer 3, November 1990. (see also page 68)

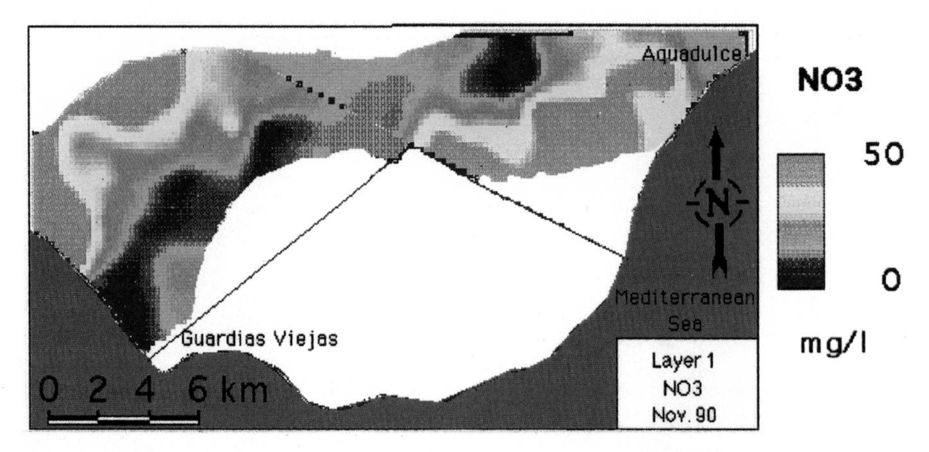

Fig. 12. Map of nitrates in Layer 1, November 1990. (see also page 72)

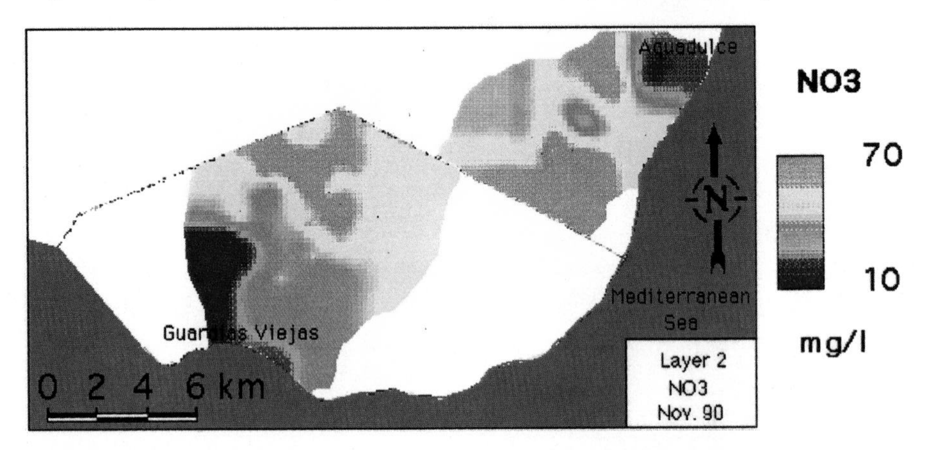

Fig. 13. Map of nitrates in Layer 2, November 1990. (see also page 72)

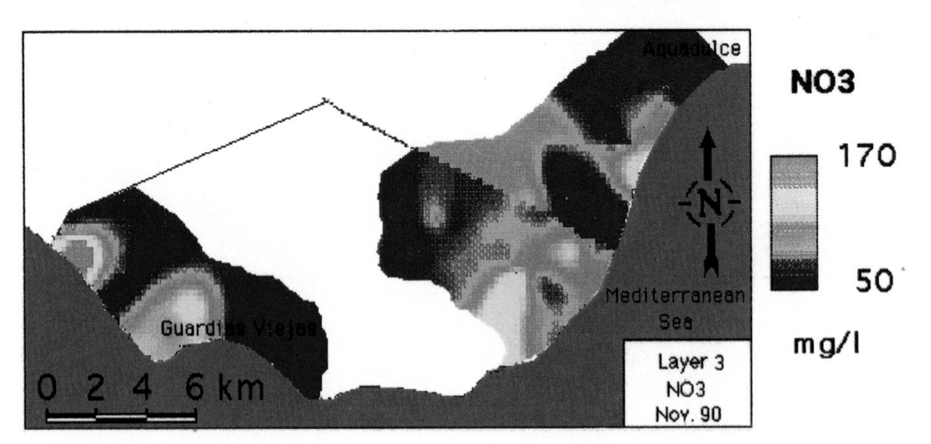

Fig. 14. Map of nitrates in Layer 3, November 1990. (see also page 72)

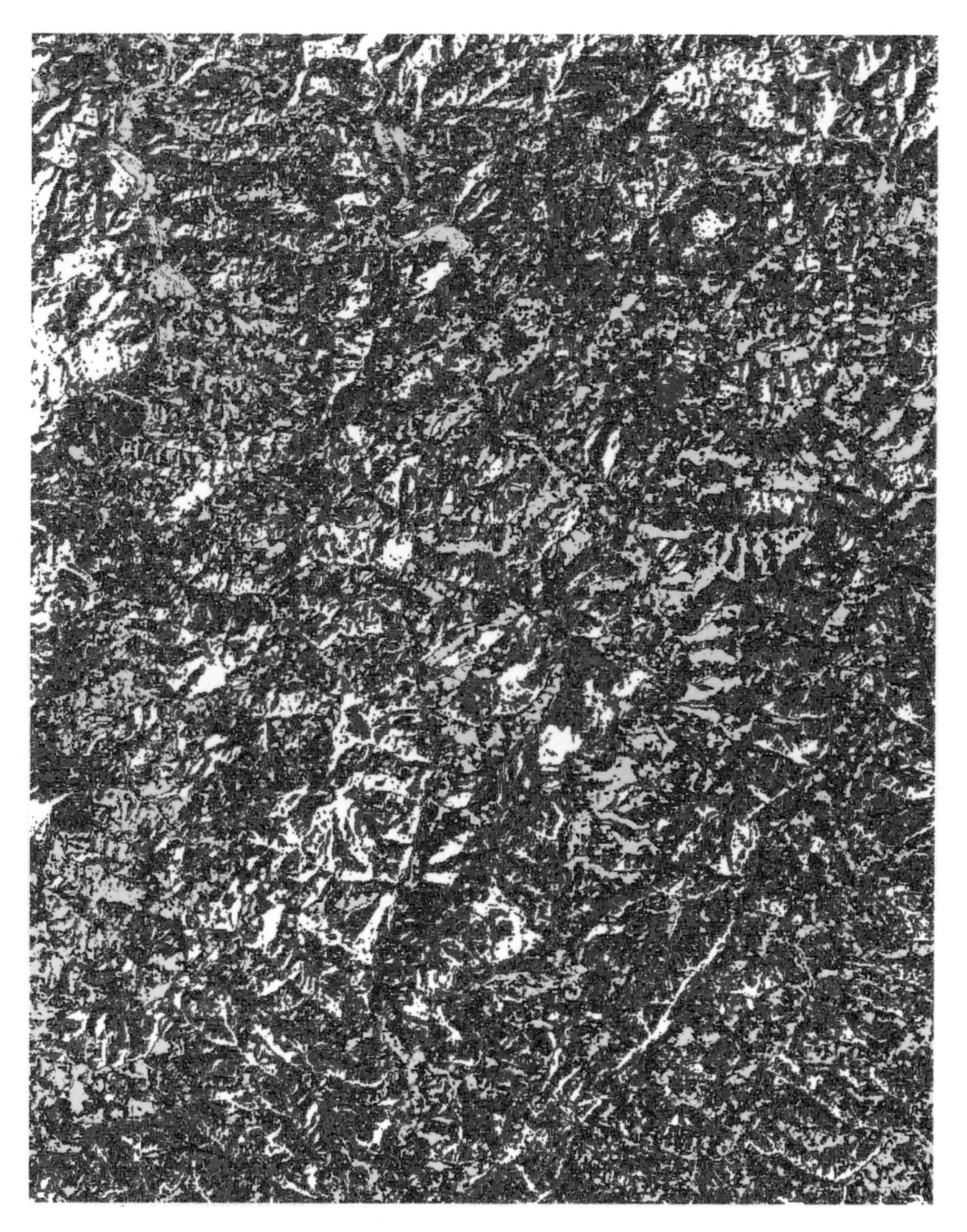

Fig. 1. The classified studied area using the results obtained by different classification algorithms: Bormida Valley land use map using Landsat 5 data. (see also page 153)

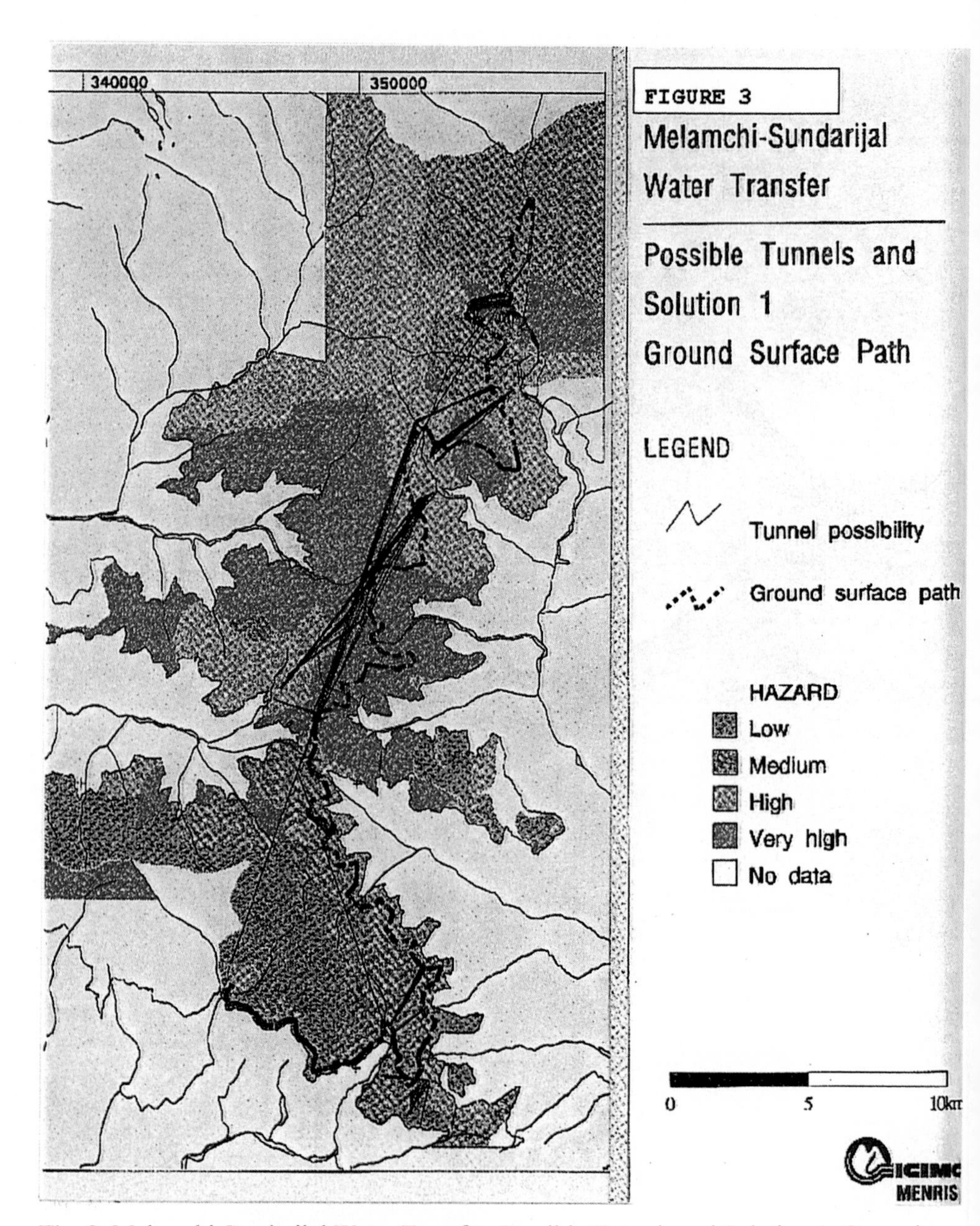

Fig. 3. Melamchi-Sundarijal Water Transfer. Possible Tunnels and Solution 1: Ground Surface Path. (see also page 169)

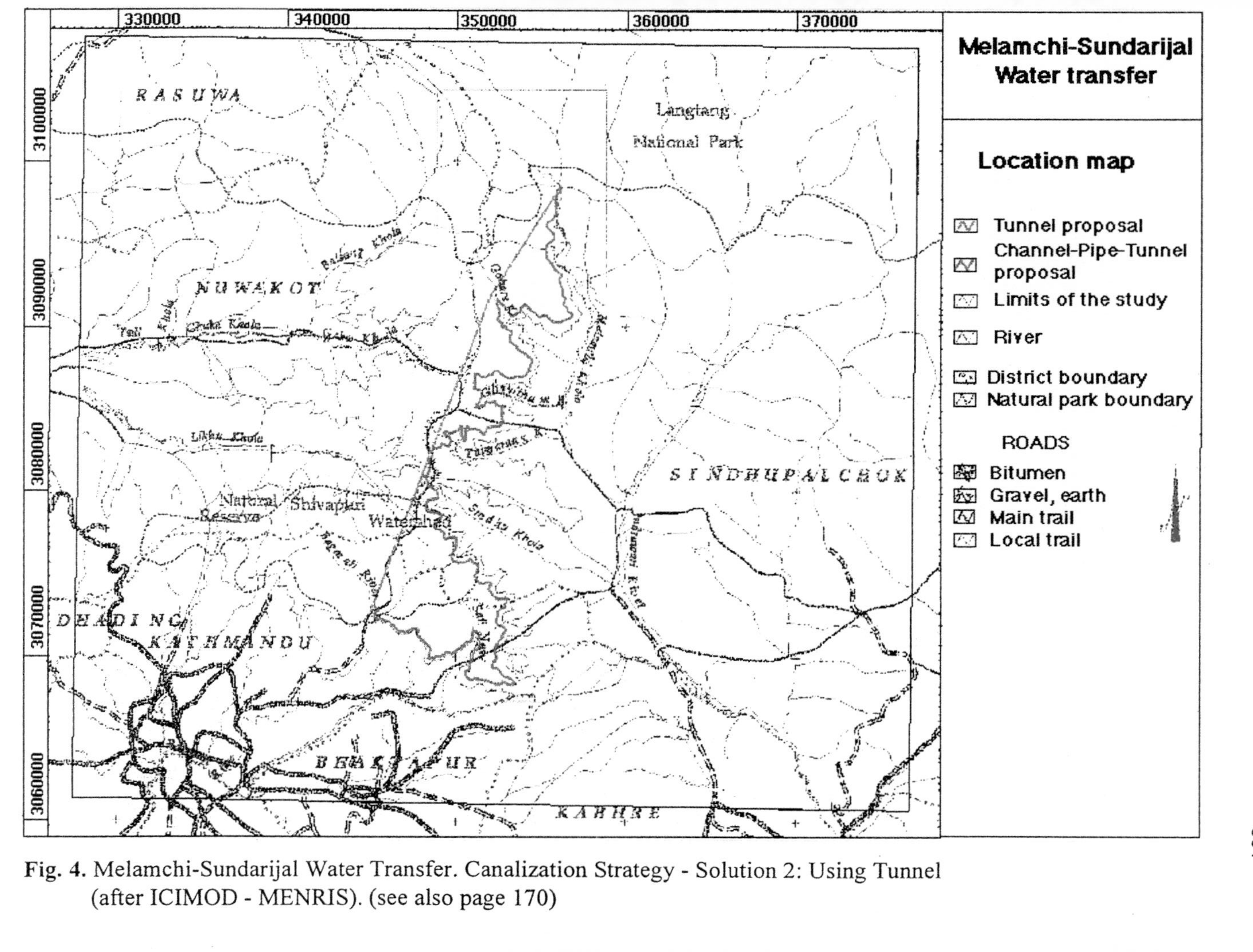

Fig. 4. Melamchi-Sundarijal Water Transfer. Canalization Strategy - Solution 2: Using Tunnel (after ICIMOD - MENRIS). (see also page 170)

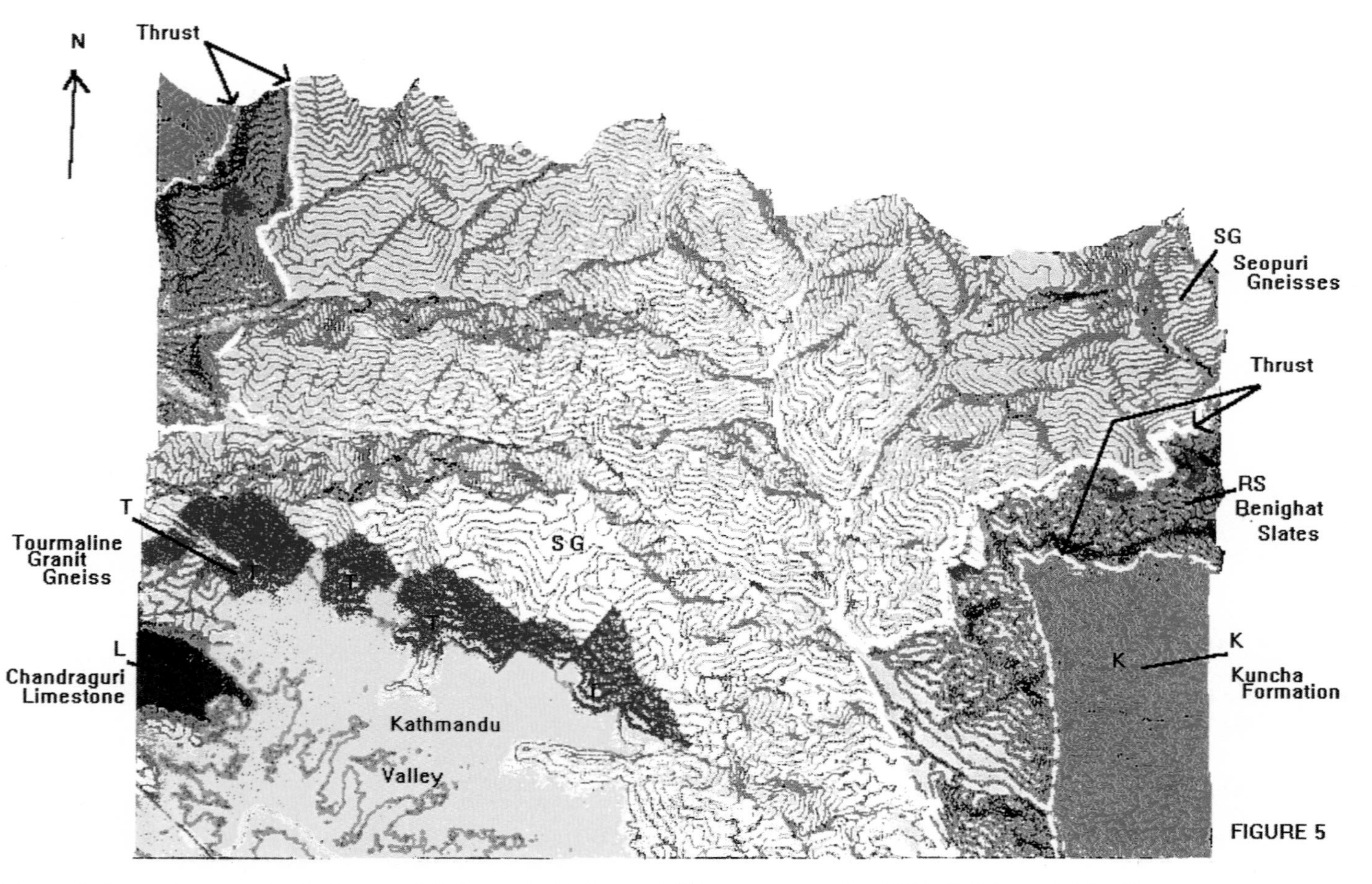

Fig. 5. 3-dimensional geological map of Kathmandu Valley and Shivapuri area: the SPOT-DEM overlaid with the geological data. (see also page 172)

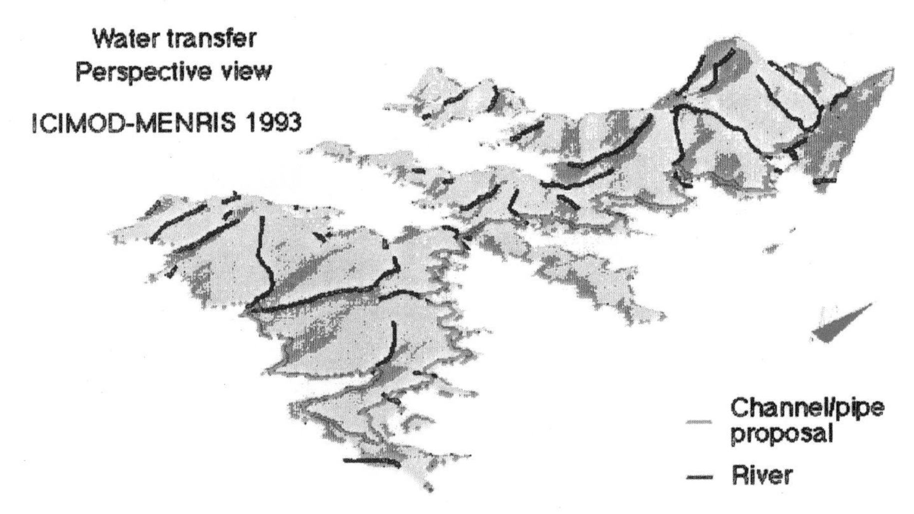

Fig. 2. Melamchi-Sundarijal Water Transfer: Spot Relief View Digital Elevation Model above 1600 m altitude overlaid by the Channel/Pipe Proposal. Sun elevation 35° and azimuth 100°, Observator elevation 20° and azimuth 150° (after ICIMOD). (see also page 168)

Fig. 1. Landsat MSS false color composite image (bands 7,5,4 in RGB) acquired on March 21, 1977. This scene represents the northern part of maritime Syria. (see also page 199)

Fig. 2. Landsat MSS false color composite image (bands 7,5,4 in RGB) acquired on April 05, 1986 (Northern part of maritime Syria). (see also page 200)

Fig. 3. Landsat MSS false color composite image (bands 7,5,4 in RGB) acquired on March 23, 1993 (Northern part of maritime Syria). (see also page 201)

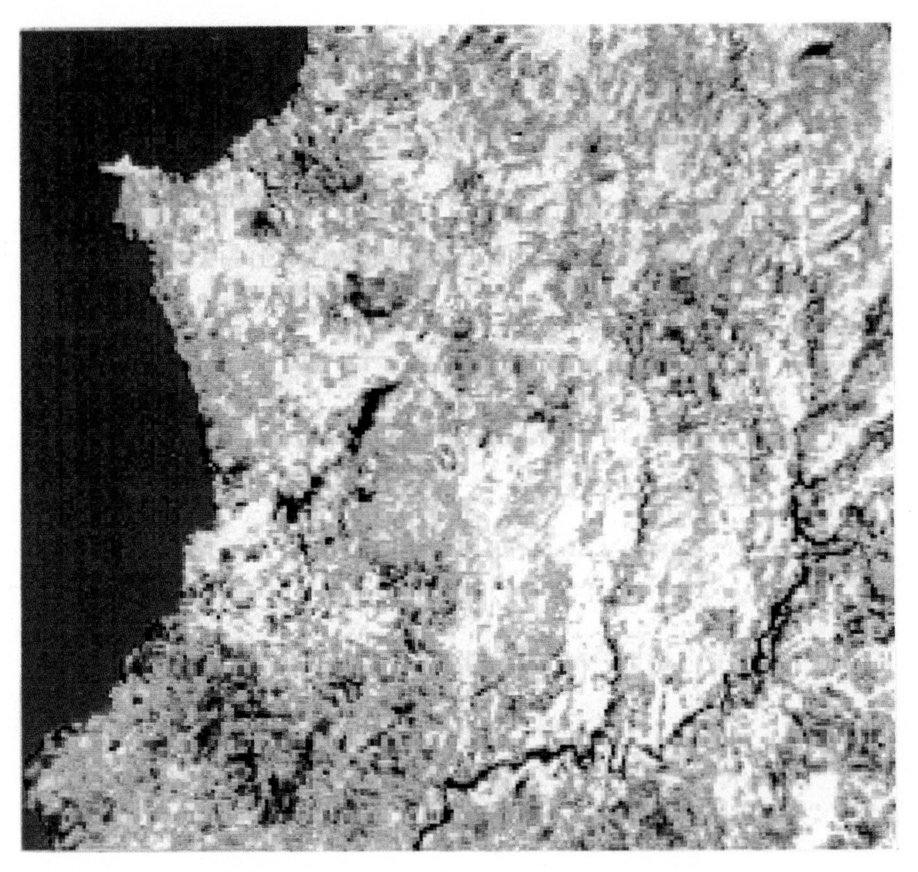

Fig. 4. NDVI image derived from the MSS scene of March 23, 1993. Light grey indicates high NDVI values, dark grey and black low NDVI values. (see also page 202)

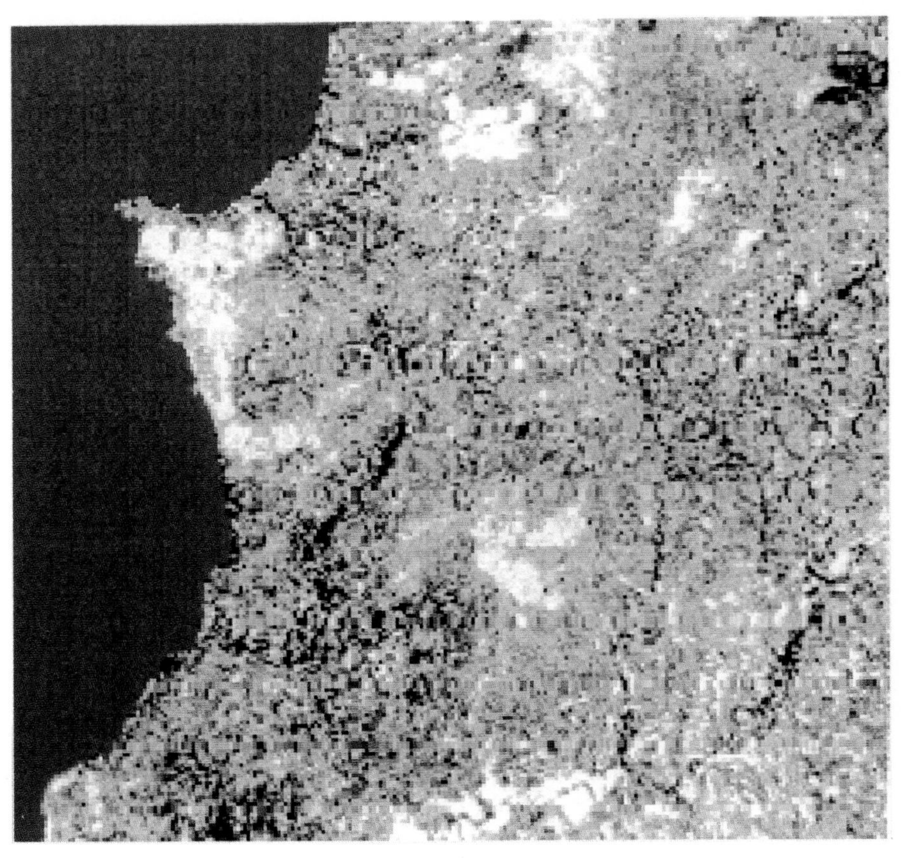

Fig. 5. Subtracted NDVI image obtained by subtracting March 1993, MSS data from March 1977 data. Very light tones indicate vegetation decrease while very dark tones indicate vegetation increase in 1993. (see also page 204)

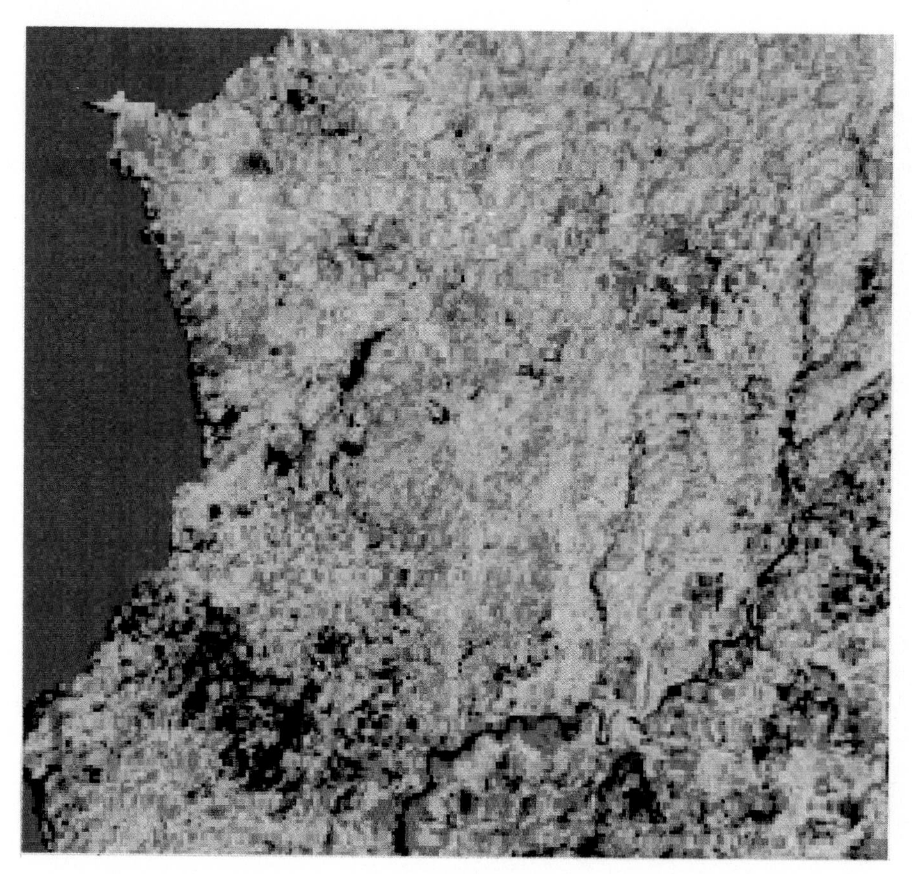

Fig. 6. RGB color composite image of the 1977, 1986, 1993 NDVI assigned to RBG. White, black and grey express no major change in NDVI values between the three dates while colors illustrate change in vegetation cover (magenta = decrease in NDVI value in 1993; green = increase in NDVI value focussed in 1993). (see also page 205)

Fig. 7. RGB color composite image of the 1997, 1993 NDVI assigned to R and BG respectively. White, black and grey express no major change in NDVI values between the two dates while light red illustrates low decrease, dark red high decrease in NDVI value in 1993. Light and dark cyan indicate an increase in NDVI value in 1993. (see also page 206)

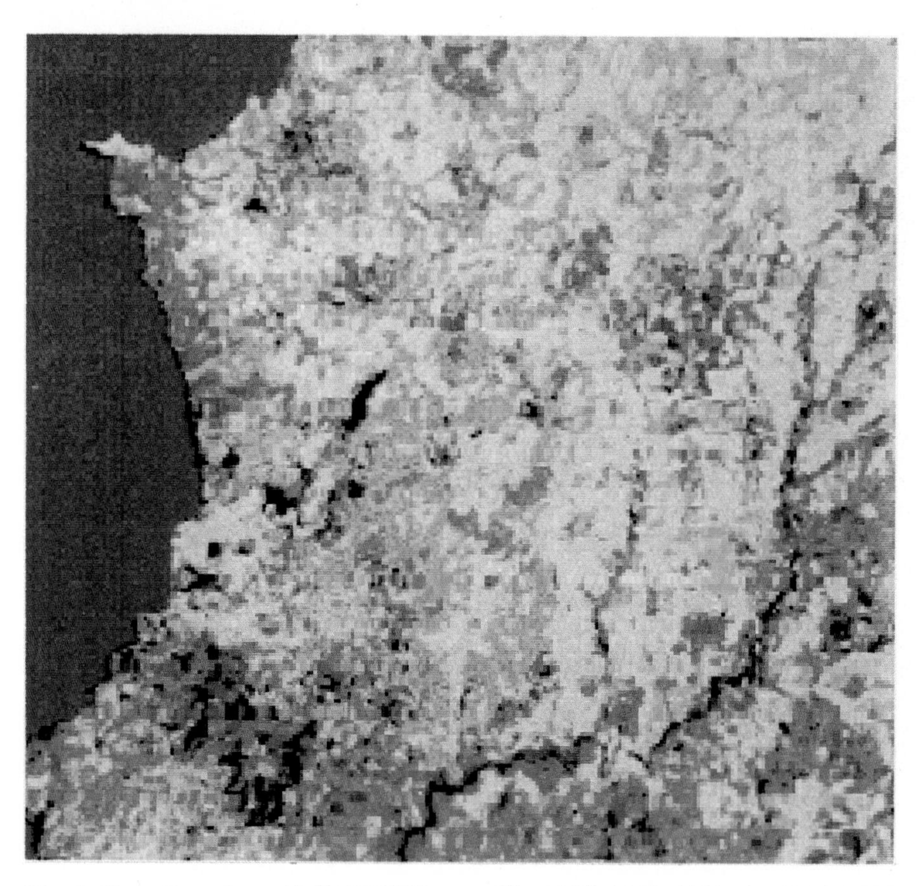

Fig. 8. Sixty-four classes RGB-NDVI image. The 1977, 1986, 1993 classified NDVI are assigned to RBG respectively. White, black and grey express no major change in NDVI values between the three dates while colors illustrate change in vegetation cover (magenta = important decrease in NDVI value in 1993; green = important increase in NDVI value in 1993; yellow = important decrease in 1986 and 1993; light blue = increase in 1986 and 1993; orange = important decrease in 1986 and slight increase in 1993). (see also page 207)

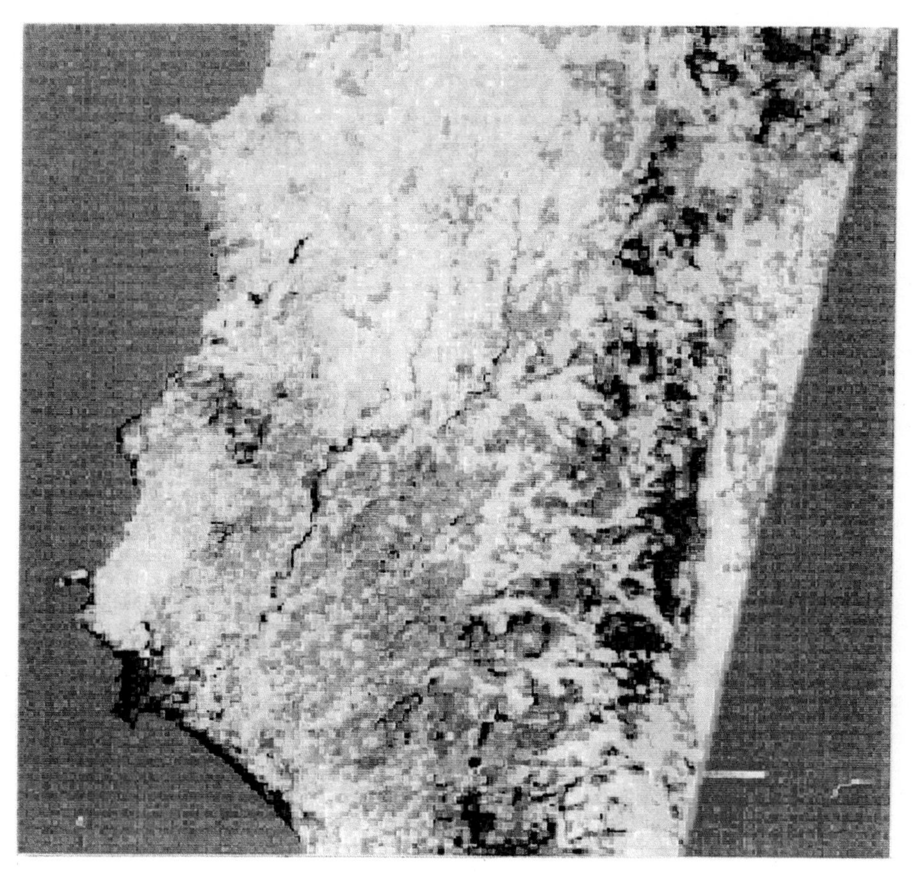

Fig. 9. Sixty-four classes RGB-NDVI image: generalization to a larger area. Same color coding as in Fig. 8. (see also page 208)

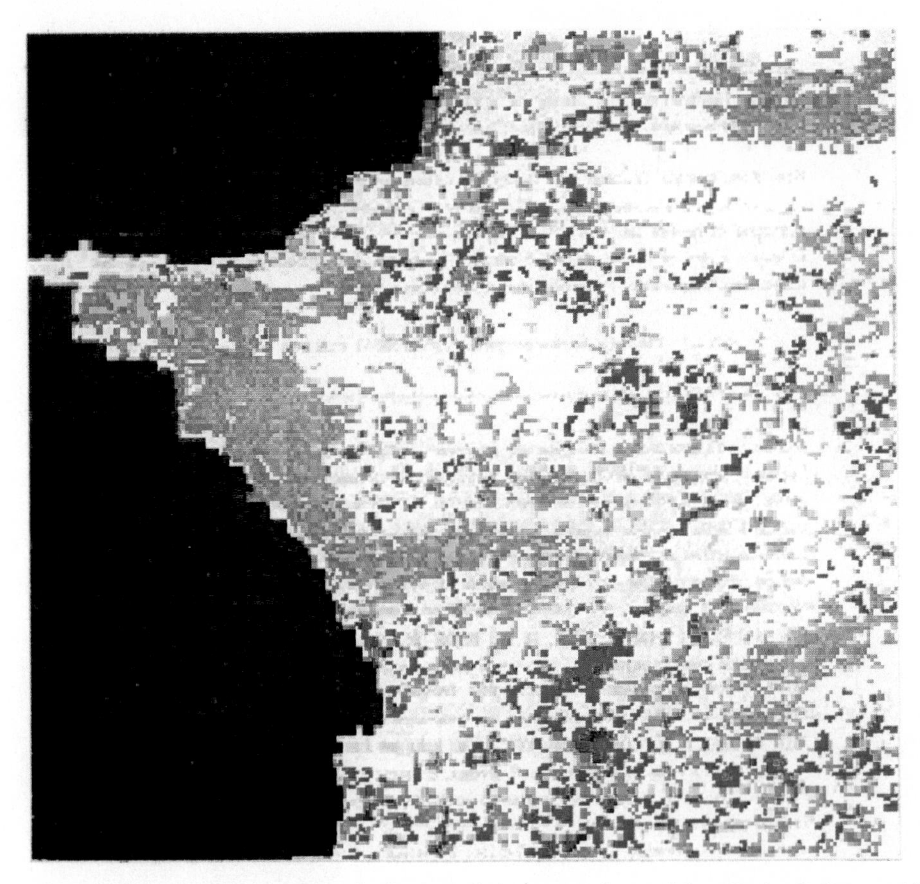

Fig. 10. Mapping the magnitued of vegetation changes in maritime Syria. Subtracted NDVI image was obtained by subtracting the 1993 classified vegetation index image from the 1977 classified NDVI image. White indicates no change in NDVI slice units. Magenta, red and yellow represent classes whose NDVI value was lowered by one, two and three slice units respectively. Cyan, blue and black (except sea surface) represent classes whose NDVI value rose by one, two and three slice units respectively. (see also page 209)

Volumes in the Series
Data and Knowledge in a Changing World

The Information Revolution:
Impact on Science and Technology
1996, 50 Figures, 17 Tables, 274 p.
Edited by J.-E. Dubois and N. Gershon
ISBN 3-540-60855-9

Modeling Complex Data for Creating Information
1996, 84 Figures, 12 Tables, 278 p.
Edited by J.-E. Dubois and N. Gershon
ISBN 3-540-61069-3

Industrial Information and Design Issues
1996, 102 Figures, 8 Tables, 294 p.
Edited by J.-E. Dubois and N. Gershon
ISBN 3-540-61457-5

Geosciences and Water Resources:
Environmental Data Modeling
1997, 95 Figures, 35 Tables, 340 p.
Edited by C. Bardinet and J.-J. Royer
ISBN 3-540-61947-X

Thermodynamic Modeling and Materials Data Engineering
1997, 38 Figures, 96 Tables, 340 p.
by J.-P. Caliste, A. Truyol and J. Westbrook (Eds.)

Springer and the environment

At Springer we firmly believe that an international science publisher has a special obligation to the environment, and our corporate policies consistently reflect this conviction.
We also expect our business partners – paper mills, printers, packaging manufacturers, etc. – to commit themselves to using materials and production processes that do not harm the environment. The paper in this book is made from low- or no-chlorine pulp and is acid free, in conformance with international standards for paper permanency.

Printing: Mercedesdruck, Berlin
Binding: Buchbinderei Lüderitz & Bauer, Berlin